INTRACTABLE SEIZURES

Diagnosis, Treatment, and Prevention

ADVANCES IN EXPERIMENTAL MEDICINE AND BIOLOGY

Recent Volumes in this Series

Volume 491
THE MOLECULAR IMMUNOLOGY OF COMPLEX CARBOHYDRATES—2
Edited by Albert M. Wu

Volume 492
NUTRITION AND CANCER PREVENTION: New Insights into the Role of Phytochemicals
Edited under the auspices of the American Institute for Cancer Research

Volume 493
NEUROIMMUNE CIRCUITS, DRUGS OF ABUSE, AND INFECTIOUS DISEASES
Edited by Herman Friedman, Thomas W. Klein, and John J. Madden

Volume 494
THE NIDOVIRUSES (CORONAVIRUSES AND ARTERIVIRUSES)
Edited by Ehud Lavi, Susan R. Weiss, and Susan T. Hingley

Volume 495
PROGRESS IN BASIC AND CLINICAL IMMUNOLOGY
Edited by Andrzej Mackiewicz, Maciej Kurpisz, and Jan Żeromski

Volume 496
NONINVASIVE ASSESSMENT OF TRABECULAR BONE ARCHITECTURE AND THE COMPETENCE OF BONE
Edited by Sharmila Majumdar, Ph.D., and Brian K. Bay, Ph.D.

Volume 497
INTRACTABLE SEIZURES: Diagnosis, Treatment, and Prevention
Edited by W. McIntyre Burnham, Peter L. Carlen, and Paul A. Hwang

Volume 498
DIABETES AND CARDIOVASCULAR DISEASE: Etiology, Treatment, and Outcomes
Edited by Aubie Angel, Naranjan Dhalla, Grant Pierce, and Pawan Singal

Volume 499
FRONTIERS IN MODELING AND CONTROL OF BREATHING
Edited by Chi-Sang Poon and Homayoun Kazemi

Volume 500
BIOLOGICAL REACTIVE INTERMEDIATES VI: Chemical and Biological Mechanisms of Susceptibility to and Prevention of Environmental Diseases
Edited by Patrick M. Dansette, Robert Snyder, Marcel Delaforge, G. Gordon Gibson, Helmut Greim, David J. Jollow, Terrence J. Monks, and I. Glenn Sipes

---

A Continuation Order Plan is available for this series. A continuation order will bring delivery of each new volume immediately upon publication. Volumes are billed only upon actual shipment. For further information please contact the publisher.

INTRACTABLE SEIZURES
Diagnosis, Treatment, and Prevention

Edited by

W. McIntyre Burnham
University of Toronto
Toronto, Ontario, Canada

Peter L. Carlen
Toronto Western Research Institute
Toronto, Ontario, Canada

and

Paul A. Hwang
University of Toronto
Toronto, Ontario, Canada

KLUWER ACADEMIC / PLENUM PUBLISHERS
NEW YORK, BOSTON, DORDRECHT, LONDON, MOSCOW

Library of Congress Cataloging-in-Publication Data

Intractable seizures: diagnosis, treatment, and prevention/edited by W. McIntyre Burnham, Peter L. Carlen, and Paul A. Hwang.
p. ; cm. — (Advances in experimental medicine and biology; v. 497)
Includes bibliographical references and index.
ISBN 0-306-46621-X
1. Epilepsy. I. Burnham, W. McIntyre, 1939– II. Carlen, Peter L. III. Hwang, Paul A. 1949– IV. Series.
[DNLM: 1. Seizures—diagnosis. 2. Seizures—prevention & control. 3. Seizures—therapy. WL 340 I614 2001]
RC372 .I495 2001
616.8′53—dc21

2001042722

ISBN 0-306-46621-X

233 Spring Street, New York, N.Y. 10013

http://www.wkap.com

10 9 8 7 6 5 4 3 2 1

A C.I.P. record for this book is available from the Library of Congress

Printed in the United States of America

To absent friends, Gerhard Fromm,
Eric Lothman, and Frank Morrell,
and to our teacher, Pierre Gloor

PREFACE

About 20% of the people who have epilepsy have seizures that are resistant to anticonvulsant medications. Another 20% have seizures which are only partially responsive to drugs. These drug-resistant seizures are called "intractable", and the patients who have them—about 1 in 250 in the general population—present a major challenge to physicians, neurologists and epilepsy associations.

The present volume describes the symptomatology of the major "intractable" syndromes, the most appropriate drugs for each, and the possibilities for surgical control. Research related to the causes and effects of intractable seizures is presented, and new directions in therapy are discussed.

ACKNOWLEDGMENTS

The present volume originated in a conference sponsored by the Bloorview Children's Hospital Foundation. The editors would like to thank the Foundation for their generous support. We would also like to thank Ms. Alla Vilner of the Bloorview Epilepsy Research Program and Mr. David Banh for their dedicated work in preparing this book.

ABBREVIATIONS

ACPD	amino-cyclopentane-1,3-dicarboxylate
ACTH	Corticotrophin
AED	Antiepileptic Drugs
AEM	Antiepileptic Medications
AHP	Afterhyperpolarization
AMPA	amino-3-hydroxy-5-methyl-4-isoxazoleproprionic acid
AVMs	Arteriovenous Malformations
BZP	Benzodiazepine
CA1	Field CA1 of the Hippocampus
CA2	Field CA2 of the Hippocampus
CA3	Field CA3 of the Hippocampus
Ca^{2+}	Calcium Ions
Cl^-	Chloride Ions
CNS	Central Nervous System
CPS	Complex Partial Seizures
CRH	Corticotropin-Releasing Hormone
CSF	Cerebrospinal Fluid
CT	Computerized Tomography
DAG	Diacyl Glycerol
EAA	Excitatory Amino Acids
EEG	Elecroencephalography
EcoG	Electrocorticography
ESES	Electrical Status Epilepticus of Sleep
FCD	Focal Cortical Dysplasia
G	G-protein
GEPR	Genetically Epilepsy Prone Rats
GSH	glutathione
HHE	Hemiconvulsions Hemiplegia Epilepsy Syndrome
IP3	inositol-1.4.5-triphosphate
IS	Infantile Spasms
KA	Kainic Acid
LGS	Lennox-Gastaut Syndrome
LTP	Long-Term Potentiation
MCT	Medium-Chain Triglyceride
mGluRs	Metabotropic Glutamate Receptors
MK801	(+)-5-methyl-10,11-dihydro-5H-dibenzo[a,d]cycloheptan-5,10-imine maleate

MR	Magnetic Resonance
MTS	Mesial Temporal Sclerosis
Na^+	Sodium Ions
NGF	Nerve Growth Factor
NMDA	N-methyl, D-aspartate
NO	Nitric Oxide
nREM	Non Rapid Eye Movement Sleep (Slow Wave Sleep)
PCP	phencyclidine
PET	Positron Emission Tomography
PDS	Paroxysmal Depolarization Shift
PIP_2	phosphatidyl inositol-4,5-diphosphate
PKC	Phosphokinase C
PLC	Phospholipase C
PNS	Peripheral Nervous System
PTE	Posttraumatic Epilepsy
REM	Rapid Eye Movement Sleep (Dream Sleep)
SE	Status Epilepticus
SPECT	Single Photon Emission Computed Tomography
SRS	Spontaneous Recurrent Seizures
VGB	Vigabatrin
VPA	Valproate
WS	West's Syndrome

TABLE OF CONTENTS

Special Topic

1

INTRACTABLE EPILEPSY IN ADULTS

Ilo E. Leppik

Clinical Professor of Neurology and Pharmacy
University of Minnesota
Director of Pharmacology Research
MINCEP Epilepsy Care, P.A.
5775 Wayzata Blvd. Suite 255
Minneapolis, Minnesota 55416
tel: (612) 525-4511
fax: (612) 525-1560

Drug-resistant or "intractable" epilepsy presents a major challenge for physicians, epilepsy associations and researchers. Dr. Leppik discusses the evaluation of patients who may be intractable, the nature of intractable epilepsy and the factors which contribute to intractability.

INTRODUCTION

Most patients with epilepsy can achieve complete control of seizures with medication. Nevertheless, a substantial number develop intractable epilepsy—and all of the attendant medical and psychosocial difficulties associated with uncontrolled seizures. Various estimates have placed the number of persons continuing to have seizures in spite of medical treatment at 17% to 31% of the population of persons with epilepsy.[14,13] In this chapter, three areas will be reviewed: 1) evaluation of a person with uncontrolled seizures; 2) definition of "intractable" epilepsy; and 3) predictors of intractability.

EVALUATION OF A PERSON WITH UNCONTROLLED SEIZURES (POSSIBLE INTRACTABLE EPILEPSY)

A person who continues to have seizures after initiation of treatment with antiepileptic medications (AEMs) must be carefully evaluated before being diagnosed as

Table 1. Evaluation of a Person with Intractable Epilepsy

1. Seizures must be correctly diagnosed
   1.1 Correctly identify seizure type
   1.2 Rule out non-epileptic events
      2.1 Physiologic, i.e. Cardiac
      2.2 Psychogenic
2. Treatment must be monitored
   2.1 For therapeutic concentrations
   2.2 For compliance
      2.1 Serial blood levels
      2.2 Asking about compliance
      2.3 Monitoring using an electronic dispensing device

"intractable". The seizure type must be correctly diagnosed, the appropriate AEMs used and compliance determined (Table 1).

In the majority of cases, physicians or trained persons do not observe initial seizures. Usually a patient has behavior which appears to have been a seizure and is brought to a medical facility for diagnosis and treatment. Many seizures are induced by sleep deprivation, alcohol or drug abuse, or some other provocation. In the face of a normal EEG, negative family history and normal structural studies, recurrent seizures are unlikely, especially if a provoking factor is identified.[4] Single seizures usually do not require initiation of therapy.

Usually after a second unprovoked seizure, a diagnosis of "epilepsy" is made and treatment with AEMs started. At this point, it is necessary to monitor the outcome of treatment. There is some evidence that prognosis of treatment is related to the rapidity with which seizures are brought under control after the onset of epilepsy. It is essential that adequate resources be expended at this point to bring seizures under complete control, both for the quality of life of the patient and reduction of expenditure later if intractability can be prevented. This will involve determination of compliance, adjustment of AEM dosage and—if seizures persist—review of the initial diagnosis.

Determination of Compliance

It is necessary to obtain three measurements of AEM concentrations approximately a few months after initiation of treatment. The absolute concentrations and their variability provides crucial information about the pharmacokinetic profile and compliance of the patient. If all three levels are within 20% to 25% of each other, compliance is likely.[6] If the variability between serial samples is greater than 25%, noncompliance must be suspected and efforts made to assist the person with this problem.[4] Compliance can be evaluated using serial blood levels, asking about medication taking habits, or using computerized monitoring devices. Non-compliance is the major cause of failure of AEM treatment.

Adjustment of AEM Dosage

If compliance is good, but the AEM concentrations low, the person is a rapid metabolizer. This may be due to genetic factors, but is most likely caused by induction of metabolism by other substances. For example, moderate alcohol use (more than 3 ounces per day) may affect the cytochrome p-450 system sufficiently to double the clearance of phenytoin. Cigarette smoking will increase the clearance of most AEMs. Many prescription medications and other AEMs also can increase clearance. If concentrations are consistently low, an upward dose change is needed. For phenytoin, which has saturable kinetics, small dose increases may result in disproportionally large increases in levels. A number of formulas have been devised for calculating the magnitude of the dose increase needed for phenytoin.[17] Carbamazepine appears to follow linear kinetics, and the dose can be increased proportionally to the desired increase in concentration. Because of autoinduction of its metabolism, carbamazepine's dose must be increased a few weeks after initiation of therapy. For valproate, because of its protein binding properties, dose increases usually result in lower than expected concentrations. Often doubling the dose increases the concentration by less than twofold.[18]

It is advisable to instruct the patient to obtain a measurement of AEM concentrations as soon as possible after a seizure. This is critical in assessing the progress of treatment. If the AEM concentration is below the target range, and if values from clinic visits were in the target range, non-compliance has been demonstrated unless there has been a significant illness at the time of the seizure.

Review of the Diagnosis

If seizures persist after compliance has been verified, and suitable blood concentrations achieved, the accuracy of the diagnosis must be reviewed. Often, the physician has not been able to observe seizures directly, but rather has relied on accounts by witnesses. These persons are usually more concerned about dealing with the crisis than observing the event, and their reports may be misleading. From descriptions alone it is difficult to distinguish between brief complex partial seizures and absence seizures. This mistake in diagnosis may lead to treatment with inappropriate AEMs. In addition, a substantial number of persons referred for video/EEG monitoring are found not to have epileptic seizures, but rather to suffer from non-epileptic events. These may be physiologic or psychogenic. There are a number of physiological phenomena which may resemble epilepsy. These include, among others, syncope with clonic activity, and hypoxic seizures from unrecognized cardiac rhythm disturbances. Psychogenic events resembling seizures are difficult to diagnose without video/EEG monitoring.[3]

Definition—Diagnosis of intractable epilepsy

Intractable is defined as "of persons and animals: not to be guided; not manageable or docile; uncontrollable; refractory; stubborn".[10]

For seizures to occur, a population of epileptic neurons must exist within the central nervous system (CNS). These neurons are subject to paroxysmal depolarization shifts (PDS), which render them hyperexcitable. Hyperexcitable neurons may be limited to a specific area of the CNS, as in localization-related epilepsies, or may represent a widely distributed pattern involving neuronal networks with diminished inhibition or excessive excitation, as in generalized epilepsies.[1]

Presently available antiepileptic medications (AEMs) do not alter the epileptic neurons permanently. Rather, they modulate neuronal excitability by many actions, including influencing the chloride channel (valproate, benzodiazepines, barbiturates) or limiting the spread of repetitive discharges by affecting the sodium channel (phenytoin and carbamazepine). These effects are present only when sufficient concentrations of AEMs occur at the specific receptor sites.

If the epileptogenic process is relatively weak, that is, the seizure threshold is high, complete seizure control can be achieved with standard doses of AEMs at concentrations in the usually effective range. On the other hand, the epileptogenic lesion may be "not manageable or docile"—intractable—and be only partially controlled by AEMs.

It is important to note, however, even with a relatively weak epileptogenic process, a temporary reduction or absence of AEMs at the receptor sites can lead to the occurrence of seizures in an otherwise controlled person. Thus, even a person well-controlled at times of adequate medication may be subject to seizures when the AEM doses are incorrect or changes in drug absorption or metabolism occur because of changes in product formulation[16] or the presence of disease states which alter metabolism.[7] A description of "intractable epilepsy" must consider the severity of the seizure type, its frequency, and the consequences of the seizure to the individual's life style. For example, one complex partial seizure every six months would be severely incapacitating to a person capable of competitive employment requiring the operation of a motor vehicle. The same seizure, with the same frequency, might be considered excellent control for a multi-handicapped person in a residential facility.

When is a case of epilepsy considered to be "intractable"? In this presentation, it is proposed that a person be defined as having "intractable" epilepsy if seizures continue to occur, while the person is documented to have a therapeutic concentration of at least one appropriate standard medication, one year after the onset of epilepsy. This definition simplifies the process of classification by providing a clear demarcation between controlled and intractable. It is dependent on the measurement of AEM concentrations at the time of the seizure. This is necessary to document that the seizure was indeed a failure of the AEM to control the epileptogenic focus and not due to either incorrect dosing by the physician, noncompliance by the patient, errors of dispensing or changes in formulation. Within this definition of intractable, however, there is still a wide spectrum of severity, with some individuals having almost daily seizures with multiple medications while others may have only one or two seizures per year.

It is also proposed that a person who has been diagnosed as "intractable" be continued in this category until a one-year seizure free period has elapsed. This time period is sufficient in most jurisdictions to permit renewal of driver's license and to begin the process of reintegrating the individual into the social and economic fabric of his or her community.

In an early definition used in epidemiological studies, Juul-Jensen equated "intractable" epilepsy in clinical practice to "severe epilepsy" and included social factors in his classification.[5] The National Association of Epilepsy Centers (NAEC) has considered as intractable a person whose seizures do not come under control after nine months of treatment under the care of a neurologist.[8] The present proposed definition incorporates both of these concepts. In order to develop health care statistics for patients with intractable epilepsy, a temporary IDC code has been assigned to persons with intractable epilepsy. It consists of adding a suffix of 1 in the second decimal place of the 345 code (Table 2).

Table 2. ICD-9-CM Diagnostic Codes for Seizures with Codes for Intractable Epilepsy

Diagnosis	Intractable	Not Intractable
Generalized convulsive epilepsy	345.11	345.10
Partial epilepsy, with impairment of consciousness	345.41	345.40
Partial epilepsy, without mention of impairment of consciousness	345.51	345.50
Other forms of epilepsy	345.81	345.80
Epilepsy, unspecified	345.91	345.90

PREDICTION OF INTRACTABILITY / POSSIBILITY OF REMISSION

It is important to be able to predict the probability of intractability or remission in a person who is beginning to have seizures. In one study, records were reviewed from two populations of patients.[15] One population consisted of unselected patients identified from a single neurology group practice. From 6,000 patients, 108 met the criteria of at least one nonfebrile epileptic seizure and at least five years of follow-up data. The other population consisted of 181 patients referred to an epilepsy center who had active epilepsy at time of referral.

Three patterns of seizure occurrence were identified (Table 3). The "burst" pattern consisted of repeated seizures at onset, followed by remission of at least two years until the time of the study. The "intermittent" pattern consisted of repeated seizures at onset, followed by a least one remission of more than 2 years, with subsequent seizures. The "continuous" pattern consisted of repeated attacks without remission to the time of the study.

In 70 of 83 patients in remission at the time of the study, the burst pattern was evident. In only 13 of 83 patients was there a remission followed by a recurrence followed by remission to the time of the study. The most common pattern for intractable epilepsy in this study, therefore, was onset of seizures with no remissions, i.e., a continuous pattern (Table 3).

Most of the remissions that did occur had occurred by the end of year three. Remission status at the end of 10 and 15 years from the onset of epilepsy could be predicted from the five year status with an 85% and 79% accuracy, respectively. Results from this study demonstrate that there may be a bimodal distribution of persons with epilepsy; one

Table 3. Longitudinal Patterns of Epilepsy (modified from Ref. 14)

Pattern	Group A (n = 108)	Group B (n = 181)	Group A & B (n = 289)
Burst	70 (65%)		70 (24%)
Intermittent	13 (12%)	39 (22%)	52 (18%)
Continuous	25 (23%)	142 (78%)	167 (58%)

A "burst" consisted of one or more seizures at onset, followed by complete control of seizures. The "intermittent" group consisted of a burst, followed by a remissions of 2 or more years, followed by more seizures. The "continuous" pattern consisted of seizures with no remissions. Group A consisted of unselected patients from a general practice. Group B consisted of patients with intractable seizures. The most common pattern for intractable epilepsy was continuous seizures from onset with no remissions.

Table 4. Presence of the Following Is Often Associated with Intractability

Seizure Type:	Complex partial seizures Multiple seizure types Frequent tonic-clonic seizures Status epilepticus
Age of Onset:	Onset in childhood (for some syndromes such as Lennox-Gastaut) Onset in adulthood
EEG Findings:	Slow background Interictal activity
Etiology:	Symptomatic epilepsy
Clinical Findings:	Abnormal neurological examination Abnormal neuropsychological testing
Treatment:	Delay from time of diagnosis to treatment[14,16]

in which the process is mild and easily treated, with early remissions likely, and the other in whom epilepsy is intractable from the onset.[14]

Thus, early response to treatment—or lack of it—is one predictor of intractability. Other predictors of intractability have been considered by a number of reports (Table 4). One group of researchers found that the presence of at least two of the following had a high degree of association with intractability: multiple neurological or psychological handicaps, different seizure types, clustering of seizures, slowing of the background EEG rhythms, and interictal EEG discharges while on AEMs.[11,12] In another study, persons who continued to seizure during optimum single-drug treatment were found to have neuropsychological handicaps, partial or mixed seizure patterns, symptomatic epilepsy, or many seizures prior to onset of treatment.[14] From a study of 1,868 persons with epilepsy in Japan, 58.3% had remissions.[9] Factors associated with favorable outcome were idiopathic seizures, onset before age 10, few seizures, seizures during sleep, treatment within one year of onset of epilepsy, single seizure type, normal EEG background, and normal neurological and psychological testing.

Some types of seizure are particularly drug resistant. Complex partial seizures seem to have the poorest prognosis of all seizure types in adults. In one study of 666 persons with "temporal lobe epilepsy", only 40% were free of seizures for one year or more.[2] In another study, only 31% of persons with complex partial seizures could be controlled with high-dose single AEM treatment.[2]

SUMMARY

Intractable epilepsy can be defined as seizures which have not been completely controlled with AEMs one year after onset, despite accurate diagnosis and carefully monitored treatment. Once intractable, there is a low probability of remission. Predictors of intractability include the presence of partial seizures, structural abnormalities on imaging studies, and abnormalities on the neurological examination. If a person continues to have seizures in spite of adequate treatment with AEMs, surgical or other options should be considered.

ACKNOWLEDGMENTS

This work was supported by the National Institute of Neurological and Communicative Disorders and Stroke (P50 NS1-16308). Word processing by Diane Rider is gratefully acknowledged.

REFERENCES

1. Commission on Classification and Terminology of the International League Against Epilepsy. Proposal for revised classification of epilepsies and epileptic syndromes, *Epilepsia*, 30 (1989) 389–399.
2. Currie, S., Healthfield, K.W.G., Henson, R.A., and Scott, D.F. Clinical course and prognosis of temporal lobe epilepsy, a survey of 666 patients, *Brain*, 94 (1971) 173–190.
3. Gumnit, R.J. Psychogenic Seizures. In: E. Wyllie (Ed.) *The Treatment of Epilepsy: Principles and Practice, 2nd ed.*, Wiliams and Wilke, Baltimore (1997) pp. 677–680.
4. Hauser, W.A., Anderson, V.E., Loewenson, R.B., and McRoberts, S.M. Seizure recurrence after a first unprovoked seizure, *N. Engl. J. Med.*, 307 (1982) 522–528.
5. Juul-Jensen, P. Epilepsy. A Clinical and social analysis of 1020 adult patients with epileptic seizures, *Act. Neurol. Scand.*, 40 (1964) 5.
6. Leppik, I.E. Compliance during the treatment of epilepsy, *Epilepsia*, Suppl 2 (1988) 79–84.
7. Leppik, I.E., Fisher, J., Kriel, R., and Sawchuk, R.J. Altered phenytoin clearance during febrile illness, *Neurology*, 36 (1986) 1367–1370.
8. The National Association of Epilepsy Centers. Recommended guidelines for diagnosis and treatment in specialized epilepsy centers, *Epilepsia*, 31 (Suppl 1) (1990) S1–S12.
9. Okuma, T. and Kumashiro, H. (Chairmen for the Group for the Study for Prognosis of Epilepsy in Japan) Natural History and prognosis of epilepsy: report of a multi-institutional study in Japan, *Epilepsia*, 22 (1981) 35–53.
10. The Oxford English Dictionary, Oxford University Press, Glasgow, New York (1971).
11. Rodin, E. *The prognosis of patients with epilepsy*, Charles C. Thomas, Springfield, Ill. (1968).
12. Rodin, E., Klutke, G., and Chayasirisobohn, S. Epileptic patients who are refractory to anticonvulsant medications, *Neurology*, 32 (1982) 1382–1384.
13. Schmidt, D. Single drug therapy for intractable epilepsy, *J. Neurol.*, 229 (1983) 221–226.
14. Shorvon, S.D. and Reynolds, E.H. Early prognosis of epilepsy, *Br. Med. J.*, 285 (1982) 1699–1701.
15. Shorvon, S.D. and Sander, J.W.A.S. Temporal patterns of remission and relapse of seizures in patients with epilepsy. In: Schmidt, D. and Morselli, P.L. (Eds.) *Intractable Epilepsy*, Raven Press, New York (1986) pp. 87–97.
16. Tyrer, J.H., Eadie, M.J., Sutehrland, J.M., and Hooper, W.D. Outbreak of anticonvulsant intoxication in an Australian City, *Br. Med. J.*, 4(1970) 271–273.
17. Welty, T.E., Robinson, F.C., and Mayer, P.R. A comparison of phenytoin dosing methods in private practice seizure patients, *Epilepsia*, 27 (1986) 76–80.
18. Yu, H.Y. Clinical implications of serum protein binding in epileptic children during sodium valproate maintenance therapy, *Ther. Drug Monit.*, 6 (1984) 414–423.

2

COMPLEX PARTIAL SEIZURES

Clinical Description and Diagnosis

Warren T. Blume

University Hospital
University of Western Ontario
London, Ontario, Canada
N6A 5A5
tel: (519) 663-3690
fax: (519) 663-3753

A few types of epilepsy are particularly drug-resistant. These include West's syndrome, Lennox-Gastaut syndrome and complex partial epilepsy. Of these, complex partial epilepsy is by far the most common.

INTRODUCTION

"Partial" or "localization-related" seizures are those whose initial clinical and electroencephalographic manifestations indicate activation of a system of neurons limited to part of one cerebral hemisphere.[6] "Complex" partial seizures are those in which consciousness is impaired. For this purpose consciousness has been defined as "that integrating activity by which Man grasps the totality of his phenomenal field".[7] The often difficult clinical assessment of consciousness has impaired somewhat the value of this distinction as receptive or expressive dysphasia, apraxia, distraction and amnesia may all appear to represent impaired consciousness but occur in a patient who is otherwise aware of his environment. Cardinal symptoms of complex partial seizures are presented in Table 1.

The diagnosis of complex partial seizures utilizes the following elements: seizure description by patient and observers, neurological examination, neurological history, interictal and ictal EEG and MRI scanning. PET scanning and functional MRI have yet to be proven cost effective in this determination.

Complex partial seizures are found in over half of the adult patients with epilepsy.[9] Their onset can occur at any time in life. They may arise from the frontal, parietal or—most commonly—temporal lobes.

Table 1. Common Symptoms of Complex Partial Seizures*

- there is impairment of interaction with the environment during the seizure
- behavior during the seizure is often a motionless stare with automatisms
- the seizure is often preceded by a simple partial seizure (aura), which may be an epigastric sensation, an affective symptom such as fear or an experiential phenomenon such as an "out of body" experience

*See also Table 3.

FUNCTIONS OF MESIAL TEMPORAL LOBE ("LIMBIC") STRUCTURES

As most complex partial seizures arise from a temporal lobe, features of limbic system activation will be present, prompting a brief review of these functions (see Table 2). Because the hippocampal formation is physiologically and structurally closely associated with the amygdala, it is difficult to clearly distinguish their functions. However, most of the following functions have been ascribed to the amygdala.

Stimulation of the amygdala in animals and humans has produced a variety of responses in the emotional, motor and vegetative spheres. Emotional effects include: fear, bewilderment, anxiety and other emotional reactions such as sadness and disgust. Arrest of spontaneous ongoing movements, swallowing, chewing, searching movements, contralateral turning of head and eyes, and inhibition or facilitation of spinal reflexes are the principal motor effects.[5] Autonomic changes may be: inhibition or activation of gastric motility and secretion, pupillary dilatation, pilo-erection and changes in respiration and cardiovascular function.

Diffuse and abdominal sensations may also result from amygdala activation.[12]

Interpretation of one's environment may also be a function of the amygdala and hippocampal region, in concert with the temporal neocortex.[3,13]

Finally, olfaction, a less important function in humans, is still represented, and olfactory sensations can be evoked with stimulation of the amygdala.[20]

As the hippocampus is closely related to memory, epileptic activation of the hippocampus involved in verbal memory (usually left) or both hippocampi may result in amnesia for the attack, a deficit which can masquerade as impaired awareness.

CLINICAL MANIFESTATIONS OF TEMPORAL LOBE COMPLEX PARTIAL SEIZURES

The characteristics of any ictal phenomenon reflect the structures activated at that time. These may or may not coincide with the site of origin. This concept applies particularly to complex partial seizures, as propagation to the mesial temporal area from extra-temporal structures occurs.

Table 2. Functions of the Limbic System

- Alimentary
- Emotional
- Interpretative, experiential
- Autonomic
- Memory

Temporal lobe complex partial seizures often have two components: 1) an "aura", which is a simple partial seizure preceding the complex partial seizure, and 2) the complex partial seizure with automatisms (see Table 3).

Simple Partial Component (Aura)

Auras (the simple partial component) have been reported in 20% to about 67% of patients with temporal lobe seizures.[18]

The most common aura can be classed as epigastric/emotional, occurring in 52% of Rasmussen's series.[23] Thus, the patient may feel one or more of the following abdominal sensations: nausea, butterflies, an emptiness, roller-coaster sensation, tightness, churning, or even pain.[26]

Fear is the most common affective symptom. One might think that fear would herald any epileptic seizure, the patient fearing its consequences. Actually, symptoms which precede tonic-clonic attacks, with possible injury, are accompanied by concern, not fear. Instead, the gripping abdominal fear or of an imaginary person approaching from behind reflects limbic system involvement, either in the temporal lobe or, more rarely, in the cingulate gyrus. Depression with crying and disgust may occur, but anger is distinctly an unusual symptom.

Olfactory and/or gustatory sensations occur in about 10% of temporal lobe auras. Olfactory sensations represent mesial temporal (uncal) activation.[22] While gustatory sensations may reflect involvement within the circular sulcus adjacent to the insula.[21] Activation of the adjacent parietal and frontal operculum may also produce gustatory sensations.[14]

Diffuse or cephalic sensations, such as a tingling, fullness, or "difficult-to-describe" feeling, may represent seizures involving many cerebral structures including the temporal lobe.[23]

More complex experiential phenomena, while fascinating, occur in a smaller percentage of temporal lobe seizure aurae. "Illusions" are faulty interpretations of present experience in relationship to the environment. These involve macropsia, micropsia, distortion of shapes, or excessively loud or muffled sounds, whose quality may be distorted. "Hallucinations", defined as life-like experiences unrelated to the present environment and reality, may be visual, auditory, or may combine both modalities with other sensory phenomena in a very complex scene.[26] Thus, hallucinations may involve scenery, persons, someone calling or an entire experience, as follows: sitting in a boat in the Thousand Islands region on a hot, humid, and overcast Saturday afternoon watching people on the shore having a bar-b-que and thinking about their misfortune in having unfavourable weather for their event! Other phenomena include the well known deja-vu or jamais-vu feelings, out-of-body sensations and dreamy states.

Complex Partial Component (Psychomotor Phase)

Proceeding from one or more aura are the complex partial or "psychomotor components" of temporal lobe seizures. Video taping of such attacks permitted Kotagal et al.[19] to chart the most common sequence of these phenomena. Typical attacks involved both a motionless stare, and automatisms.

"Automatisms" are repetitive, more or less coordinated, involuntary motor acts which occur during a state of clouded consciousness for which the patient is amnesic.[8,9] They may be spontaneous or reactive. "Spontaneous" automatisms are stereotyped, and

Table 3. Temporal Lobe Complex Partial Seizures: Common Symptoms

Simple Partial Component (aurae)
- Emotional: fear, depression
- Epigastric
- Olfactory, gustatory
- Experiential:
  - Visual
  - Auditory
  - Corporeal
- Diffuse sensations

Complex Partial Component
- Impaired awareness
- Motionless stare
- Automatisms:
  - Oroalimentary: chewing, swallowing
  - Mimetic: smiling, pouting
  - Hand: fumbling, grasping
    - bilateral, ipsilateral
- Autonomic
- Motor: head turning
  - dystonic posturing

may be a constant feature of the patient's seizure, irrespective of his or her environment. "Reactive" automatisms may occur if the patient continues an activity underway at seizure onset such as washing dishes or making change. In Kotagal's study, a behavioral arrest with motionless stare appeared first, followed by oroalimentary automatisms, which included lip and tongue movements (smacking), masticatory movements, gulping and/or swallowing.[19] (Curiously, Gates and Gumnit[10] found restlessness to appear early in such seizures.) "Mimetic" automatisms, which consist of facial expressions including smiling or pouting, often correlated with the behavioral arrest.[19] Following oroalimentary automatisms, were repetitive hand automatisms, consisting of fumbling or grasping movements. Subsequently, patients may look about and then may walk about. Hand automatisms may have lateralizing value if unilateral, and if the contralateral limb becomes involved in dystonic posturing permitting only the limb ipsilateral to seizure involvement to perform the manual automatism.

Arrest of speech may signal dominant temporal lobe seizure involvement (usually left); while non-sensical, but non-garbled, whole-word speech may accompany non-dominant temporal seizures. Lateralizing significance of head turning during a complex partial seizure of temporal lobe origin is controversial.[18,1] Dystonic posturing, in contrast, may be a more reliable sign of contralateral temporal lobe involvement.

Tachycardia, sinoatrial arrest, respiratory irregularities, pallor, pilomotor erection, and mydriasis are some autonomic phenomena which may accompany seizures of the temporal lobe.

Post-Ictal Phase

In contrast to absence attacks, patients emerge *gradually* from temporal lobe seizures. Examination of language, once the patient is sufficiently mentally clear to understand instructions, is important as dysphasia may reflect significant involvement of the temporal lobe "dominant" for language. Such does not necessarily indicate side of seizure origin, however. Therefore, once a patient is able to carry out a simple non-verbal instruction, naming function could be performed as well as saying days of the week and counting forwards and backwards.

FRONTAL LOBE COMPLEX PARTIAL SEIZURES

Frontal lobe complex partial seizures occur frequently, are brief (30 seconds or less), begin and end abruptly, occur nocturnally, and often exhibit explosive symptoms. In the middle of the night, a patient may emit a blood curdling scream, may thrash about in bed, may exhibit bicycling or peddling, repetitive automatisms, may gesture wildly, may dance or run about and then may become alert with minimal postictal features. Complicating the analysis is the occasional existence of both temporal and frontal, principally orbital frontal, epileptogenesis. Such seizures may spread to the supplementary motor area, creating postural type seizures. Secondary generalization may occur slightly more often with frontal lobe attacks than with temporal lobe seizures.

OCCIPITAL LOBE COMPLEX PARTIAL SEIZURES

Manifestations of occipital lobe-originating seizures may not only reflect involvement there, but also spread to the mesial temporal region and to the premotor cortex.

Hints that complex partial seizures may originate posteriorly include overt visual symptoms, the patient describing his attacks as "a blackout", visual experiential phenomena, or simply a visual flavour to a description of the symptoms. Ictal signs raising this possibility include: ocular deviation, a pulling ocular sensation, rapid and forced blinking, and ictal headache or peri-ocular pain. A permanent or postictal visual field deficit may also suggest an occipital origin.

INCOMPLETELY GENERALIZED SEIZURES

Several aspects of absence attacks distinguish them from complex partial seizures: sudden loss of consciousness without warning, bilaterally symmetrical motor phenomena such as peri-ocular of facial myoclonus, vertical ocular rotation, a usually briefer duration, and sudden offset without postictal features.

If the absence attack (manifested on EEG as bilaterally synchronous 3 Hertz spike-waves) incompletely involves the cortex, permitting some preservation of consciousness, unusual phenomena may occur and these have been well described by Howell.[15] These include a feeling of diffuse paralysis, diffuse tremulousness, adversive movements with tonic unilateral or bilateral forearm flexion, loss of use of a limb, blinking, staring and automatisms as remembered phenomena, vertigo, a high-pitched whistling noise, blindness or blurred vision, a feeling of bradycardia, unusual cephalic or epigastric sensations, and weird or "far away" feelings. The EEG will help to distinguish this process from the more common temporal lobe complex partial seizures. Such distinction may alter therapeutic preferences, as medications for absence attacks would be more appropriate (see Table 4).

Table 4. Incompletely Generalized Absence Seizures

- Impaired consciousness (not lost)
- Disparate localizations
- Bilateral EEG phenomena

THE NEUROLOGICAL EVALUATION

The Neurological History

Results of a population-based case-control study of risk factors for complex partial seizures by Rocca et al.[24] provide valuable guides to history taking in seeking an etiology for complex partial seizures. Their data indicate that febrile seizures, neonatal convulsions, cerebral palsy, significant head trauma (loss of consciousness, post traumatic amnesia, or evidence of skull fracture) and viral encephalitis occurred more commonly in patients with complex partial seizures than in control subjects. A history of epilepsy or febrile seizures in the patient's mother was also related. Unclear is whether the febrile convulsions cause the complex partial seizures, share the same risk, or represent only an early manifestation of a seizure disorder. Pre-natal and peri-natal factors were not associated with complex partial seizures (see Table 5).

Characteristics of febrile convulsions may reveal the side of temporal lobe complex partial seizure onset; a forme fruste of the Hemiconvulsions Hemiplegia Epilepsy Syndrome (HHE) of Gastaut may be present. In this syndrome, lateralized febrile status epilepticus (Hemiconvulsions) may be followed by Hemiplegia or Hemiparesis of varying degrees and durations, and later by Epilepsy. The epilepsy is most commonly complex partial of temporal lobe origin, originating on the side implicated by the previous hemiconvulsions or the motor deficit. Less common are focal motor seizures emanating from the involved Rolandic region. Minimal evidence of the HHE syndrome may be revealed by such history taking and mild lateralizing motor findings (see below).

The role of minor head injury in creating complex partial seizures remains incompletely determined. For example, a 32-year old patient has had a 12-year history of complex partial seizures beginning two weeks after colliding head-on with another non-helmeted patient at hockey. Consciousness was not impaired and there was no peri-traumatic amnesia, although the patient had a headache and felt unwell. No other antecedent for the patient's left mesial temporal sclerosis has been determined by history, including no febrile convulsions. Interictal spikes appear prominently in both temporal convexities, but the clinically typical seizures of fear and sadness with impairment of consciousness and automatisms all arise from the left mesial temporal area.

Table 5. History and Examination

- Neonatal convulsions
- Febrile seizures
- Hemiconvulsions Hemiplegia Epilepsy Syndrome (HHE)
- Cerebral palsy
- Head trauma
- Encephalitis

Neurological Examination

Some clues as to the nature of epileptogenesis can be discerned by a neurological examination, seeking particularly the residua of the HHE syndrome (see above) and for evidence of extratemporal onset of complex partial seizures. A mild hemiparesis in a patient with febrile convulsions may suggest the HHE syndrome with its implications for epileptogenesis (see above). A visual field deficit would suggest an occipital origin for complex partial seizures, as visual field impairments are exceedingly rare in temporal lobe processes, despite the presence of Meyer's Loop. Theoretically, frontal lobe signs could be found in patients with frontal lobe complex partial seizures, but these are rare. If the cingulate gyrus were considered to be involved in producing fear and impairment of consciousness, a search for upper motor neuron signs in the feet would help to determine whether the process extended posteriorly to the Rolandic area.

The foregoing only lists some of the more prominent examination features to be considered, and does not imply that a complete neurological examination would not be necessary.

Central Nervous System Functional Enquiry

Patients with language-dominant temporal lobe complex partial seizures will usually complain of verbal or generalized memory impairment reflecting a hippocampal dysfunction. Non-dominant temporal lobe seizure disorders less commonly involve *overt* memory difficulties, although such may be disclosed by neuropsychological testing. Reading difficulties either to the right or left may reflect occipital dysfunction and, therefore, suggest this region as a source of complex partial seizures. Personality disorders and impaired judgement could represent frontal lobe dysfunction.

Electroencephalography

The interictal EEG can support or alter the clinical impression that complex partial seizures arise from the temporal lobe by the distribution of spike discharges and non-paroxysmal abnormalities. Bitemporal spikes are noted in 25–33% of patients with temporal lobe seizures.[11] Conversely, about 90% of children with temporal lobe spikes have seizure disorders.[17] Such spikes are best expressed by mandibular notch electrodes (M1,M2),[25] while the mid temporal positions (T3,T4) detect such spikes better than the anterior temporal-inferior frontal electrodes (F7,F8), which still record the majority of such discharges. These discharges are almost always accompanied by regional excess theta and delta activity.[24]

Interictal temporal spikes will predict the side of seizure origin in over 95% of patients if at least three times as many spikes appear in one temporal lobe as the other.[2] Similarly, unilateral focal temporal delta activity correlates with seizure origin in 95% of patients.[2]

The EEG may be normal in 30–40% of patients with temporal lobe epilepsy and, therefore, multiple EEGs, including a sleep EEG, may be required.[18]

A recorded clinically typical seizure would establish the diagnosis; this consists of sequential rhythmic waves and/or spikes whose morphology evolves over several seconds and which arises distinctly from the temporal-lobe interictal "background" activity.

In assessing interictal EEG, one must be mindful of normal spikes appearing in temporal lobe electrode positions such as: small sharp spikes, wicket spikes, 14 and 6/second positive spikes, and even "psychomotor variant".

Magnetic Resonance Imaging

MRI is of considerable use in the evaluation of patients with complex partial seizures. Two questions are addressed: 1) is there a structural abnormality in the area implicated by the clinical and EEG data, and 2) are the other structures normal? MRI is particularly valuable in detecting mesial temporal sclerosis, which presents as a decreased grey-white matter differentiation, an increased T_2 signal and a reduced size of the mesial temporal lobe. Such diagnostic MRI abnormalities were demonstrated in 25 of 27 patients with pathologically proven mesial temporal sclerosis.[16]

Additionally, MRI can detect cortical dysplastic lesions, gangliogliomata, dysembryoplastic neuroendoepithelial tumour (DNET) and other lesions either in the temporal lobe or in the extra-temporal areas.

It must be emphasized that any neuroimaging abnormality must correlate with the clinical data for significance, as irrelevant findings may appear.

Positron Emission Tomography (PET) and Single Photon Emission Computed Tomography (SPECT), while disclosing neurobiological information, have not proven to be cost effective in evaluation of patients with complex partial seizures.

DISTINCTION FROM NON-EPILEPTIC CONDITIONS

Episodic Microsleep

Episodic diurnal microsleep may be represented by impaired concentration, decreased performance and irritability. More pronounced episodes, however, may share some symptoms with complex partial seizures, such as intermittent confusion, lapses of attention, staring and even semi-automatic actions. Prior to such episodes, however, the patient may feel fatigued and acknowledge that he is sleepy. An afferent stimulus will instantly alert the patient from microsleep, but would unlikely abort a complex partial seizure. A rapid descent into non-REM sleep on one or more EEGs would suggest the diagnosis of episodic microsleep. Such microsleep episodes usually represent poor sleep habits; less commonly they may represent narcolepsy. The latter might be diagnosed by the presence of other members of the narcolepsy tetrad, such as cataplexy. A REM onset sleep during EEG would help confirm the diagnosis.

Panic Attacks

Panic attacks last minutes to hours, during which the patient becomes pre-occupied by anxiety or fear of a horrible personal calamity. Sweating, trembling, palpitations of the heart, and abdominal distress may accompany these attacks. The patient may feel breathless and suffocated, and may hyperventilate and thus become light headed with possible tingling of the lips and extremities. Hyperventilation may cause confusion, tetany and even loss of consciousness.

Transient Global Amnesia

In this rare syndrome afflicting middle-aged and elderly persons, a sudden loss of memory for current and recent events renders the patient bewildered for several hours. Consciousness is preserved, intellectual activity (aside from memory) is intact and there is no language disorder. The attacks end abruptly but may recur in 20–25% of patients. The neurophysiology remains unknown and the EEG changes, if any, are non-specific.

Night Terrors and Somnambulism

Night terrors occur in neurologically normal children about 30 minutes after falling asleep. They occur during stage 3–4 non-REM sleep. The child sits up in bed, screams and breathes deeply with tachycardia. Visceral automatisms, such as chewing and swallowing, are absent. Somnambulism may accompany such attacks. Automatisms, such as turning on or off a light or rotating a door knob, may occur. Facial expression is blank, or may be fearful if a component of night terror is present.

Hypoglycemia

Episodic confusion may rarely represent hypoglycemia, but this is accompanied by premonitory weakness, sweating and pallor.

REFERENCES

1. Blume, W.T. Focal motor seizures and epilepsia partialis continua. In: E. Wyllie (Ed.) *The treatment of epilepsy: principles & practice*, Lea & Febiger, Philadelphia (1993) pp. 393–400.
2. Blume, W.T., Borghesi, J.L., and Lemieux, J.F. Interictal indices of temporal seizure origin, *Ann. Neurol.*, 34 (1993) 703–709.
3. Blume, W.T., Girvin, J.P., and Stenerson, P. Temporal neocortical role in ictal experiential phenomena, *Ann. Neurol.*, 33 (1993) 105–107.
4. Blume, W.T. and Kaibara, M. *Atlas of Adult Electroencephalography*, Raven Press, New York (1994) pp. 212–214.
5. Brodal, A. *Neurological Anatomy in Relation to Clinical Medicine*, Oxford University Press, New York (1981) pp. 660–663.
6. Commission on Classification and Terminology of the International League Against Epilepsy. Proposal for revised clinical and electroencephalographic classification of epileptic seizures, *Epilepsia*, 22 (1981) 489–501.
7. Evans, P. Henri Ey's concepts of the organization of consciousness and its disorganization: an extension of Jacksonian theory, *Brain*, 95 (1972) 413–440.
8. Gastaut, H. *Dictionary of Epilepsy*, World Health Organization, Geneva, Switzerland (1973).
9. Gastaut, H., Gastaut, J.L., Goncalves e Silva, E., and Fernandez Sanchez, G.R. Relative frequency of different types of epilepsy: a study employing the classification of the international league against epilepsy, *Epilepsia*, 16 (1975) 457–461.
10. Gates, J.R. and Gumnit, R.J. Partial seizures of temporal lobe origin. In: M. Dam and L. Gram (Eds.) *Comprehensive Epileptology*, Raven Press, New York (1990) pp. 187–195.
11. Gibbs, F.A., Gibbs, E.L., and Lennox, W.G. Epilepsy: a paroxysmal cerebral dysrhythmia, *Brain*, 60 (1937) 377–388.
12. Gloor, P. Temporal lobe epilepsy: its possible contribution to the understanding of the functional significance of the amygdala and of its interaction with neocortical-temporal mechanisms. In: B.E. Eleftheriou (Ed.) *The Neurobiology of the Amygdala*, Plenum Press, New York (1972) pp. 423–457.
13. Gloor, P., Olivier, A., Quesney, L.F., Andermann, F., and Horowitz, S. The role of the limbic system in experiential phenomena of temporal lobe epilepsy, *Ann. Neurol.*, 12 (1982) 129–144.

14. Hausser-Hauw, C. and Bancaud, J. Gustatory hallucinations in epileptic seizures. Electrophysiological, clinical and anatomical correlates, *Brain*, 110 (1987) 339–359.
15. Howell, D.A. Unusual centrencephalic seizure patterns, *Brain*, 78 (1955) 199–208.
16. Jackson, G.D., Berkovic, S.F., Tress, B.M., Kalnins, R.M., Fabinyi, G.C., and Bladin, P.F. Hippocampal sclerosis can be reliably detected by magnetic resonance imaging, *Neurology*, 40 (1990) 1869–1875.
17. Kellaway, P. The incidence, significance and natural history of spike foci in children. In: C. Henry (Ed.) *Current Clinical Neurophysiology*, Elsevier, Amsterdam (1981) pp. 151–175.
18. Kotagal, P. Psychomotor seizures: clinical and EEG findings. In: E. Wyllie (Ed.) *The Treatment of Epilepsy: Principles and Practices*, Lea & Febiger, Philadelphia (1993) pp. 378–392.
19. Kotagal, P., Luders, H., Williams, G., Wyllie, E., Nichols, T., and McPherson, J. Temporal lobe complex partial seizures: analysis of symptom clusters and sequences, *Epilepsia*, 29 (1988) 661.
20. Mark, V.H., Ervin, F.R., and Sweet, W.H. Deep temporal lobe stimulation in man. In: B.E. Eleftheriou (Ed.) *The Neurobiology of the Amygdala*, Plenum Press, New York (1972) pp. 485–507.
21. Penfield, W. and Jasper, H. *Epilepsy and the Functional Anatomy of the Human Brain*, Little, Brown and Company, Boston (1954) pp. 149.
22. Penfield, W. and Jasper, H. *Epilepsy and the Functional Anatomy of the Human Brain*, Little, Brown and Company, Boston (1954) pp. 524.
23. Rasmussen, T. Localizational aspects of epileptic seizure phenomena. In: R.A. Thompson and J.R. Green (Eds.) *New Perspectives in Cerebral Localization*, Raven Press, New York (1982) pp. 177–203.
24. Rocca, W.A., Sharbrough, F.W., Hauser, W.A., Annegers, J.F., and Schoenberg, B.S. Risk factors for complex partial seizures: a population-based case-control study, *Ann. Neurol.*, 21 (1987) 22–31.
25. Sadler, R.M. and Goodwin, J. Multiple electrodes for detecting spikes in partial complex seizures, *Can. J. Neurol. Sci.*, 16 (1989) 326–329.
26. So, N.K. Epileptic auras. In: E. Wyllie (Ed.) *The Treatment of Epilepsy: Principles and Practices*, Lea & Febiger, Philadelphia (1993) pp. 369–377.

3

DRUG THERAPY FOR COMPLEX PARTIAL SEIZURES

Joseph Bruni

Division of Neurology
St. Michael's Hospital, Wellesley Central Site
Jones Building, Suite 318
160 Wellesley Street East
Toronto, Ontario, Canada M4Y 1J3
tel: (416) 926-7707
fax: (416) 926-5176

Complex partial seizures present a major challenge for medical management. Dr. Bruni reviews the available drugs, including the novel anticonvulsants which have recently become available.

INTRODUCTION

Optimal medical management of the patient with complex partial seizures should take into account a number of factors, including the efficacy of a drug, its adverse effects profile, its pharmacokinetics, its ease of use, the potential for drug interactions, the age and sex of the patient, emergency use and cost. Only recently have data from controlled clinical trials become available comparing the relative efficacy and adverse effects of the standard antiepileptic drugs. It has become apparent that, because of a lack of significant differences in efficacy, other considerations play an important role in the choice of a specific antiepileptic drug.

In the initial therapy of patients with complex partial seizures, monotherapy should be the prime goal. If there is a lack of satisfactory response, and/or if the patient develops unacceptable adverse effects after an optimal dose has been achieved, alternate monotherapy should be considered. Only after failure of two or three major antiepileptic drugs as monotherapy should adjunctive or combination therapy be considered.

Although data are available on the relative efficacy of single antiepileptic drugs, data on "rational polytherapy"—as to which drug combinations would be preferable—are lacking. Often the choice of drug combinations is made empirically. Factors that might influence this decision, include the (different) mechanism of action of the antiepileptic drugs, a lack of drug interactions and a lack of synergistic or additive adverse effects. Combination therapy may be more effective than monotherapy in up to 30 percent of patients.

EFFICACY AND SAFETY: STANDARD DRUGS

Only a few comparative monotherapy trials have been performed with the standard antiepileptic drugs, phenytoin, carbamazepine, primidone, phenobarbital, and valproic acid.[7,14,19,20,23,34] Ramsay et al.[23] found no significant difference in efficacy between phenytoin and carbamazepine as initial therapy in adults. The overall rate of side effects was the same. Control was attained in 85 percent of patients on monotherapy. In a multicenter Veterans' Administration Study, 622 patients were treated with phenytoin, carbamazepine, phenobarbital or primidone monotherapy.[19] Patients were followed up to a mean of 36 months. Control of secondarily generalized tonic-clonic seizures did not differ significantly. Carbamazepine provided complete control of partial seizures more often than primidone or phenobarbital. The study recommended that overall carbamazepine and phenytoin are the first drugs of choice for monotherapy of adults with partial or secondarily generalized tonic-clonic seizures. Therapy with phenobarbital or primidone was less successful because of a higher incidence of adverse effects. Turnbull et al.[34] found no differences in efficacy between sodium valproate and phenytoin for previously untreated patients. There was no difference in time to a two-year remission or time to first seizure. Generally, patients with partial seizures did significantly less well than patients with tonic-clonic seizures alone. Fifty-one percent of patients treated with valproate achieved a two-year remission versus a 44 percent rate of two-year remission with phenytoin (no significant difference). This study found no difference in toxicity between the two drugs. Callaghan et al.[7] reported that there was no significant difference in efficacy between carbamazepine, phenytoin and sodium valproate in the treatment of partial seizures. Mattson et al.[20] compared the efficacy of carbamazepine and valproate and found that carbamazepine was superior to valproate in efficacy against complex partial seizures in some outcome measures. No differences, however, were detected in the number of patients that became seizure free. Heller et al.[23] found no significant differences in the efficacy of phenytoin, carbamazepine, phenobarbitone or sodium valproate in newly diagnosed adult patients with epilepsy. More patients had to be withdrawn from therapy with phenobarbitone because of side effects than patients treated with one of the other antiepileptic drugs. Thus, there seem to be few differences in efficacy among the standard antiepileptic drugs, although there are clear differences in adverse effects.

Adverse effects, predominantly central nervous system side effects, are relatively common with all the standard antiepileptic drugs especially during the early stages of treatment. Allergic skin reactions can be observed in 5 to 10 percent of patients. These are most frequently observed with phenytoin, carbamazepine, and the barbiturates. The barbiturates produce a higher incidence of sedation, cognitive side effects and sexual dysfunction. Cosmetic side effects have been most frequently reported for phenytoin. Idiosyncratic reactions such as pancreatitis have been reported with carbamazepine and valproate. Carbamazepine may rarely cause agranulocytosis. Valproate may rarely cause hepatic toxicity, especially in young children under the age of two with neurologic handicaps and on polytherapy. The adverse effects of antiepileptic drugs have been recently reviewed.[22]

All of the currently used antiepileptic drugs are associated with a higher incidence of congenital anomalies and malformations, although there does not appear to be a specific syndrome associated with any of them. Valproate[37] and carbamazepine,[27] however, have been associated with a 1–2 percent risk and a 0.5–1 percent risk respectively of neural-tube defects.

PHARMACOKINETIC CONSIDERATIONS

Pharmacokinetic considerations that influence the choice of an antiepileptic drug include the absorption features, the route and rate of elimination and the potential for drug interactions. Because of its saturation kinetics and poor water solubility, changes in the absorption of phenytoin may be observed with generic formulations. Carbamazepine generics may also have variable bioavailability, because of the poor water solubility of carbamazepine. Drugs with short half-lives, such as valproate and carbamazepine, generally have to be given in two or three divided daily doses.

Initiation of therapy should be done with small doses for drugs such as valproate, carbamazepine and the barbiturates. Dose increases should be gradual with phenytoin because of its saturation kinetics of elimination.

Drug interactions become an important consideration when the standard antiepileptic drugs are used concurrently. Phenytoin, carbamazepine and the barbiturates are generally enzyme inducers. Interactions with valproate are less common, although valproate may result in a higher phenobarbital concentration if they are used concurrently, requiring a reduction in dosage of phenobarbital.

COMBINATION THERAPY

Although approximately 60 to 70 percent of patients can be satisfactorily controlled with monotherapy, some patients benefit from combination or adjunctive therapy. The Veterans' Administration trial found that 40 percent of patients improved with the addition of a second drug and 9 percent achieved complete seizure control. Dean and Penry found that valproate added to carbamazepine improved seizure control in 39 percent of patients and 16 percent were completely controlled.[8] An earlier study[16] found that phenytoin or carbamazepine added to primidone or phenobarbital controlled 16 percent of patients not previously controlled. Rodin et al.[26] found that primidone added to carbamazepine or phenytoin controlled 17 percent of patients.

The use of adjunctive therapy has recently been stimulated by the introduction of new antiepileptic drugs for the treatment of complex partial seizures.[13] Clobazam, vigabatrin, gabapentin, lamotrigine and topiramate are recently introduced antiepileptic drugs that are playing a greater and greater role in the treatment of complex partial seizures.

THE ROLE OF THE NEW ANTIEPILEPTIC DRUGS

Clobazam

Clobazam is a 1,5 benzodiazepine that was introduced in clinical trials in 1977. It has been shown to be effective adjunctive therapy in a variety of seizure types, including both primary generalized and partial seizures. Eight double blind and many open studies have demonstrated the efficacy of clobazam.[30] A summary of the data from the open studies suggests that 30 percent of all patients became seizure free and that improvement in seizure frequency occurs in up to 95 percent of patients. The results of controlled studies are less satisfactory.

Clobazam is generally well tolerated, but side effects include sedation, irritability, mood swings, depression, dizziness, fatigue, and weight gain. Adverse effects are reported in 10 to 50 percent of patients, although generally these are mild and do not require discontinuation of therapy.

Clobazam can be recommended as adjunctive therapy when there is a lack of satisfactory control with the major antiepileptic drugs. A limiting factor in its use is the development of tolerance, which can occur in up to 50 percent of responding patients. Clobazam is not available in the United States.

Vigabatrin

Vigabatrin, a GABA-T inhibitor, has been evaluated in over 22 controlled and numerous open studies.[4,12,13,15,24,25,31,33] A greater than 50 percent reduction in seizure frequency has been reported in approximately 50 percent of refractory patients. Approximately 5–10 percent of refractory patients become seizure free. In monotherapy trials,[15] 60 percent of newly diagnosed patients were equally successfully treated with either vigabatrin or carbamazepine.

In almost all clinical trials, vigabatrin has been well tolerated. Allergic reactions are rare. Dose-related adverse effects include drowsiness, headaches and ataxia. No adverse effects on cognition have been reported. Behavioural or psychiatric side effects have been reported in only a small number of patients. Recently there has been concern about the retinal toxicity of vigabatrin with constriction of visual fields. At present the number of pregnancies reported is small and no firm conclusion can be drawn concerning teratogenicity.

Vigabatrin can be recommended as adjunctive therapy when there is a lack of satisfactory control with the major antiepileptic drugs or the other new agents.

Gabapentin

Gabapentin has been assessed as adjunctive therapy in the treatment of refractory partial seizures. Monotherapy trials are currently ongoing, and preliminary data indicate that some patients can achieve monotherapy. Gabapentin's mechanism of action remains to be defined. The U.S. Gabapentin Study Group[36] found that gabapentin-treated patients had a significantly greater reduction in seizure frequency as compared to placebo. Eight percent of placebo-treated patients, versus 18 to 26 percent of gabapentin-treated patients, had at least a 50 percent reduction in seizure frequency. The International Gabapentin Study Group[1] reported a −17.8 to −21.8 percent change in seizure frequency on gabapentin as compared to a −0.3 percent change in the placebo group. The U.K. Gabapentin Study Group[35] reported a response rate of 25 percent for patients receiving gabapentin, compared with 9.8 percent for patients who received placebo. Long-term studies have demonstrated that tolerance does not develop and that gabapentin is well tolerated. In an open label study involving less refractory patients, 71 percent of patients had a 50 percent or greater reduction in seizures and 46 percent became seizure free.[6]

In virtually all clinical trials, gabapentin has been found to be a safe antiepileptic drug. Serious toxicity is exceedingly rare. The most frequently observed side effects include drowsiness, dizziness, fatigue, and ataxia. Gabapentin does not appear to have a negative effect on cognitive function. The teratogenic effects are not fully known. To date, 10 pregnancies have been reported: 6 were electively terminated and 4 pregnancies resulted in normal births. Similar to clobazam and vigabatrin, gabapentin can be recommended as adjunctive therapy when there is a lack of satisfactory control with the major antiepileptic drugs. Drug interactions do not occur.

Lamotrigine

Lamotrigine, a compound that stabilizes presynaptic neuronal membranes by blocking voltage-dependent sodium channels, has demonstrated efficacy against refractory partial seizures in 10 double-blind placebo-controlled clinical trials.[17] A 17 to approximately 60 percent decrease in seizure frequency compared to placebo has been reported. The use of lamotrigine in children has to be further explored. More recent data from open clinical trials suggest that lamotrigine may have broad-spectrum activity against various types of generalized seizure disorders.[9,21] A recent study comparing lamotrigine monotherapy to carbamazepine monotherapy in 260 newly diagnosed patients with partial or tonic-clonic seizures suggested similar efficacy with lamotrigine being better tolerated.[3] A recent study comparing lamotrigine monotherapy to phenytoin monotherapy also suggested similar efficacy.[32]

The commonest reported adverse effects of lamotrigine include nausea and vomiting, headaches, dizziness, tremor, and ataxia. Sedation is uncommon. Cognitive dysfunction does not appear to occur. Rash occurs in approximately 4 to 5 percent of patients, and represents the most frequent reason for discontinuation of therapy. Rarely, a Stevens-Johnson syndrome can occur. Skin reactions are more likely to occur in children, with high initial doses, or with rapid titration of dose. Valproic acid co-therapy appears to be an added risk factor.

Lamotrigine has favourable pharmacokinetics with a long half-life of approximately 30 hours. Phenytoin, the barbiturates and carbamazepine shorten its half-life, and valproate lengthens the half-life. To date, approximately 50 pregnancies have been reported. There does not appear to be a specific teratogenic effect, although additional data are required.

Topiramate

Topiramate is a new compound with multiple mechanisms of action. These include enhancement of GABA currents, blockage of AMPA/kainate glutamate receptors, and limitation of repetitive firing by blocking voltage-dependant sodium channels.[29] Double-blind placebo controlled studies in children and adults have demonstrated the efficacy and safety of topiramate in patients with refractory partial seizures.[5,10,18] More recent data suggest that topiramate has broad spectrum activity against primary generalized tonic-clonic seizures.[2]

To date there has been no evidence of serious systemic side effects. Renal stones were reported in 1.5 percent of patients. The most frequently observed side effects are CNS in nature and include dizziness, ataxia, fatigue, nervousness, abnormal thinking and language disturbances. Weight loss may be observed in 10 to 20 percent of patients. The teratogenic effects are not full known. Similar to the other new antiepileptic drugs,

Table 1. Medical Treatment of Complex Partial Seizures

Standard Drugs	*New Anticonvulsants*
Carbamazepine	Clobazam
Valproate	Vigabatrin
Phenytoin	Gabapentin
	Lamotrigine
	Topiramate

topiramate can be recommended as adjunctive therapy when there is lack of satisfactory control with the major antiepileptic drugs. Hepatic enzyme inducing antiepileptic drugs stimulate the metabolism of topiramate and topiramate may increase contraceptive failure of oral contraceptives.

Felbamate

Although felbamate has demonstrated efficacy as adjunctive therapy and as monotherapy in the treatment of partial seizures, it should only be used with extreme caution, when all other therapies fail. Aplastic anemia and hepatic failure occurred at a rate of approximately 1 in 2,000.[11,28]

CONCLUSION

Complex partial seizures remain a challenging problem for medical management. Although a significant number of patients can be satisfactorily controlled, patients with this seizure type are some of the most refractory to medical treatment. Phenytoin, carbamazepine and valproate are major antiepileptic drugs in the treatment of this seizure type. In choosing one drug over another, factors such as potential adverse effects and drug interactions also have to be taken into account along with clinical efficacy. The role of the new antiepileptic drugs, such as clobazam, vigabatrin, gabapentin, lamotrigine and topiramate has to be further refined. More comparative data are required before definite conclusions can be reached concerning the relative merits and efficacy of these compounds as initial monotherapy. It is hoped that the medical management of complex partial seizures will continue to improve, and it is with interest that we will follow the clinical development of newer compounds such as oxcarbazepine, tiagabine, and remacemide.

REFERENCES

1. Anhut, H., Ashman, P., Feuerstein, T.J., Sauermann, W., Saunders, M., and Schmidt, B. Gabapentin as add-on therapy in patients with partial seizures: A double-blind placebo-controlled study, *Epilepsia*, 35 (1994) 795–801.
2. Biton, V., Montouris, G.D., Ritter, F., Riviello, J.J., Reife, R., Lim, P., and Pledger, G. A randomized placebo-controlled study of topiramate in primary generalized tonic clonic seizures, *Neurology*, 53 (1999) 1330–1337.
3. Brodie, M.J., Richens, A., Yuen, A.W.C., and the UK Lamotrigine/Carbamazepine Monotherapy Trial Group. Double-blind comparison of lamotrigine and carbamazepine in newly diagnosed epilepsy, *Lancet*, 345 (1995) 476–479.

4. Browne, T.R., Mattson, R.H., and Penry, J.K. Multicenter long-term safety and efficacy study of vigabatrin for refractory complex partial seizures: An update, *Neurology*, 41 (1991) 363–364.
5. Bruni, J. Efficacy of topiramate, *Can. J. Neurol. Sci.*, 25 (1998) 56–57.
6. Bruni, J. Outcome evaluation of gabapentin as add-on therapy for partial seizures, *Can. J. Neurol. Sci.*, 25 (1998) 134–140.
7. Callaghan, N., Kenny, R.A., O'Neill, B., Crowley, M., and Gogginl T. A prospective study between carbamazepine, phenytoin and sodium valproate as monotherapy in previously untreated and recently diagnosed patients with epilepsy, *J. Neurol. Neurosurg. Psychiatry*, 48 (1985) 639–644.
8. Dean, C. and Penry, J.K. Carbamazepine valproate therapy in 100 patients with partial seizures failing carbamazepine monotherapy: Long term follow up, *Epilepsia*, 29 (1988) 687.
9. Dulac, O. The use of lamotrigine in children, *Rev. Contemp. Pharmacother.*, 5 (1994) 133–139.
10. Elterman, R.D., Glauser, T.A., Wyllie, E., Reife, R., Wu, S.C., and Pledger, G. A double-blind, randomized trial of topiramate as adjunctive therapy for Daihol-onset seizures in children, *Neurology*, 52 (1998) 1338–1334.
11. Faught, E., Sachdeo, R., Remler, M., Chayasirisobhon, S., Iragui-Madoz, V.J., Ramsay, R.E., Sutula, T.P., Kanner, A., Harner, R.N., Kuzniecky, R., Kramer, L.D., Kamin, M., and Rosenberg, A. Felbamate monotherapy for partial onset seizures: An active control study, *Neurology*, 43 (1993) 688–692.
12. Gram, L., Klosterskov, P., and Dam, M. Gamma-vinyl GABA: A double-blind placebo controlled trial in partial epilepsy, *Ann. Neurol.*, 17 (1985) 262–266.
13. Guberman, A., Bruni, J., Deforges, C., Jones, M., Lee, M.A., McLaughlan, R., Ogunyemi, A., Pillay, N., Purves, S., Sadler, M., Starreveld, E., Sussman, N., Vachon, L., and Weaver, D. Vigabatrin in uncontrolled complex partial seizures in adult patients with epilepsy: A double-blind, placebo-controlled, dose-ranging study, *Can. J. Neurol. Sci.*, 21 (1994) Suppl. 2: S17–S18.
14. Heller, A.J., Chesterman, P., Elwes, R.D.C., Crawford, P., Chadwick, D., Johnson, A.L., and Reynolds, E.H. Phenobarbitone, phenytoin, carbamazepine, or sodium valproate for new diagnosed adult epilepsy: A randomized comparative monotherapy trial, *J. Neurol. Neurosurg. Psychiatry*, 58 (1995) 44–50.
15. Kalviainen, R., Aikia, M., Saukkonen, A.M., Mervaala, E., and Riekkinen, P.J. Sr. Vigabatrin vs carbamazepine monotherapy in patients with newly diagnosed epilepsy, *Arch. Neurol.*, 52 (1995) 989–996.
16. Kosteljanetz, M., Christiansen, J., Dam, A.M., Hansen, B.S., Lyon, B.B., Pedersen, H., and Dam, M. A controlled clinical trial in focal motor and generalized epilepsy, *Arch. Neurol.*, 36 (1979) 22–24.
17. Leach, J.P. and Brodie, M.J. Lamotrigine: Clinical use. In: R.H. Levy, R.N. Mattson, and B.S. Meldrum (Eds.) *Antiepileptic Drugs*, Raven Press (1995) 889–895.
18. Marson, A.G., Kadir, Z.A., and Chadwick, D.W. New antiepileptic drugs: a systematic review of their efficacy and tolerability, *BMJ*, 313 (1996) 1169–1174.
19. Mattson, R.H., Cramer, J.A., Collins, J.F., Smith, D.B., Delgado-Escueta, A.V., Browne, T.R., Williamson, P.D., Treiman, D.M., McNamara, J.O., McCutchen, C.B., Homan R.W., Crill, W.E., Lubozynski, M.F., Rosenthal, N.P., and Mayersdorf, A. Comparison of carbamazepine, phenobarbital, phenytoin and primidone in partial and secondarily generalized tonic-clonic seizures, *N. Engl. J. Med.*, 313 (1985) 145–151.
20. Mattson, R.H., Cramer, J.A., and Collins, J.F. A comparison of valproate with carbamazepine for the treatment of complex partial seizures and secondarily generalized tonic-clonic seizures in adults, *N. Engl. J. Med.*, 327 (1992) 765–771.
21. Panayiotopoulos, C.P., Ferrie, C.D., Knott, C., and Robinson, R.O. Interaction of lamotrigine with sodium valproate, *Lancet*, 341 (1993) 445.
22. Plaa, G.L. and Willmore, L.J. General Principles: Toxicology. In: R.H. Levy, R.H. Mattson, and B.S. Meldrum (Eds.) *Antiepileptic Drugs*, Raven Press (1995) pp. 51–60.
23. Ramsay, R.E., Wilder, B.J., Berger, J.R., and Bruni, J. A double-blind study comparing carbamazepine with phenytoin as initial seizure therapy in adults, *Neurology*, 33 (1983) 904–910.
24. Reynolds, E.H., Ring, H.A., Farr, I.N., Heller, A.J., and Elwes, R.D. Open double-blind and long term study of vigabatrin in chronic epilepsy, *Epilepsia*, 32 (1991) 530–538.
25. Rimmer, E.M. and Richens, A. Double-blind study of gamma-vinyl GABA in patients with refractory epilepsy, *Lancet*, 1 (1984) 189–190.
26. Rodin, E., Rim, C.S., Kitano, H., Lewis, R., and Rennick, P.M. A comparison of the effectiveness of primidone versus carbamazepine in epileptic patients, *J. Nerv. Ment. Dis.*, 163 (1976) 41–46.
27. Rosa, F.W. Spina bifida in infants of women treated with carbamazepine during pregnancy, *N. Engl. J. Med.*, 324 (1991) 674–677.

28. Sachdeo, R., Kramer, L.D., et al. Felbamate monotherapy: Controlled trial in patients with partial onset seizures, *Annals of Neurology*, 32 (1992) 386–392.
29. Schneiderman, J.H. Topiramate: Pharmacokinetics and pharmacodynamics, *Can. J. Neurol. Sci.*, 25 (1998) 53–55.
30. Shorvon, S.D. Benzodiazepines: Clobazam. In: R.H. Levy, R.H. Mattson, and B.S. Meldrum (Eds.) *Antiepileptic Drugs*, Raven Press (1995) pp. 763–777.
31. Sivenius, J., Ylinen, A., Murros, K., Mumford, J.P., and Riekkinen, P.J. Vigabatrin in drug resistant partial epilepsy: A 5 year follow-up study, *Neurology*, 41 (1991) 562–565.
32. Steiner, T.J., Dellaportas, C.I., Findley, L.J., Gross, M., Gibberd, F.B., Perkin, G.D., Park, D.M., and Abbott, R. Lamotrigine monotherapy in newly diagnosed untreated epilepsy: a double-blind comparison with phenytoin, *Epilepsia*, 40 (1999) 601–607.
33. Tassinari, C.A., Michelucci, R., Ambrosetto, G., and Salvi, F. Double blind study of vigabatrin in the treatment of drug resistant epilepsy, *Arch. Neurol.*, 44 (1987) 688–692.
34. Turnbull, D.M., Howel, D., Rawlins, M.D., and Chadwick, D.W. Which drug for the adult epileptic patient: Phenytoin or valproate? *Lancet*, 290 (1985) 815–819.
35. UK Gabapentin Study Group. Gabapentin in partial epilepsy, *Lancet*, I (1990) 1114–1117.
36. US Gabapentin Study Group. Gabapentin as add-on therapy in refractory epilepsy: A double-blind, placebo-controlled, parallel-group study, *Neurology*, 43 (1993) 2292–2298.
37. Valproic acid and spina bifida. Preliminary reports, *M.M.W.R.*, 31 (1982) 565–566.
38. Wilder, B.J. (Ed.) Rational polytherapy in the treatment of epilepsy, *Neurology*, 45 (1995) Suppl 2.

4

SURGERY FOR COMPLEX PARTIAL SEIZURES

John P. Girvin

London Health Sciences Centre
University Campus
339 Windermere Road
London, Ontario, Canada
N6A 5A5
tel: (519) 663-3664
fax: (519) 663-3398

Dr. Girvin discusses the technique of surgery for complex partial seizures, some of the questions and controversies related to this technique and the probability of a successful outcome.

INTRODUCTION

In this paper, my intention is to cover many of the generic aspects of the surgery for complex partial seizures (CPSs), as opposed to specific results. Results will be referred to, but the paper will not consist of an analysis of these results in detail. The rationale for the resective surgery of epilepsy is embodied in the methodology of that resection, i.e., the use of the so-called technique of subpial dissection. The rationale for this, the implications of it, and the varying techniques which may be used in carrying it out will be discussed. The surgical anatomy of the anterior temporal lobe, the commonest resection for CPSs, will be referred to specifically, because of the implications in the removal of seizure foci in the temporal limbic structures with respect to the consequent relief of seizures. Finally, there are a number of controversies that exist within the field of surgery for CPSs and these will be referred to briefly.

TECHNIQUE OF RESECTION

The technique of resection in epilepsy surgery differentiates epilepsy surgery from general neurosurgery involving the same areas of brain resection, e.g., removal of tumours, arteriovenous malformations (AVMs), etc. This is not to say that the same technique may not be used in general neurosurgery, but only that this is not the rule.

The technique of subpial dissection has as its rationale: 1) the removal of the epileptogenic cortex; 2) the minimization of the morphological (gliotic) and physiological (deafferented neurons) surgical scar left by the subpial removal; and 3) the preservation of the blood supply to the cortex which is left intact. The technique itself consists of the removal only of cortex which is below the pia mater, i.e., "subpial". Providing that the narrow subarachnoid spaces in the sulci are not entered, then blood vessels within the sulci are left intact. Hence, there should be no compromise of the blood supply which these "en passage" vessels carry to other parts of the cortex which may be left intact. The technique involves, where possible, the use of sulci as boundaries to the resection, as a resection line at the base of a sulcus will result in the smallest scar and the smallest likelihood of leaving behind deafferented neurons. Obviously, some parts of the resection line will be perpendicular to the axes of the sulci or fissures, and, in these cases, the white matter is cut across as cleanly as possible, usually with bipolar coagulation forceps on very low current. The "cleaner" such a resection line can be made, the smaller will be the residual scar on the intact part of the resection line which is left behind. This residual scar will be kept to a minimum by the absence of undermining of intact cortex, the absence of mechanical manipulation of the resection line and by the use of low current in cauterization.

Adherence to the foregoing principles of resection will lead to the greatest benefit from the point of view of removal of epileptic scar, at the lowest cost, i.e., the residual scar left behind. As with any trauma to the brain, however, one must be aware that there will be an irreducible minimum of "success" with such surgery. That is to say, no matter how much the surgery is optimized or how little scar is left behind, the fact is that there *is* scar left behind and this scar can give rise to future seizures. At a minimum, this is probably little more than 5% or thereabouts.

The other point which is worth emphasizing—which is more practical, but certainly of interest to the non-surgical neuroscientist as well as the surgeon—is the question of how much tissue is available from such a resection for the pathologist to look at and what is the quality of that tissue. There have been a number of technical advances in neurosurgery, one being the ultrasonic aspirator (CUSA). This instrument is now used by a number of epilepsy surgeons in carrying out their subpial resections. The resection line, of course, is not made by a scalpel beneath the pial surface. Traditionally this has been carried out by the use of a combination of suction and electrocautery. With fine suction at relatively low power, this is done with probably no more than the loss of a millimetre or two of tissue along the resection line. The use of the ultrasonic aspirator may increase this considerably. This becomes particularly pertinent with respect to the removal of the mesial temporal structures. That is to say, if these structures are removed by mechanically separating the parenchyma from the underlying leptomeningeal coverings, then there will be a relatively completely intact tissue of good quality submitted for histological examination by the pathologist. Once again, reference must be made to the ultrasonic aspirator, for, when this is used, there may be a loss of tissue which is much greater than with other methods of dissection.

If the ultrasonic aspirator is used to remove all of the tissue from the leptomeninges without there being intact pathology, or if the cortex is extremely adherent to the leptomeninges, then the removal may not leave an ideal specimen for morphological examination. Under these circumstances, as pointed out to me by Dr. Rasmussen many years ago, the surgeon must record carefully his or her intraoperative findings for these will represent the "gross" morphological abnormalities of the resected tissue. As most surgeons know, the subtle gross abnormalities of the mesial temporal structures are much more difficult to discern with any degree of certainty then are abnormalities of the neocortex.

CORTICECTOMY FOR COMPLEX PARTIAL SEIZURES OF EXTRATEMPORAL ORIGIN

As already indicated in the foregoing, foci in the temporal lobe are the commonest causes of CPSs. Surgery for CPSs of extratemporal origin usually involves some type of removal of neocortex. This may involve a whole lobe, e.g., lobectomy, or a portion of the lobe, e.g., corticectomy or topectomy. The surgical principle involved in corticectomies is the removal of the gray matter, or "neocortical mantle". In such resections, there is an attempt to: 1) preserve all "en passage" vessels coursing through the area of resection (particularly important in the case of inferior rolandic and parieto-occipital resections); and 2) limit the depth of the resection to the depths of the associated sulci (particularly important in the parieto-occipital or occipital resections).

ANTERIOR TEMPORAL LOBECTOMY

Surgical Anatomy

At most of the international conferences dealing with epilepsy surgery, there have been sessions devoted to a discussion of the surgical anatomy of the temporal lobe. A fact which is well known to the comprehensive epilepsy programs, pertains to the efficacy of an anterior temporal lobectomy with respect to the removal of certain structures in the temporal lobe. The complexity of the anatomy of the mesial temporal lobe, the importance of the recognition of this complex surgical anatomy by the operating surgeon, and the fact that the majority of CPSs of temporal lobe origin originate in these structures account for the importance of this consideration. The removal of the neocortex of the temporal lobe is really not dissimilar, with respect to complexity, to the removal of one of the other lobes of the brain. However, unlike other such removals where the primary removal is devoted to neocortex, the removal of the mesial, or limbic, temporal lobe structures requires intimate knowledge of the surgical anatomy of this area.

The complexity of the surgical anatomy and the importance of familiarity with the anatomy of this area can be appreciated to some extent by a consideration of the structures which remain in the resection bed following a temporal lobectomy, and which are separated from the previously removed mesial temporal structures only by the leptomeninges. These include: 1) the basal vein of Rosenthal; 2) the anterior choroidal artery; 3) the posterior communicating and posterior cerebral arteries; 4) the third cranial nerve; 5) the cerebral (midbrain) peduncle; and 6) very occasionally the 4th cranial nerve and/or the optic tract.

The reason for the emphasis on the mesial anatomy of the temporal lobe is that the investigation of intractable seizures following a failed so-called anterior temporal lobectomy frequently results in the demonstration of the retention of various parts of the amygdala and the hippocampus. It has been the experience of this author that this can very commonly happen in those centres where very few cases are being done and/or where the surgeons have not had the benefit of training with individuals who have had a large experience in the radical removal of infero-mesial temporal lobe structures.

Perhaps the most striking personal example of this was in a young man who had had such a temporal lobectomy and who was referred some nine years later with continuing intractable seizures. This was in the days prior to the magnetic resonance (MR) scan and hence it was very difficult to know how much of the infero-mesial temporal lobe might or might not have been intact. In any case, the focus was determined to be in the temporal area on the same side as the previously removed temporal lobe. An operation was undertaken with the view of perhaps extending the resection line more posteriorly. Intraoperatively, however, it was found that of at least two-thirds of the medial amygdala was intact and the hippocampus was nearly completely intact. Further, the hippocampus contained a small benign tumour, the removal of which resulted in a permanent seizure-free status.

In the removal of the infero-mesial structures, the amygdala can only be radically and safely removed by determining not only the easily exposed (superficial) superior aspect of the temporal stem at the insula, but also by clearly defining the anterior extent of the temporal stem. Dissection around the anterior stem onto its underside will then allow easy definition of the deep (inferior) aspect of the temporal stem. This latter subpial dissection around the anterior part of the stem to its underside exposes the junction of the most superior and medial part of the amygdala with the stem. (The reader will be aware that the uncus can really be considered as the anterior part of the parahippocampal gyrus topographically.) Once having identified both the superior and the inferior aspects of the temporal stem, this allows the surgeon to remove the temporal stem and the medial part of the amygdala by joining these two lines by whatever dissection technique is being used, usually bipolar coagulation forceps. As this resection is tediously carried posteriorly it will merge with the anterior part of the hippocampus and uncus, and the anterior part of the choroidal fissure inferiorly.

The hippocampus similarly can only be removed by the clear identification of the medial extent of the resection, which will be the choroidal fissure. Once this is identified, then a purely subpial resection carried laterally from this fissure, beginning with the fimbria, will be naturally carried to the hippocampal sulcus (medial side). A similar resection carried from the lateral part of the posterior resection line will lead to the lateral lip of the hippocampal sulcus. This will complete the posterior resection line and thus separate the hippocampus from its posterior projections. Following this, the hippocampus can be removed from its bed by subpial mechanical separation of the hippocampus from its investing leptomeninges. Coagulation of a few small hippocampal vessels entering the sulcus from the anterior choroidal artery will complete the removal.

"Tailoring" of the Temporal Lobectomy?

There are a number of controversies related to the investigation and the surgery of CPSs. One of those controversies pertains to whether one carries out a typical "standard" temporal lobectomy or whether one considers "tailoring" the temporal lobectomy. In the standard lobectomy, the surgery consists of a 3.5–5 cm removal of the lateral neocortex, as measured from the tip of the temporal pole, plus the amygdala and varying extents of the hippocampus. The proponents of the standard lobectomy claim: 1) that there is really no difference in the results between a standard and a "tailored" lobectomy; 2) that such a lobectomy can be carried out under general anesthesia; and 3) that a lobectomy carried out under general anesthesia may be more complete and qualitatively better than one carried out under local anesthesia.

The so-called "tailored" lobectomy utilizes intraoperative electrocorticography (ECoG). The epileptogenic activity identified on the ECoG guides the extent of resection. Thus, the lobectomy is "tailored" in that it is designed to remove only that parenchyma which has been identified by the ECoG as being involved with epileptogenic electrical discharge. Usually such a lobectomy is carried out under local anesthesia and, in the dominant hemisphere, involves mapping of the temporal speech area, and, in some cases, other neuropsychological functions by electrocortical stimulation.

MEMORY AND TEMPORAL LOBECTOMY

Beginning in 1980, temporal lobectomies were carried out when a clear focus could be determined, even in patients who exhibited bilateral failure of the carotid amytal test for memory. Initially this was undertaken with a great deal of trepidation and with extensive intraoperative stimulation of the hippocampus during its removal in stages. More recently, surgery has been undertaken without such stimulation, because there has not been any evidence that a global amnestic deficit results from this practice.

RESULTS OF SURGERY FOR COMPLEX PARTIAL SEIZURES

Complex partial seizures may arise from any part of the brain. The most common CPS disorders, however, arise from the temporal lobe. The resective surgery for this is the anterior temporal lobectomy. In the temporal lobectomy, there are varying degrees of removal of the neocortex and infero-mesial temporal (limbic) structures. In general, the results of resective surgery carried out in comprehensive epilepsy units are not really dissimilar. In individuals in whom seizures are not abolished, however, there are obviously some differences of opinion with respect to what is a "satisfactory" result and what should be deemed an "unsatisfactory" result. My own personal view is that the operation should be considered truly successful only when the patient has been seizure free for two years off of all anticonvulsant medication. It is only this type of result which leaves the individual with no stigmata of epilepsy. This is not to say that such individuals may not go on to have a seizure in the future or that they may not have to be placed on anticonvulsant medication or, in some circumstances, even become intractable again. This is certainly the exception, however, and not the rule in those who have been seizure free and off medication for at least two years.

A case can certainly be made for an "excellent" result in individuals in whom only the presence of some anticonvulsant medication is required in order to ensure the absence

Table 1. Results of Surgery for Complex Partial Seizures

Origin of Seizures	Probability of Successful Outcome
Temporal Lobe	80–85%
Frontal Lobe	50–75%
Posterior Hemisphere	25–50%
Rolandic	5–10%

of seizures or in those few cases in which only insignificant auras (simple partial seizures) exist, with or without medication. When any seizures persist post-operatively, however, it then becomes a matter of philosophy as to how much such individuals are helped. In general, under such circumstances, the individual remains with the stigmata of the epileptic. That is to say, they are unable to drive. They are unable to hold certain jobs. They must take all precautions, no matter how slight, which pertain to what can and cannot be done alone during the course of leisure-time activities. This begs the question, therefore, as to how much such patients are helped.

Putting aside the foregoing, if one reads the literature and accepts the fact that "good", or "satisfactory", results include individuals who may have auras, the "odd" seizure, or "better than a 95% reduction" in their preoperative seizure frequency, then in most good comprehensive epilepsy units temporal-lobe surgery will result in something of the order of an 80–85% "success rate".

There is no doubt that the surgery for complex partial seizures of origin other than the temporal lobe is not as successful. In a qualitative estimate of our own results, the success of seizure surgery in seizures of frontal-lobe origin would probably be something of the order of 50–75%, in seizures of posterior hemisphere origin 25–50%, and in those unusual cases which appear to have a rolandic origin probably less than 5–10%.

5

WEST'S SYNDROME (INFANTILE SPASMS)

Clinical Description and Diagnosis

Richard A. Hrachovy

Baylor College of Medicine
Department of Neurology
One Baylor Plaza
Houston, TX 77030
tel: (713) 790-3716
fax: (713) 793-1574

West's syndrome is a serious epileptic syndrome which usually begins in the first year of life. It involves seizures known as "infantile spasms". Dr. Hrachovy describes the disorder and its symptoms.

INTRODUCTION

Infantile spasms are a unique disorder, peculiar to infancy and early childhood. These seizures, which have been recognized as an epileptic phenomenon since they were first described by William West in 1841,[53] are referred to by a variety of names, including "flexion spasms", "jackknife seizures", "massive spasms" and "infantile myoclonic seizures". In 1952, Gibbs and Gibbs[16] described the interictal EEG pattern, hypsarrhythmia, which was noted to occur in a large number of infantile spasms patients. Most patients with infantile spasms have some degree of mental and developmental retardation, and the triad of infantile spasms, hypsarrhythmia and retardation has become known as "West's syndrome". In 1958, a major breakthrough in the treatment of this disorder was reported by Sorel and Dusaucy-Bauloye[48] who noted improvement in the EEG and cessation of spasms in patients treated with adrenocorticotrophic hormone (ACTH).

The incidence of infantile spasms in the United States has been estimated to be 1/4,000–6,000 live births.[50] Onset of spasms is usually between 4 and 8 months, although they may appear as early as the first week of life, or as late as 3 years of age (Fig. 1). A familial occurrence of infantile spasms is rare, and appears to be related not to epilepsy, but to the basic disease processes that are associated with this disorder (e.g., tuberous

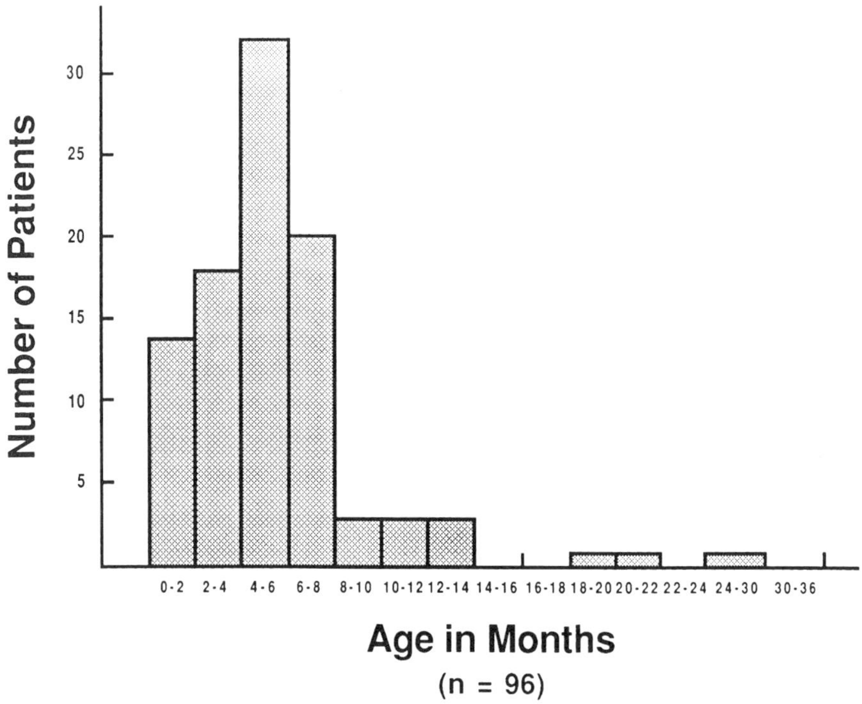

Figure 1. Age at onset of infantile spasms in 96 infants. (Reproduced from ref. 19 by permission)

sclerosis). Infantile spasms have been reported to occur in both monozygotic and dizygotic twins.[6,43,46] There is no clear evidence to indicate that one sex is afflicted with this disorder more than the other.[36]

Over the past four decades, a considerable world literature concerning infantile spasms has accumulated.[36] The literature has been replete with disagreements and controversies, however, largely because clinical descriptions of the seizures and evaluation of the effects of various treatment modalities have relied primarily on routine bedside observations of parents and other caretakers. With the introduction of long-term polygraphic/video monitoring techniques in the 1970s,[12] our understanding of the clinical manifestations of this disorder was greatly enhanced. Such monitoring techniques have provided an objective method of evaluating the acute effects of therapy on the EEG and seizure frequency.

Since 1976, our group has performed long-term polygraphic/video monitoring studies on more than 200 infantile spasm patients. In this chapter, I will review some of the data concerning the clinical and EEG features of this disorder that we have accumulated over the past two decades.

CLINICAL MANIFESTATIONS

An infantile spasm is a brief contraction involving muscles of the neck, trunk, and extremities, usually bilaterally and symmetrically. Polygraphically, the muscle action in an infantile spasm consists of an initial phasic contraction lasting less than 2 seconds,

Table 1. Cardinal Symptoms/Diagnostic Criteria for Infantile Spasms

Clinical Spasm	A brief synchronous contraction of the flexor and/or extensor muscles of the head, trunk and extremities. The spasm consists of an initial phasic contraction usually lasting less than two seconds, followed by a less intense, but more sustained, tonic contraction lasting less than 10 seconds. Only the phasic component may be seen.
	There is marked variability in the intensity of contractions. Minimally intense or fragmentary spasms are common.
	Spasms usually occur in clusters upon arousal from sleep, but many patients exhibit isolated spasms as well.
Interictal EEG	A variety of interictal EEG patterns may be seen, including focal and diffuse slowing and/or generalized, focal or multifocal epileptiform discharges. Rarely, the interictal EEG may be normal.
	The most common interictal EEG finding is hypsarrhythmia or one of its variations.
Ictal EEG	Ictal EEG patterns consist of attenuation episodes, generalized slow-wave transients, generalized sharp- and slow-wave transients, and rhythmic faster frequencies, which may occur singly or in various combinations.
	The most common ictal EEG change is a high-voltage generalized slow-wave transient, followed by attenuation of the background activity in all regions.
Age at Onset	1 week to 3 years, usually between 4–8 months.

followed by a less intense, but more sustained, tonic contraction lasting from about 2 to 10 seconds. The latter tonic phase may not be present, and the spasm may consist only of a brief phasic contraction. The character of the spasm depends on whether the flexor or extensor muscles are predominantly affected, and on the distribution of the muscle groups involved.[12,35] Positions of the body (e.g., sitting vs supine) may influence the type of spasm seen. The intensity of the spasm may vary from a massive contraction of all muscle groups to an isolated contraction of an individual muscle (e.g., abdominal recti).

There are three types of motor spasms: flexor, extensor, and mixed flexor/extensor. In our monitoring experience, the mixed type is most frequent (about 42%), followed by flexor spasms (about 34%), while extensor spasms are the least common (about 23%).

The fully expressed flexor spasm consists of flexion of muscles of the neck, trunk and extremities. The contraction of the abdominal muscles may be so forceful as to cause the torso to jackknife at the waist (i.e., "jackknife seizure"). Involvement of the muscles of the arms may result in either adduction (self-hugging motion) or abduction of the arms to either side of the head with the arms flexed at the elbow. The force of the contraction varies, as does the extent of the musculature involved. If the contraction is confined to the neck muscles, the patient may only bob his head (head nodding attack). Similarly, the spasms may involve only the upper limb girdle, with elevation and adduction of the shoulders in a shrug-like movement.

Extensor spasms ("cheerleading seizures") are consequent to predominate muscle contractions of extensor muscles of the neck and trunk, with extensor adduction or abduction of the arms, legs, or both.

Mixed flexor/extensor spasms usually involve flexion of the neck, trunk, and arms with extension of the legs, or, less commonly, flexion of the legs and extension of the arms with varying degrees of flexion of the trunk and neck.

Asymmetrical spasms, with greater involvement over one side of the body, occur rarely, even in patients who show focal neurological, EEG or neuroradiological features.

Most patients experience more than one type of motor spasm. Also, periods of akinesia may occur following a motor spasm, or may occur as independent phenomena in association with ictal EEG changes (see below).

About two-thirds of motor spasms are associated with eye movements consisting of deviation alone or deviation followed by rhythmic nystagmoid movements. In some patients with minimal spasms, eye movement may be the only clinically evident change. Respiratory pauses occur with about two-thirds of spasms, but heart rate changes are rare (<1%). Patients will frequently cry following a spasm or cluster of spasms, but crying does not appear to occur as an ictal phenomenon.

Infantile spasms are not elicited by photic stimulation, but rarely may be induced by tactile or auditory stimulation (e.g., hand claps).

In our monitoring experience, about 80% of infantile spasms occur in clusters. Although most spasms occur in clusters, however, infants typically experience both clusters of spasms and isolated spasms. During long-term polygraphic/video monitoring studies, the number of spasms per cluster has ranged from 2–138, with succession rates of up to 15 spasms per minute. In the longer clusters, the intensity of the spasms may initially wax and then wane.

Approximately the same number of spasms occur during the night as during the day; however, infantile spasms rarely occur when the infant is actually asleep (<3%), and, instead, usually occur immediately after arousal (Fig. 2).[20] Although there is relatively little variation in spasm frequency during consecutive 24-hour monitoring periods, there is marked variation in spasm frequency when patients are monitored at two-week intervals (Fig. 3).[19] The number of spasms recorded during 24-hour monitoring periods has ranged from 3 to 763.

EEG FEATURES

Interictal EEG

A variety of interictal EEG patterns may be seen. These include diffuse slowing, focal slowing, undifferentiated background activity, focal or multifocal spikes and sharp waves, generalized slow-spike-and-slow-wave activity and poorly-defined sleep characteristics (e.g., vertex transients and sleep spindles).[36] These abnormalities may occur in isolation or in various combinations. Rarely, the interictal background EEG activity may be normal.

The most common interictal pattern seen in infantile spasm patients, however, is hypsarrhythmia (Fig. 4). As originally described by Gibbs & Gibbs,[16] this pattern consists of extremely high voltage (500–1000 μV), random and disorganized bilaterally asynchronous slow-wave activity mixed with spikes and sharp waves in all cortical regions. The prototypical pattern is usually seen during the early stages of this disorder. Prolonged monitoring studies have revealed that hypsarrhythmia is a highly dynamic and variable EEG pattern, changing constantly throughout the day and night.[24] During NREM sleep, there is an increase in the amplitude of the background EEG activity, and in many patients there is a tendency for the multifocal spikes and sharp waves to group, resulting in a pseudo-periodic appearance. Attenuation episodes frequently occur during this stage of sleep. In some patients, the hypsarrhythmic pattern may be seen only during NREM

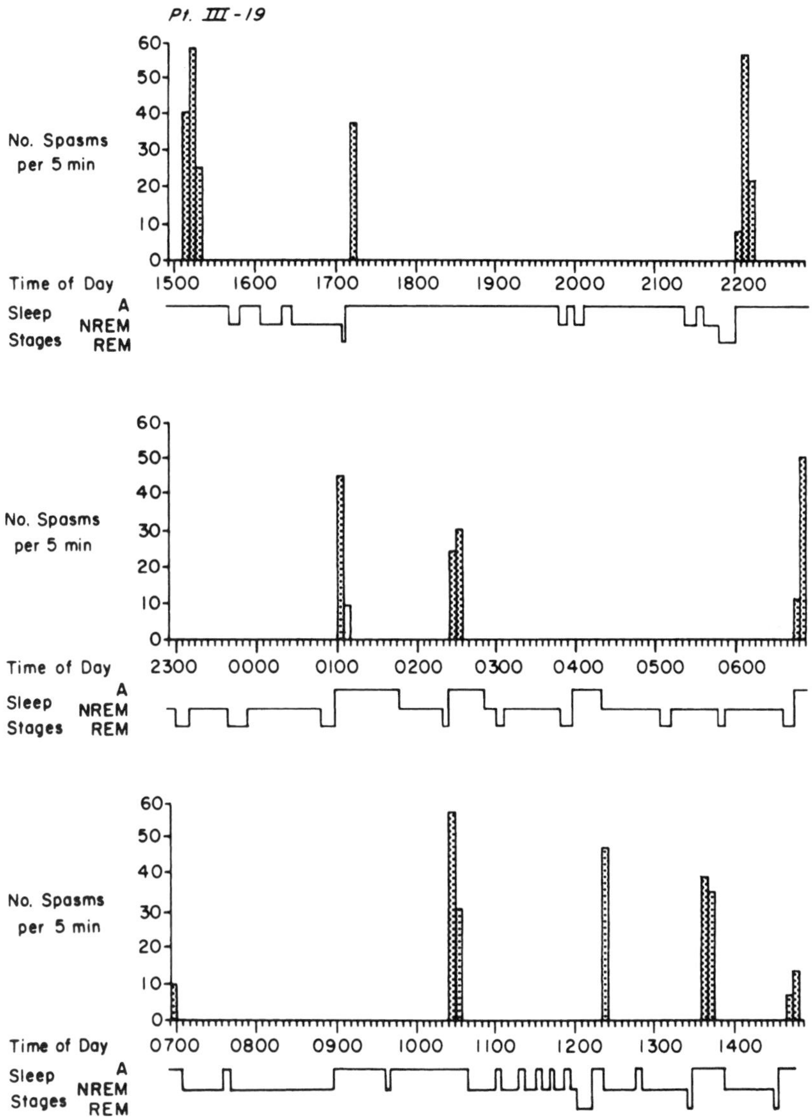

Figure 2. Relation of spasm occurrence to the sleep-awake cycle. A = awake; NREM = slow-wave sleep; REM = rapid-eye-movement sleep. (Reproduced from ref. 20 by permission)

sleep. During REM sleep, which is markedly reduced in infantile spasm patients, there is usually a marked reduction or complete disappearance of the hypsarrhythmic pattern (Fig. 5).[25] In many patients, the hypsarrhythmic pattern is also markedly reduced immediately after arousal from sleep. This relative "normalization" may last from a few seconds to many minutes before the hypsarrhythmic pattern reappears. Finally, relative normalization of the hypsarrhythmic pattern may also occur during a cluster of spasms.[24]

In addition to the prototypical hypsarrhythmic pattern, certain variations of this pattern have been identified.[24] These variations, which are frequently referred to as "modified hypsarrhythmia", include the following:

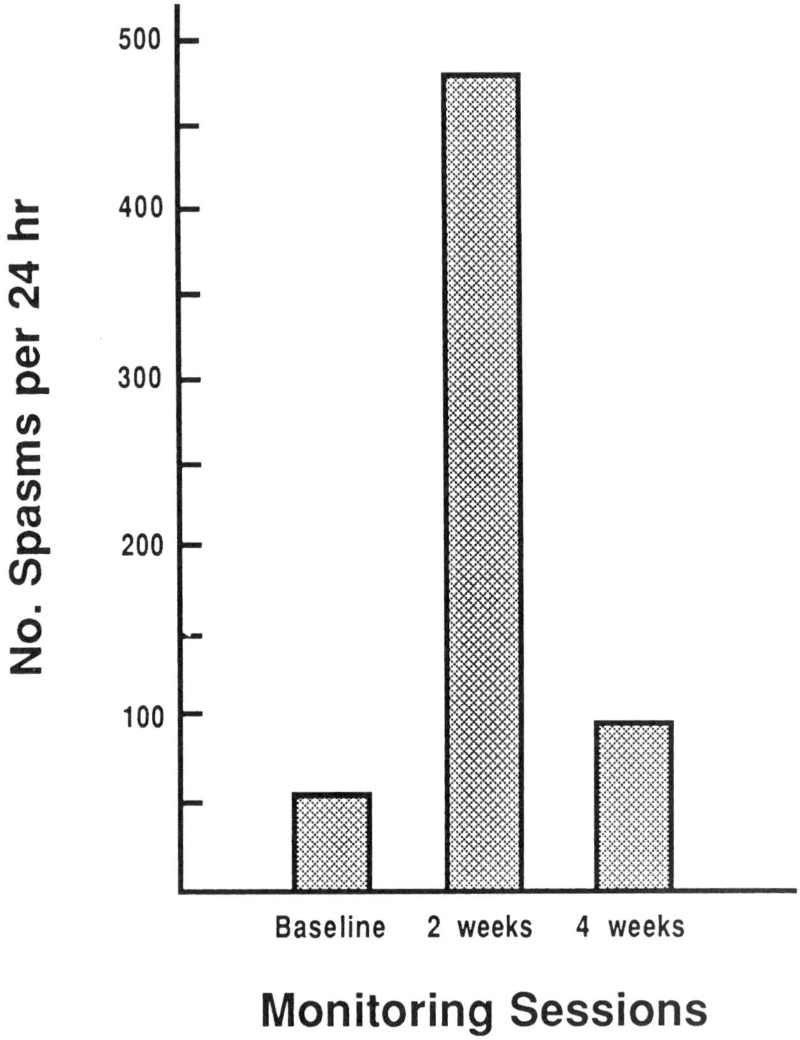

Figure 3. Variation in spasms frequency in a patient monitored at 2-week intervals. (Reproduced from ref. 19 by permission)

(a) Hypsarrhythmia with increased interhemispheric synchronization (Fig. 6).[9,30,36,51] The increased interhemispheric synchronization may be evidenced as bursts of generalized spike-and-wave activity or increased synchronization of the background activity.
(b) Hypsarrhythmia with a consistent focus of abnormal discharge (Fig. 7). In this variant, a consistent focus of spike-and-wave or sharp-wave activity is clearly distinguishable from the other multifocal discharges present, and on occasion focal electrographic seizures may be recorded (see below).
(c) Asymmetrical hypsarrhythmia (Fig. 8)—This variant refers to a hypsarrhythmic pattern in which there are consistent voltage asymmetries between the two sides. The terms "hemihypsarrhythmia" or "unilateral hypsarrhythmia" have

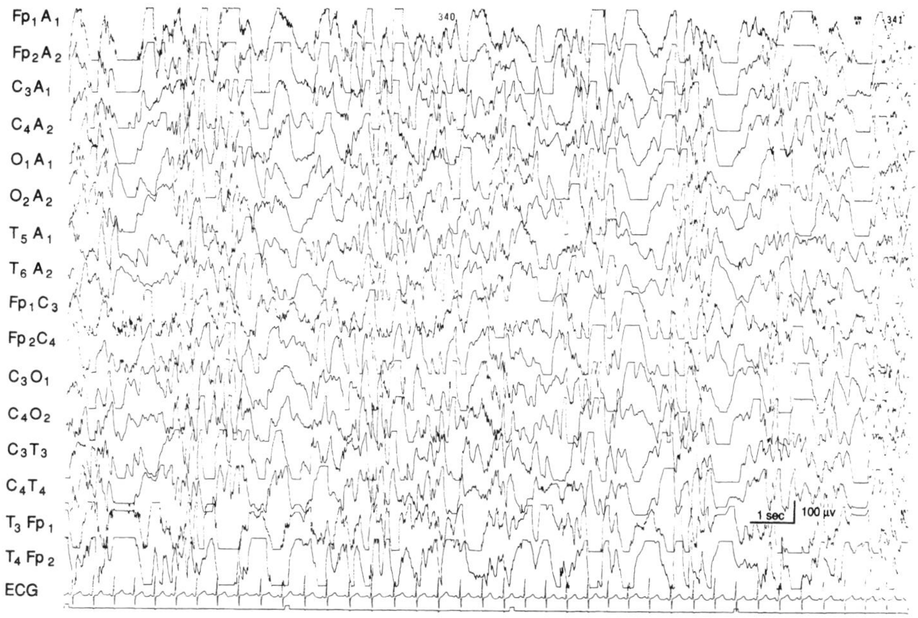

Figure 4. Hypsarrhythmic EEG pattern in an 8-month-old infant. (Reproduced from ref. 19 by permission)

been used to describe this variant.[7,10,37,41,49] Various pathological entities have been associated with this pattern, including porencephaly, megalocephaly, and various developmental anomalies.

(d) Hypsarrhythmia comprised primarily of high voltage asynchronous slow activity with very little sharp wave or spike activity (Fig. 9).

(e) Hypsarrhythmia with episodes of generalized, regional or localized attenuation (Fig. 10). In this variation, the hypsarrhythmic activity is interrupted by recurrent episodes of voltage attenuation that may be generalized, regional or localized. In their maximal expression, the attenuation episodes occur every few seconds in a semiperiodic manner resulting in a suppression-burst pattern (Fig. 11).[11,18,40] Recognition of the suppression-burst variant is important because of the poor prognosis for long-term outcome associated with this pattern. Attenuation episodes may also be the sole ictal manifestation of clinical infantile spasms (see below); however, long-term monitoring studies have shown that patients may have hundreds of random attenuation episodes that are not accompanied by clinical spasms.

Ictal EEG

A variety of ictal EEG patterns have been associated with infantile spasms.[35] These patterns consist of attenuation episodes, generalized slow wave transients, generalized sharp-and slow-wave transients, and rhythmic faster frequencies (usually alpha or beta) which may occur singly or in various combinations. The most common EEG signature of an infantile spasm is a high-voltage generalized slow-wave transient, followed by attenuation of the background activity in all regions that may last from one to many seconds

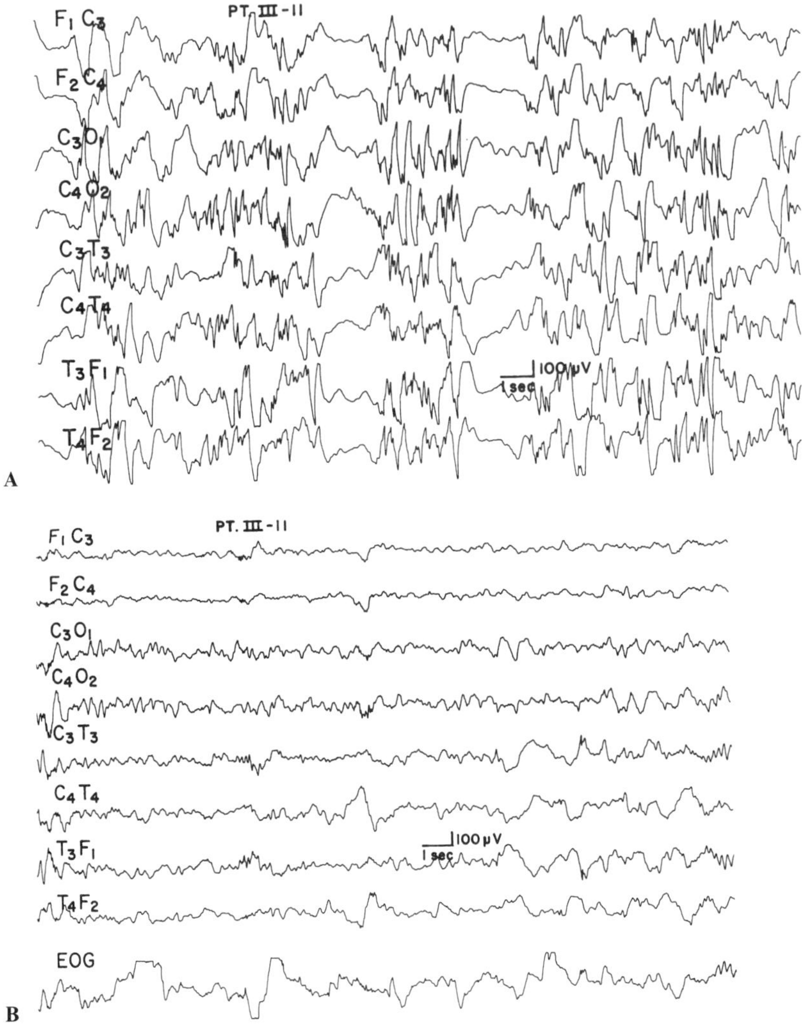

Figure 5. EEG samples of a 4-month-old infant during sleep. A: NREM sleep. B: REM sleep. (Reproduced from ref. 24 by permission)

(Fig. 12). Monitoring studies have revealed no close correlation between the character of the ictal EEG events and the various types of motor spasms.

The major diagnostic criteria for infantile spasms are summarized in Table 1.

OTHER SEIZURES OR BEHAVIORS WHICH MAY MIMIC INFANTILE SPASMS

The diagnosis of this condition is often delayed for weeks or months because the parents, caretakers or physician do not recognize the motor events as being infantile spasms. Other seizures that may be confused with infantile spasms include tonic seizures,

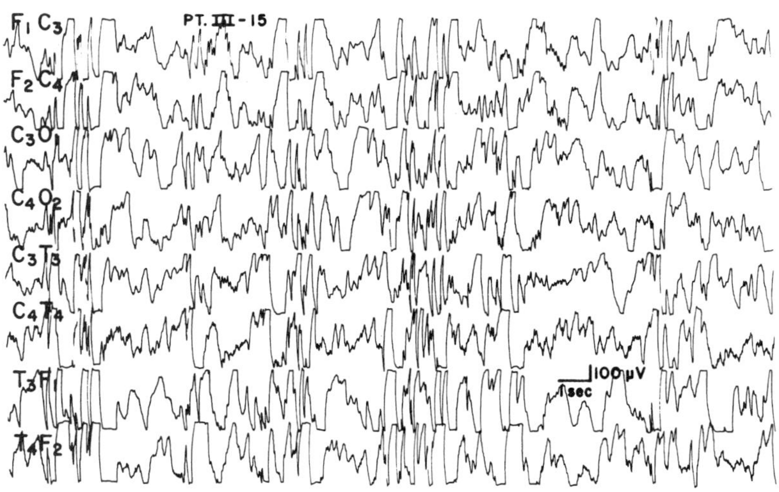

Figure 6. EEG of an 8-month-old infant, showing a modified hypsarrhythmia characterized by the presence of synchronous and symmetrical frontal-dominant spike-and-slow-wave and sharp-and-slow-wave activity. (Reproduced from ref. 24 by permission)

atonic seizures and other types of myoclonic seizures (e.g., benign myoclonic epilepsy).[39] Non-epileptic movements that are sometimes suspected of being infantile spasms include moro reflexes, arousal movements, startles, hypnagogic jerks, phasic muscle twitches occurring during REM sleep, and transient flexor or extensor posturing of non-epileptic origin. In many instances, video/EEG monitoring is required to document the character of a suspected motor event. This is especially true in patients in whom the diagnosis of infantile spasms is suspected, but whose routine EEG does not show hypsarrhythmia.

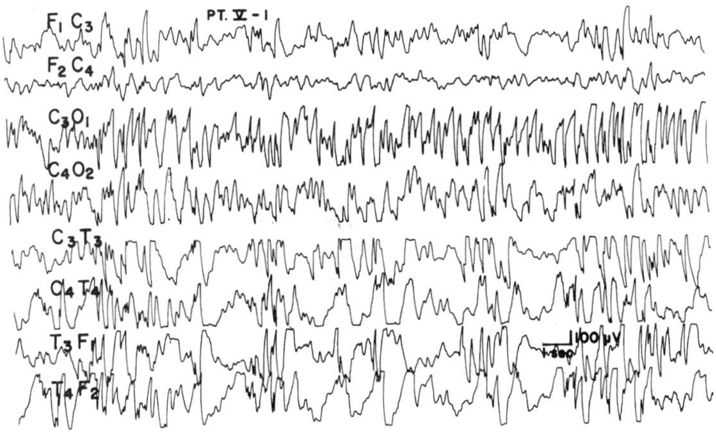

Figure 7. EEG of a 3-month-old infant, showing a modified hypsarrhythmic pattern with a distinguishable focus of irregular spike-and-slow-wave and sharp-and-slow-wave activity in the left occipital region. (Reproduced from ref. 24 by permission)

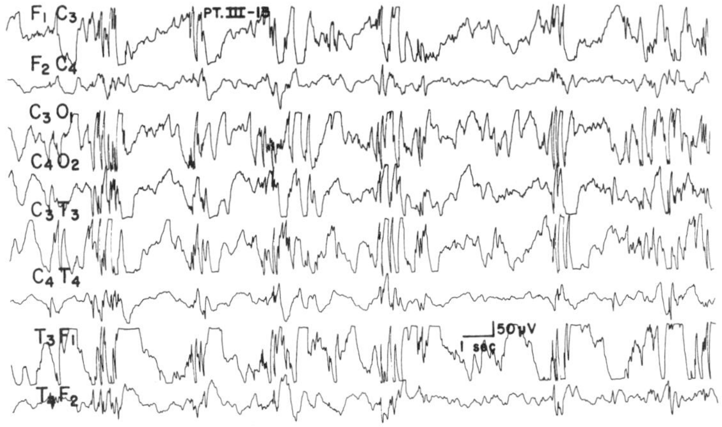

Figure 8. EEG of a 6-month-old infant, showing asymmetrical hypsarrhythmia. A CT scan revealed a porencephalic cyst in the right hemisphere. (Reproduced from ref. 24 by permission)

PROBLEMS WITH PARENTAL ESTIMATION OF SPASM FREQUENCY

Polygraphic/video monitoring studies have shown that parents often underestimate the number of spasms their child is experiencing (Fig. 13).[21,35] In the extreme, the parents may report that no spasms are occurring when, in fact, tens or hundreds of spasms may be present per day. Conversely, and less commonly, some parents may report the presence of spasms in a child who is, in fact, seizure free. Several factors probably account for these discrepancies between parents' estimation of seizure frequency and seizure

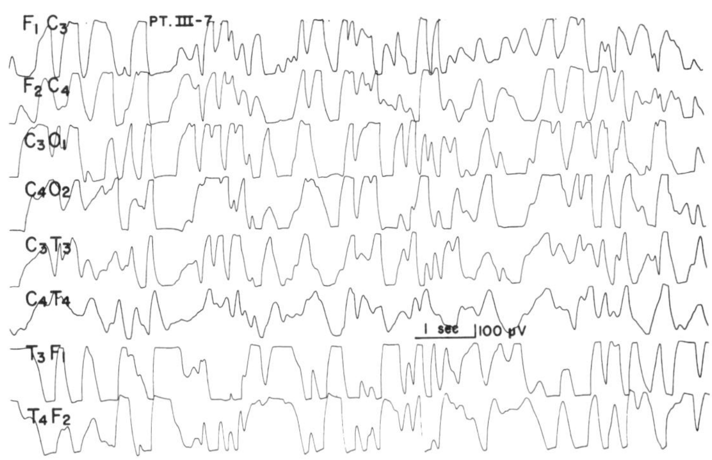

Figure 9. EEG of a 13-month-old infant, showing a modified hypsarrhythmia characterized by the presence of high-voltage, primarily asynchronous, slow activity with little spike or sharp-wave activity. (Reproduced from ref. 24 by permission)

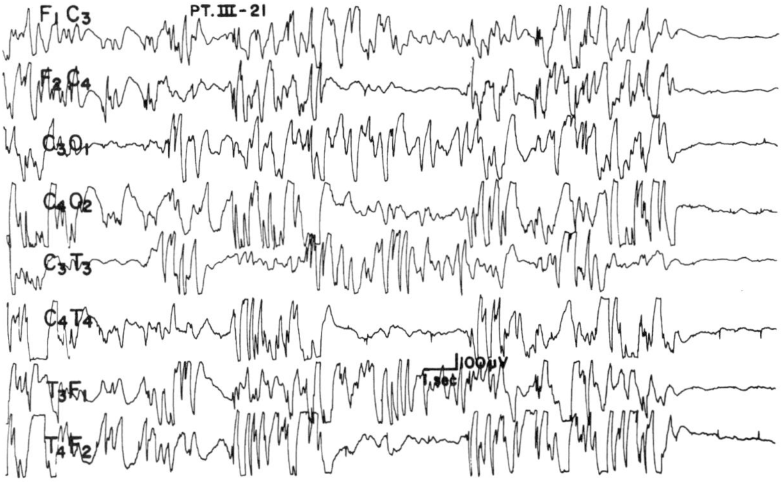

Figure 10. EEG of a 9-month-old infant, showing a modified hypsarrhythmic pattern with episodes of lateralized and generalized voltage attenuation. (Reproduced from ref. 24 by permission)

frequency as documented by long-term monitoring. First, spasms occur in relatively equal numbers throughout the day and night, thus making it difficult for parents to see all spasms that are occurring. Also, most spasms occur in brief clusters immediately after arousal from sleep, often before the infant starts to cry. Thus, by the time the parent hears the child crying and attends to the child's needs, the spasms may have ceased. In addition, minimal or less intense motor spasms (e.g., head nods, shoulder shrugs, etc.) are often extremely difficult for parents or physicians to differentiate from other normal or abnormal movements. Finally, once therapy for the disorder has been initiated, many parents are biased to see improvement in seizure frequency, considering the dismal prognosis of this disorder.

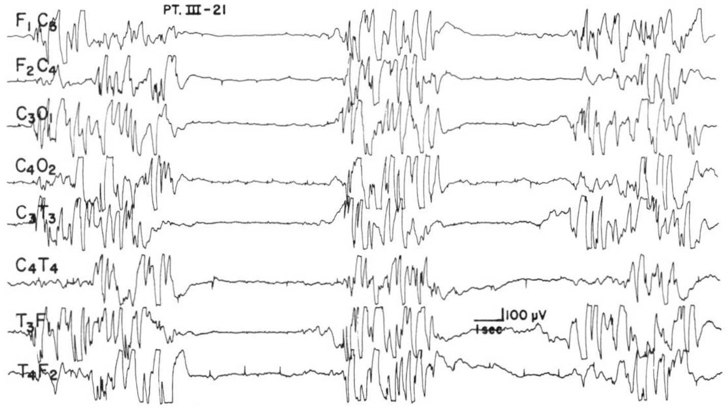

Figure 11. EEG of a 9-month-old infant, showing the suppression-burst variation of hypsarrhythmia. (Reproduced from ref. 24 by permission)

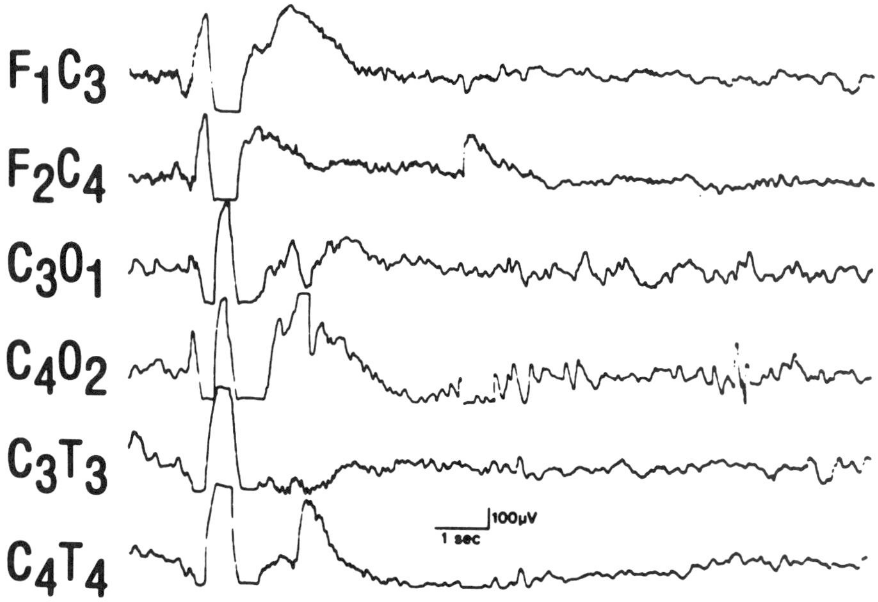

Figure 12. Most common ictal EEG change associated with infantile spasms, a generalized slow-wave transient followed by attenuation of background activity in all regions.

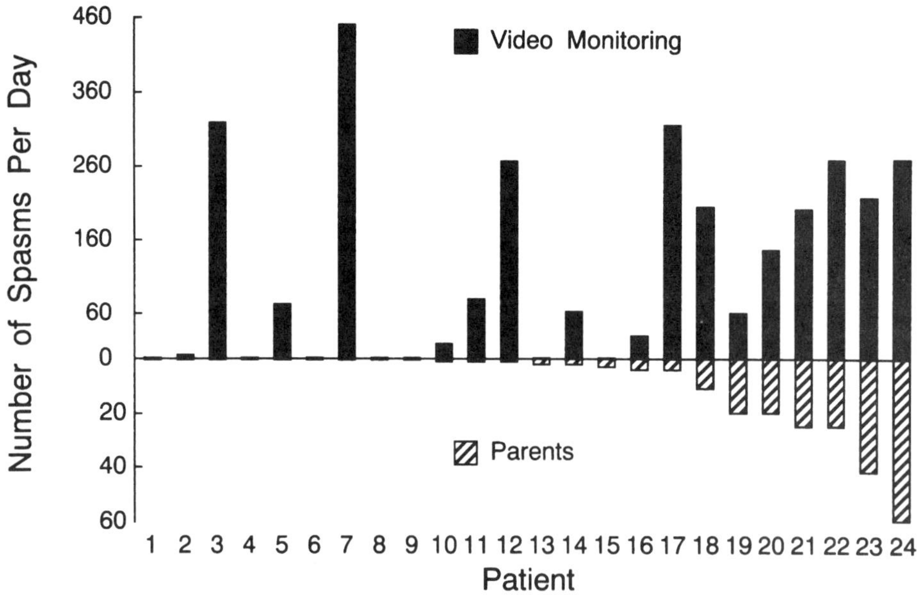

Figure 13. Spasm frequency after institution of ACTH or prednisone therapy; comparison of parents' estimates with results of 24-hour polygraphic/video monitoring. The 24-hour monitoring studies were performed 2–4 weeks after initiation of hormonal therapy.

ASSOCIATED ETIOLOGIES & LABORATORY TESTING

In about 40% of cases, no associated etiological factors can be identified.[36] In the remaining 60%, various pre-, peri-, and postnatal factors have been identified. These include cerebral dysgenesis, intrauterine infection, prematurity, intracranial hemorrhage, stroke, hypoxia-ischemia, genetic disorders (e.g., tuberous sclerosis), traumatic delivery, inborn errors of metabolism (e.g., non-ketotic hyperglycinemia), head injury and postnatal CNS infection. About 50% of patients have major neurological defects, and another 40% have minor ones.[34] These consist primarily of focal motor deficits, microcephaly and blindness. About 85–90% of patients show some degree of mental and developmental retardation.[14,36]

All infants diagnosed with infantile spasms require a thorough evaluation including neurological and physical examinations, developmental assessment and various laboratory studies. CT or MRI scans should be performed on all patients, because these studies can provide crucial information concerning etiology, and, thus, can aid in classification of patients (see below). Neuroimaging studies reveal abnormalities in between 60–90% of cases, most commonly diffuse atrophy.[15,17,47] In those patients whose clinical histories (e.g., CNS infection, intraventricular hemorrhage), physical examination (e.g., tuberous sclerosis) or neuroradiological studies (CT or MRI) fail to identify an associated cause, metabolic studies (serum quantitative amino acids, urine organic acids, urine metabolic screen), chromosomal studies and formal ophthalmologic evaluations should be performed. In selected cases, additional diagnostic testing (e.g., serum lactate, cerebrospinal fluid lactate levels) may be indicated. More recently, positron emission tomography (PET) has been promoted as a means of detecting brain malformations which may not be seen on CT or MRI scans in infants with infantile spasms.[4] However, the value of PET in the routine evaluation of infantile spasm patients is still being assessed, and its restricted availability limits its usefulness. Other more readily accessible functional imaging techniques (e.g., NMR spectroscopy) may provide similar information.

PATIENT CLASSIFICATION

In the past, the classification of infantile spasm patients has been variable and inconsistent.[36] Currently, our group believes patients are best classified on the basis of past medical history, neurological examination, developmental history and CT or MRI findings. Patients with no abnormality on neurological examination, no associated etiological factor, normal development prior to spasm onset and normal CT or MRI scans prior to institution of hormonal therapy are classified as "cryptogenic". Utilizing these criteria, about 10–15% of patients can be classified as cryptogenic. The remaining patients are classified as "symptomatic".[14,23,26–28] The use of functional imaging technologies (e.g., PET or NMR Spectroscopy) will probably reduce the percentage of patients classified as cryptogenic.

COUPLING OF INFANTILE SPASMS WITH OTHER SEIZURES

The co-existence of partial seizures in infantile spasm patients has been recognized for decades.[9,13,31,34,36,38] Partial seizures may occur prior to the onset of infantile spasms, concurrently with spasms, or after spasms have ceased, either spontaneously or following therapy. Recently, interest has centered around the observation that focal electrical seizure discharges (FS) may be tightly coupled with infantile spasms (i.e., occur in a close temporal association). Our group was the first to describe this phenomenon in 1984,[24] and several additional reports of this phenomenon in small numbers of patients followed.[1,2,3,8,44,54] The question remains as to the significance of this association. Some investigators have used these observations to support the hypothesis that a primary cortical generator interacts with subcortical structures, resulting in infantile spasms.[3,5] To further investigate the significance of coupling of FS and infantile spasms, we analyzed the polygraphic/video monitoring studies performed on 96 consecutive patients newly diagnosed with infantile spasms and hypsarrhythmic EEGs.[22] Ten patients demonstrated FS. In only 5 patients, however, was there an apparent coupling of some FS and infantile spasms. Further analysis revealed that in only 3 patients (3% of the entire population) was the observed coupling of FS and infantile spasms statistically significant. Three types of couplings were observed in these patients: FS could precede a cluster of spasms, FS could follow a cluster of spasms, or FS could occur during a cluster of spasms. Also of importance were the observations that the FS always arose independently from multiple sites, and that in some patients coupling of FS and infantile spasms always occurred on arousal from sleep.

We concluded from this study that coupling of FS and infantile spasms occurs rarely. We also proposed three possible hypotheses for the coupling phenomenon. In some instances, apparent coupling of FS and infantile spasms may best be explained by chance coincidence. Another explanation is that FS may facilitate or induce the appearance of infantile spasms (or vice versa), as has been suggested by others. Finally, the coupling of FS and infantile spasms may result from the effect of some "critical factor" which simultaneously affects the seizure thresholds of neuronal systems involved in the generation of FS and those involved in the generation of infantile spasms. According to this hypothesis, in the presence of the critical factor, the seizure thresholds for FS and infantile spasms are concurrently altered resulting in the simultaneous or near simultaneous occurrence of both seizure types. In some of our patients the "critical factor" appeared to be the arousal mechanism, a known potent activator of infantile spasms and other seizures.

Patients with infantile spasms may also experience tonic seizures, and these two seizures can be difficult to differentiate. The clinical appearance and ictal EEG changes seen with these seizures are very similar. The major distinguishing clinical features between the two are that tonic seizures usually lack the intense initial phasic component seen with infantile spasms and tonic seizures are more prolonged. An interesting observation is that a cluster of infantile spasms may be immediately preceded by a tonic seizure. I reviewed the polygraphic/video monitoring studies performed on 57 consecutive infantile spasm patients and found coupling of tonic seizures with infantile spasms in 3 (5%) cases (personal observations). Because the clinical and EEG features of tonic seizures and infantile spasms are so similar, it is possible that these two seizures are generated by a similar mechanism and arise from the same region of the brain, probably the brain stem.

SPONTANEOUS REMISSION

The natural history of infantile spasms is incompletely understood, and descriptions of the duration of this disorder have been infrequent and imprecise. In 1973, Jeavons et al.,[32] reported that 28% of patients were free of spasms before age 1, 49%

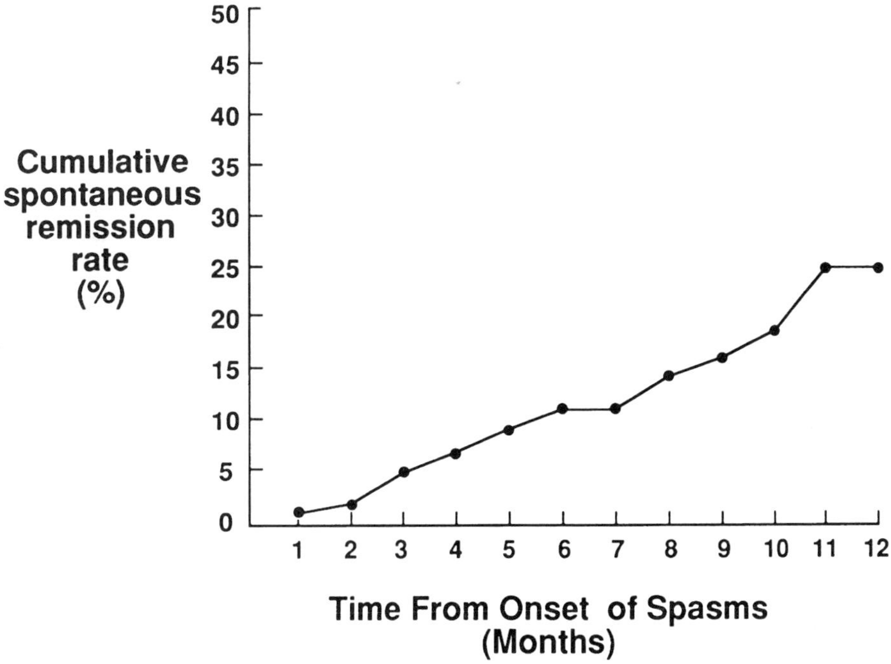

Figure 14. Spontaneous remission rate during the first year after onset of infantile spasms in patients who did not receive hormonal therapy (N = 44). (Reproduced from ref. 29 by permission)

before age 2, 65% before age 3, and 74% by age 4. However, some of their patients were treated with steroids, and EEG changes were not discussed. To further investigate the spontaneous remission rate in this disorder, we retrospectively studied 44 patients who had not been treated with hormonal drugs.[29] The results of this study revealed that spontaneous remission of infantile spasms and disappearance of the hypsarrhythmic EEG pattern could begin within 1 month of the onset of the disorder, and, that 25% of patients would experience spontaneous remission within 1 year of spasm onset (Fig. 14). The phenomenon of spontaneous remission must always be remembered when interpreting the results of any therapeutic trial in this disorder, and also must be strongly considered in any discussion of any pathophysiological mechanism(s) underlying infantile spasms.

OTHER SEIZURE TYPES WHICH FOLLOW SPASMS

The occurrence of other types of seizures in infantile spasms patients who have stopped having spasms ranges from 35–60%.[32,33,36,45] In our own prospective study of long-term outcome,[14] other seizures occurred in 53% of patients. No cryptogenic patients who responded to hormonal therapy progressed to other types of seizures. The most common types of seizures observed after spasms are partial, tonic and generalized tonic/clonic. Roughly one-third to one-half of cases reportedly develop the clinical and electroencephalographic features of the Lennox-Gastaut syndrome (LGS).[42,52] However, the time at which the transition from infantile spasms to the Lennox-Gastaut syndrome occurs is very difficult to determine in most patients, because by clinical observations alone, it may be impossible to differentiate the seizures associated with these syndromes. In some patients, the clinical and EEG features of both syndromes coexist, suggesting that infantile spasms and the Lennox-Gastaut syndrome actually represent a continuum.

REFERENCES

1. Bour, F., Chiron, C., Dulac, O., and Plouin, P. Caractéres électrocliniques des crises dans le syndrome d'Aicardi, *EEG Neurophysiol. Clin.*, 16 (1986) 341–353.
2. Carrazana, E.J., Barlow, J.K., and Holmes, G.L. Infantile spasms provoked by partial seizures, *J. Epilepsy*, 3 (1990) 97–100.
3. Carrazana, E.J., Lombroso, C.T., Mikati, M., Helmers, S., and Holmes, G.L. Facilitation of infantile spasms by partial seizures, *Epilepsia*, 34 (1993) 97–109.
4. Chugani, H.T. and Conti, J.R. Classification of infantile spasms in 139 cases: The role of positron emission tomography, *Epilepsia*, 35 (1994) 19.
5. Chugani, H.T., Shewmon, D.A., Sankar, R., Chen, B.J., and Phelps, M.E. Infantile spasms: II. Lenticular nuclei and brain stem activation on positron emission tomography, *Ann. Neurol.*, 31 (1992) 212–219.
6. Constad, H., Wagner, R.S., and Caputo, A.R. Aicardi syndrome in one dizygotic twin, *Pediatrics*, 76 (1985) 450–453.
7. de Jong, J.G.Y., Delleman, J.W., Houben, M., Manschot, W.A., de Minjer, A., Mol, J., and Slooff, J.L. Agenesis of the corpus callosum, infantile spasms, ocular anomalies (Aicardi's syndrome): clinical and pathologic findings, *Neurology*, 26 (1976) 1152–1158.
8. Donat, J.F. and Wright, F.S. Simultaneous infantile spasms and partial seizures, *J. Child Neurol.*, 6 (1991) 246–250.
9. Druckman, R. and Chao, D. Massive spasms in infancy and childhood, *Epilepsia*, 4 (third series) (1955) 61–72.
10. Drury, I., Beydoun, A., Garofalo, E.A., and Henry, T.R. Asymmetric Hypsarrhythmia: Clinical electroencephalographic and radiological findings, *Epilepsia*, 36 (1995) 41–47.
11. Fariello, R.G., Chun, R.W., Doro, J.M., Buncic, J.R., and Prichard, J.S. EEG recognition of Aicardi's syndrome, *Arch. Neurol.*, 34 (1977) 563–566.
12. Frost, J.D., Jr., Hrachovy, R.A., Kellaway, P., and Zion, T. Quantitative analysis and characterization of infantile spasms, *Epilepsia*, 19 (1978) 273–282.
13. Fukuyama, Y. Studies on the etiology and pathogenesis of flexor spasms, *Adv. Neurol. Sci. (Tokyo)*, 4 (1960) 861–867.
14. Glaze, D.G., Hrachovy, R.A., Frost, J.D., Jr., Kellaway, P., and Zion, T.E. Prospective study of outcome of infants with infantile spasms treated during controlled studies of ACTH and prednisone, *J. Pediatr.*, 112 (1988) 389–396.
15. Glaze, D.G., Hrachovy, R.A., Frost, J.D., Jr., Zion, T.E., and Bryan, R.N. Computed tomography in infantile spasms: effects of hormonal therapy, *Pediatr. Neurol.*, 2 (1986) 23–27.
16. Gibbs, F.A. and Gibbs, E.L. *Epilepsy*, (Atlas of electroencephalography; vol. 2), Addison-Wesley, Cambridge, Massachusetts, 1952.
17. Gastaut, H., Gastaut, J.L., Regis, H., Bernard, R., Pinsard, N., Saint-Jean, M., Roger, J., and Dravet, C. Computed tomography in the study of West's syndrome, *Dev. Med. Child Neurol.*, 20 (1978) 21–27.
18. Hoefer, P.F.A., de Napoli, R.A., and Lesse, S. Periodicity and hypsarrhythmia in the EEG: a study of infantile spasms, diffuse encephalopathies, and experimental lesions of the brain, *Arch. Neurol.*, 9 (1963) 424–436.
19. Hrachovy, R.A. and Frost, J.D., Jr. Infantile spasms, *Pediatr. Clin. North Am.*, 36(2) (1989) 311–329.
20. Hrachovy, R.A. and Frost, J.D., Jr. Intensive monitoring of infantile spasms. In: D. Schmidt and P.L. Morselli (Eds.), *Intractable Epilepsy: Experimental and Clinical Aspects*, Raven Press, New York (1986) pp. 87–97.

21. Hrachovy, R.A. and Frost, J.D., Jr. Severe encephalopathic epilepsy in infants: infantile spasms. In: W.E. Dodson and J.M. Pellock (Eds.), *Pediatric Epilepsy: Diagnosis and Therapy*, Demos, New York (1993) pp. 135–145.
22. Hrachovy, R.A., Frost, J.D., Jr., and Glaze, D.G. Coupling of focal electrical seizure discharges with infantile spasms: incidence during long-term monitoring in newly diagnosed patients, *J. Clin. Neurophysiol.*, 11(4) (1994) pp. 461–464.
23. Hrachovy, R.A., Frost, J.D., Jr., and Glaze, D.G. High dose/long duration vs. low dose/short duration corticotropin therapy in infantile spasms: A single blind study, *J. Pediatr.*, 124 (1994) 803–806.
24. Hrachovy, R.A., Frost, J.D., Jr., and Kellaway, P. Hypsarrhythmia: Variations on the theme, *Epilepsia*, 25 (1984) 317–325.
25. Hrachovy, R.A., Frost, J.D., Jr., and Kellaway, P. Sleep characteristics in infantile spasms, *Neurology*, 31 (1981) 688–694.
26. Hrachovy, R.A., Frost, J.D., Jr., Kellaway, P., and Zion, T. A controlled study of ACTH therapy in infantile spasms, *Epilepsia*, 21 (1980) 631–636.
27. Hrachovy, R.A., Frost, J.D., Jr., Kellaway, P., and Zion, T. A controlled study of prednisone therapy in infantile spasms, *Epilepsia*, 20 (1979) 403–407.
28. Hrachovy, R.A., Frost, J.D., Jr., Kellaway, P., and Zion, T.E. Double-blind study of ACTH vs prednisone therapy in infantile spasms, *J. Pediatr.*, 103 (1983) 641–645.
29. Hrachovy, R.A., Glaze, D.G., and Frost, J.D., Jr. A retrospective study of spontaneous remission and long-term outcome in patients with infantile spasms, *Epilepsia*, 32 (1991) 212–214.
30. Jeavons, P.M. and Bower, B.D. Infantile spasms. In: P.J. Vinken and G.W. Bruyn (Eds.), *The epilepsies*, (Handbook of clinical neurology; vol 15), Elsevier, New York (1974) pp. 219–234.
31. Jeavons, P.M. and Bower, B.D. *Infantile spasms: a review of the literature and a study of 112 cases*, London: Spastics Society and Heineman (Clinics in developmental medicine no. 15) (1964).
32. Jeavons, P.M., Bower, B.D., and Dimitrakoudi, M. Long-term prognosis of 150 cases of "West syndrome", *Epilepsia*, 14 (1973) 153–164.
33. Jeavons, P.M., Harper, J.R., and Bower, B.D. Long-term prognosis of infantile spasms: A follow-up report on 112 cases, *Dev. Med. Child Neurol.* 12 (1970) 413–421.
34. Kellaway, P. Neurologic status of patients with hypsarrhythmia. In: F.A. Gibbs (Ed.), *Molecules and Mental Health*, J.B. Lippincott, Philadelphia (1959) pp. 134–149.
35. Kellaway, P., Hrachovy, R.A., Frost, J.D., Jr., and Zion, T. Precise characterization and quantification of infantile spasms, *Ann. Neurol.*, 6 (1979) 214–218.
36. Lacy, J.R. and Penry, J.K. *Infantile spasms*, Raven Press, New York (1976).
37. Lerman, P. Hypsarrhythmia, *Electroencephalogr. Clin. Neurophysiol.*, 27 (1969) 216.
38. Lombroso, C.T. A prospective study of infantile spasms: clinical and therapeutic considerations, *Epilepsia*, 24 (1983) 135–158.
39. Lombroso, C.T. and Fejerman, N. Benign myoclonus of early infancy, *Ann. Neurol.*, 1 (1977) 138–143.
40. Maheshwari, M.C. and Jeavons, P.M. The prognostic implications of suppression-burst activity in the EEG in infancy, *Epilepsia*, 16 (1975) 127–131.
41. Ohtahara, S. Electroencephalographic studies in infantile spasms, *Dev. Med. Child Neurol.*, 7 (1965) 707.
42. Ohtahara, S. Seizure disorders in infancy and childhood, *Brain Dev.*, 6 (1984) 509–519.
43. Pavone, L., Mollica, F., Incorpora, G., and Pampiglione, G. Infantile spasms syndrome in monozygotic twins, *Arch. Dis. Child*, 55 (1980) 870–872.
44. Plouin, P., Jalin, C., Dulac, O., and Chiron, C. Enregistrement ambulatoire de l'EEG pendant 24 heures dans les spasmes infantiles épileptiques, *Rev. EEG Neurophysiol. Clin.*, 17 (1987) 309–318.
45. Riikonen, R. A long-term follow-up study of 214 children with the syndrome of infantile spasms, *Neuropediatrics*, 13 (1982) 14–23.
46. Senga, P., Mayanda, H.F., and Yidika, M. Spasmes infantiles chez deux jumeaux monozygotiques: une novelle observation, *Presse Med.*, 15 (1986) 485.
47. Singer, W.D., Haller, J.S., Sullivan, L.R., Wolpert, S., Mills, C., and Rabe, E.F. The value of neuroradiology in infantile spasms, *J. Pediatr.*, 100 (1982) 47–50.
48. Sorel, L. and Dusaucy-Bauloye, A. A propos de 21 cas d'hypsarhythmia de Gibbs. Son traitement spectaculaire par l'ACTH, *Acta Neurol. Psychiatr. Belg.*, 58 (1958) 130–141.
49. Tijiam, A.T., Stefanko, S., Schenk, V.M.D., and de Vlieger, M. Infantile spasms associated with hemihypsarrhythmia and hemimegalencephaly, *Dev. Med. Child Neurol.* 20 (1978) 779–798.
50. Van den Berg, B.J. and Yerushalmy, J. Studies on convulsive disorders in young children. I. Incidence of febrile and nonfebrile convulsions by age and other factors, *Pediatr. Res.*, 3 (1969) 298–304.

51. Watanabe, K., Iwase, K., and Hara, K. The evolution of EEG features in infantile spasms: a prospective study, *Dev. Med. Child Neurol.*, 15 (1973) 584–596.
52. Weinman, H.M. Lennox-Gastaut Syndrome and Its Relationship to Infantile Spasms (West Syndrome). In: E. Niedermeyer and R. Degen (Eds.), *The Lennox-Gastaut Syndrome*, Alan R. Liss, Inc., New York (1988) pp. 301–316.
53. West, W.J. On a peculiar form of infantile convulsions, *Lancet*, 1 (1841) 724–725.
54. Yamamoto, N., Watanabe, K., Negoro, I., Furune, S., Takahashi, I., Nomura, K., and Matsumoto, A. Partial seizures evolving to infantile spasms, *Epilepsia*, 29 (1988) 34–40.

6

DRUG THERAPY FOR WEST'S SYNDROME

Catherine Chiron and Olivier Dulac

Neuropediatric dpmt Hospital Saint Vincent de Paul
INSERM Unit 29
82 Ave. Deufert Rochereau
76574 Paris Cedex 14
France
tel: 011-331-4048-8045
fax: 011-331-4048-8046

West's syndrome resists most of the standard anticonvulsants. It may respond, however, to steroids or to some of the newer drugs like vigabatrin.

INTRODUCTION

West's syndrome (WS) is an age-related epilepsy syndrome that is considered one of the most refractory epilepsies of childhood. Although up to 25% of the cases may exhibit spontaneous recovery in the first year from the seizures,[14] nearly half the patients are left with intractable epilepsy and mental retardation.[12]

Most of the conventional antiepileptic drugs are ineffective at controlling infantile spasms (IS), and the primary treatment has consisted of steroid therapy since the first report of the efficacy of ACTH in WS.[27] A wide variety of steroid-related compounds can be used, including natural ACTH,[26] synthetic ACTH,[10] oral steroids, synthetic steroids like prednisone[24] and natural corticosteroids like hydrocortisone. They all, however, induce frequent—and often severe—side effects. Among the classical antiepileptic drugs, only some benzodiazepines, such as nitrazepam and clonazepam,[11] and high-dose valproate[23] have shown some efficacy in IS. Pyridoxine and immunoglobulins are considered as adjuvants in IS. The drug armamentarium has been enlarged over the five past years with some promising new drugs, mainly vigabatrin, a well tolerated and efficient compound,[7] which tends to be more and more used as a first-line drug in the countries in which it has been approved (Table 1).

My purpose is not to present an exhaustive review of drug therapy in IS (for review, see 12 and 25), but rather to focus on some specific studies that summarize the

Table 1. Drugs for Infantile Spasms (West's Syndrome)

First Choice
Vigabatrin
Second Choice
ACTH or Corticosteroids (short course due to toxicity)
Promising
Lamotrigine
Felbamate
Adjunct Drugs
Benzodiazepines (e.g., nitrazepam, clonazepam)
Valproate (high dose)
Occasional Use
Pyridoxine
Immunoglobulins
Thyrotropin-Releasing Hormone

experience of the Parisian Hospital Saint Vincent de Paul in this topic during the past ten years.

STEROIDS

Despite a huge literature, numerous questions remain concerning the use of steroid therapy in IS: 1) what is the most efficient and least toxic compound? 2) What is the optimal dose? 3) What is the adequate duration for treatment? 4) Is etiology a factor? 5) Does the time lag to initiation of treatment influence the outcome? Answering these questions is difficult, because the large clinical series lack a standardized protocol[20] and studies with standardized protocols include only a small number of patients. According to large series, initial seizure control ranges from 45% to 77%, relapse rate from 33% to 56%, and favorable mental outcome from 14% to 58%. In a prospective, randomized, blinded trial, the efficacy of prednisone was inferior to that of corticotropin given for 2 weeks at 150 IU,[5] but equal when corticotropin was given at lower doses.[13]

Whatever the steroid compound used, the duration of treatment is debated. Although there has been a general tendency to increase the duration of treatment over the years, there is no evidence that a prolonged course of treatment gives better results. The effect of treatment on long-term prognosis is also controversial. Because the neurological and intellectual sequelae in symptomatic patients are mainly due to preexisting brain lesions, and because favorable mental outcome after steroids is better in cryptogenic than in symptomatic cases, some authors do not advise steroid treatment in symptomatic patients.[1]

To determine the effect of etiology within the symptomatic group—and the effect of brief versus extended treatment—Schlumberger and Dulac[21] designed a standardized schedule that was systematically applied to all patients referred to them between 1986 and 1990. They excluded patients known to be unresponsive to steroids (Aicardi's syndrome, lissencephaly) or better controlled by another drug (tuberous sclerosis). A large population of 94 IS infants, aged less than 14 months, was collected. Thirty-three had cryptogenic IS, and the other 61 had symptomatic IS, including 10 with periventricular leukomalacia, 9 with term asphyxia, 5 with porencephaly, 6 with malformations, 5 with postnatal events and 26 with undetermined etiology. The treatment schedule began with a combination of sodium valproate (VPA) (40 mg/kg/d) and hydrocortisone (HC) (15 mg/kg/d) for 2 weeks. If spasms and spikes on the EEG had disappeared, HC was tapered off over a two-week period, and VPA was maintained until the age of 12 to 18 months. The non-responders were switched to tetracosactrin (TA) (synthetic ACTH, 1 mg is pharmacologically equivalent to 100 IU ACTH) 0.1 mg/kg/d for 2 weeks, followed by HC 10 mg/kg/d in decreasing doses over a one-month period while VPA was maintained as described in responders.

Initial efficacy of HC was observed in 75% of the IS patients. Only 10% of them relapsed with spasms and 45% had a favorable mental outcome. Two-thirds of the non-responders to HC were secondarily controlled by TA, and none of them relapsed. Severe side effects were found in 17% of the population. They occurred in only 12% of the patients treated with HC alone, as against 33% of those having received TA. The authors concluded that the majority of the patients with IS had a lasting response to a short treatment schedule, resulting in a better tolerance to the drug.

As far as etiology was concerned, 96% of the patients with cryptogenic IS became initially spasm-free, 90% after a short course of HC. In contrast, only 65% of the patients with symptomatic IS became spasm-free after a short course of HC, the rate rising to 78% when TA was added. Relapse rates were comparable in both groups: 23% in cryptogenic, 20% in symptomatic patients. The initial response was different according to etiology, however: it was excellent in periventricular leucomalacia (9/10), porencephaly and postnatal brain damage (9/10), but it was bad in patients with undetermined etiology (1/26), and intermediate in patients with malformations and term neonatal hypoxia (9/15). Two years later, the responders were left with only the motor and sensorial sequelae due to the preexisting brain damage, whereas the non-responders were all severely developmentally delayed.

Steroids seem, therefore, to be useful in special etiologies of IS, particularly in anoxo-ischemic lesions happening during the last trimester of pregnancy or postnatally.

OTHER CONVENTIONAL DRUGS

Among the classical antiepileptic drugs, the benzodiazepines, nitrazepam and clonazepam, are the most effective in treating IS. According to the clinical series, 30% to 77% of the patients become initially seizure-free. The relapse rate is comparable to that observed with steroids, however, and side-effects can be severe, especially with high doses. In a randomized multicentric study comparing ACTH and nitrazepam on 48 patients, Dreifuss et al.[11] failed to find any significant difference in efficacy. Both drugs induced side effects with the same frequency, but the side effects were more severe with ACTH.

Valproate in very high doses (ranging from 45 to 100 mg/kg/d) has also been proposed for treating IS.[23] It has been reported to be effective in two-thirds of cases. It is not well tolerated, however, with side effects in up to 30% of cases.

Several other therapeutic approaches have occasionally been tried in the treatment of IS. They are still considered as experimental or only adjuvant therapy. They include pyridoxine,[18] immunoglobulins[4] and thyrotropin-releasing hormone.[17]

NEW DRUGS

Vigabatrin

Among the new drugs, vigabatrin (gamma-vinyl GABA) is the most important for the treatment of IS. It is an irreversible inhibitor of GABA transaminase, the enzyme that breaks down GABA in the brain. Vigabatrin (VGB) has antiepileptic efficacy in partial epilepsy in both adults and children. A pilot multicenter open study suggested that VGB might also be useful in severe generalized epilepsies.[16] A further open study on 70 patients with IS refractory to other drugs, including steroids, reported a proportion of 68% of responders, with 43% seizure-free.[8] The effect was observed within a few days, and maintained long-term in half the cases. VGB was very well tolerated. Only 25% of the patients showed side effects, and these were usually moderate and transient. They mostly consisted of drowsiness or hyperexcitability. The recommended VGB dose ranged from 100 to 150 mg/kg/d.

Because of its efficacy—and much less toxicity than steroids—VGB was quickly proposed as first-line therapy in IS.[3] Several open trials followed in Europe, where the drug was then available,[29,6,22] with reports of 44% of patients becoming spasm-free. A retrospective European survey on 192 infants with IS, who were treated with VGB as first-line monotherapy, confirmed the high efficacy of this strategy, since 68% of them experienced initial suppression of spasms and 79% of the responders maintained seizures control in the long term.[2]

Efficacy of VGB is different according to the etiology of IS. In the first open study dedicated to IS,[8] the initial results were better in symptomatic IS (70% of cases were spasm-free) than in cryptogenic IS (25% were spasm-free), and the best results were achieved in tuberous sclerosis (85% were spasm-free). Long-term efficacy was also etiology-related. Spasms reappeared in 40% of the cryptogenic cases, but in none of the patients with tuberous sclerosis, after more than 6 years of follow-up.

Because of these dramatic results, a randomized multicentre trial was performed, limited to IS due to tuberous sclerosis. VGB and hydrocortisone were compared as first-line monotherapy.[9] The randomized drug was crossed over if spasms persisted after the first month. At 2 months, the results demonstrated a significantly better and quicker efficacy for VGB, as compared to HC (all the patients treated with VGB became initially spasm-free). Moreover, VGB was much better tolerated than HC. The preliminary long-term data, however, showed several cases of relapse, as previously reported in the studies involving VGB as first-line monotherapy.[3,29,6] Because no case of relapse of spasms had been reported in patients previously treated by VGB as add-on therapy, it is recommended to wait for more data before using VGB as prolonged monotherapy in naive patients.

Follow-up of patients treated with VGB has shown that 40% of those with IS due to tuberous sclerosis exhibit ongoing partial seizures, whereas the spasms themselves are fully controlled.[8] This outcome is associated with a dramatic improvement in the intellectual condition of the patients.

Preliminary results have recently confirmed the efficacy of VGB compared to placebo in a blinded-controlled trial.[19]

Lamotrigine

Some interesting results have been reported concerning the use of lamotrigine in IS.[28] Thirty patients with refractory IS were entered in a single-blind, placebo-controlled, add-on study. Ages ranged from 1 month to 11 years. At 3 months, 9 patients were responders, 5 of them with complete cessation of spasms. Among the responders, the proportion of symptomatic cases (8/9), and of VPA co-medication, was significantly higher than in the group of non-responders. At a mean follow-up of two years, the good response was maintained, and the 5 seizure-free patients were receiving a combination of lamotrigine and VPA. Three patients in the series exhibited adverse events, including one cutaneous rash.

CONCLUSIONS

Vigabatrin represents a dramatic improvement in the therapy of IS and is now considered the first-line drug in this epilepsy syndrome in most countries in the world. Response is rapid and clearly dependent on etiology, with efficacy being the highest in tuberous sclerosis. Further data, however, still need to be collected about potential effects on the visual field in order to optimize the use of this new compound.

Efficient, but highly toxic, steroids have now become the drugs of second choice for IS. Oral steroids are better tolerated than ACTH, but may be less efficient.

Benzodiazepines and valproate are now only considered as adjunct drugs, and the preliminary data obtained with lamotrigine and with felbamate[15] need to be confirmed.

REFERENCES

1. Aicardi, J. *Epilepsy in Children*. Raven Press, New York (1986) pp. 17–38.
2. Aicardi, J., Sabril, I.S., Mumford, J., Dumas, C., and Wood, S. Investigator and peer review groups. Vigabatrin as initial therapy for infantile spasms: a European retrospective survey, *Epilepsia*, 37 (1996) 638–642.
3. Appleton, R.E. and Monteil-Viecca, F. Vigabatrin in infantile spasms. Why add-on? *Lancet*, 341 (1992) 962.
4. Ariizumi, M., Baba, K., Hibio, S., Shiihara, H., Michihiro, N., Ogawa, K., and Okubo, O. Immunoglobulin therapy in the West syndrome, *Brain Dev*, 9 (1987) 422–425.
5. Baram, T.Z., Mitchell, W.G., Tournay, A., Snead, O.C., Hanson, R.A., and Horton, E.J. High dose corticotropin (ACTH) versus prednisone for infantile spasms: a prospective, randomized, blinded study, *Pediatrics*, 97 (1996) 375–379.
6. Buti, D., Lini, M., Bardini, R., Marvulli, I, Tromboli, P., and Nencioli, L. Early treatment of infantile spasms with vigabatrin monotherapy: results in 4 patients, *Epilepsia*, 32 (1991) 89.
7. Chiron, C., Dulac, O., Luna, D., Palacios, L., Mondragon, S., Beaumont, D., and Mumford, J. Vigabatrin in infantile spasms, *Lancet*, 1 (1990) 363–364.
8. Chiron, C., Dulac, O., Beaumont, D., Palacios, L., Pajot, N., and Mumford, J. Therapeutic trial of Vigabatrin in refractory infantile spasms. *J Child Neurol*, 6 (1991) S52–59.
9. Chiron, C., Dumas, C., Jabaqué, I., Mumford, J., and Dulac, O. Randomized trial comparing vigabatrin and hydrocortisone in infantile spasms due to tuberous sclerosis, *Epilepsy Research*, 26 (1997) 389–395.
10. Cret, L., David, M., Macabeo, V., Frederich, A., and Jeune, M. Maladie des spasmes en flexion: troubles cliniques et métaboliques observés en cours de traitement par le tetracosactide zinc. *Pediatrie*, 31 (1976) 33–48.
11. Dreifuss, F., Farwell, J., Holmes, G., Joseph, C., Lockman, L., Madsen, J.A., Minarcik, C.J. Jr., Rothner, A.D., and Shewmon, D.A. Infantile spasms, comparative trial of nitrazepam and corticotropin. *Arch Neurol*, 43 (1986) 1107–1110.
12. Dulac, O. and Schlumberger, E. Treatment of West syndrome. In: *The treatment of epilepsy: principles and practice*. E. Wyllie (Ed.) Lea and Febiger, Philadelphia (1993) pp. 595–604.
13. Hrachovy, R.A., Frost, J.D., Kellaway, P., and Zion, T.E. Double-blind study of ACTH vs prednisone therapy in infantile spasms, *J Pediatrics* (1983) 641–645.

14. Hrachovy, R.A., Glaze, D.G., and Frost, J.D. A retrospective study of spontaneous remission and long-term outcome in patients with infantile spasms, *Epilepsia*, 32 (1991) 212–214.
15. Hurst, D.L., and Rolan, T.D. The use of felbamate to treat infantile spasms, *J Child Neurol*, 10 (1995) 134–136.
16. Livingston, J.H., Beaumont, D., Arzimanoglou, A., and Aicardi, J. Vigabatrin in the treatment of epilepsy in children, *Br J Clin Pharmac*, 27 (1989) S109–S112.
17. Matsumoto, A., Kumagai, T., Takeuchi, T., Miysxski, S., and Watanabe, K. Clinical effects of thyrotropin-releasing hormone for severe epilepsy in childhood: a comparative study with ACTH therapy, *Epilepsia*, 28 (1987) 49–55.
18. Ohtsukay, Y., Mtasuda, M., Ogino, T., Kobayashi, K., and Ohtahara, S. The treatment of the West syndrome with high-dose pyridoxal phosphate, *Brain Dev*, 9 (1987) 418–421.
19. Peters, A.C., Appleton, R.E., Roi, L., and Thornton, J.L. Vigabatrin as first-line monotherapy in newly diagnosed infantile spasms: a placebo-controlled double-blind study, *Epilepsia*, 37 (1996) 118 (Abst.)
20. Riikonen, R. A long term follow-up study of 214 children with the syndrome of infantile spasms, *Neuropediatrics*, 13 (1982) 14–23.
21. Schlumberger, E. and Dulac, O. A simple, effective and well-tolerated treatment regimen for West Syndrome, *Dev Med Child Neurol*, 36 (1994) 863.
22. Schmitt, B., Wohlrab, G., and Bolthauser, E. Vigabatrin in newly diagnosed infantile spasms, *Neuropediatrics*, 25 (1994) 54.
23. Siemes, H., Spohr, H.L., Michael, T., and Nau, H. Therapy of infantile spasms with valproate: results of a prospective study. *Epilepsia*, 29 (1988) 553–560.
24. Snead, O., Benton, J.W., and Myers G.J. ACTH and prednisone in childhood seizure disorders, *Neurology*, 33 (1983) 966–970.
25. Snead, O.C., and Chiron, C. Medical treatment. In: Dulac, O., Chugani, H.T., Dalla Bernardina, B. (Eds.), *Infantile spasms and West syndrome, IV. Etiology*, W.S. Saunders, London (1994) pp. 244–256.
26. Snead, O.C., Hosey, L.C., Swann, J.W., Spink, D., Martin, D.M., and Rej, R. Treatment of infantile spasms with high dose ACTH: efficacy and plasma levels of ACTH and cortisol. *Neurology*, 39 (1989) 1027–1031.
27. Sorel, L. and Ducausy-Bauloye, A. A propos de 21 cas d'hypsarythmie de Gibbs. Son traitement spectaculaire par l'ACTH, *Acta Neurologica Belgica*, 58 (1958) 130–141.
28. Veggiotti, P., Cieuta, C., Rey, E., and Dulac, O. Lamotrigine in infantile spasms. *Lancet*, 44 (1994) 1375–1376.
29. Vles, B.I., Van der Heyden, A.M., Ghils, A., and Troost, J. Vigabatrin in the treatment of infantile spasms, *Neuropediatrics*, 24 (1993) 230–231.

7

SURGERY FOR WEST'S SYNDROME

Harold J. Hoffman

Division of Neurosurgery
The Hospital for Sick Children
Professor, Department of Surgery
University of Toronto

Many cases of West's syndrome are inoperable, because there is no clear epileptic focus. In some cases, however, there is a structural abnormality which can be surgically removed. Dr. Hoffman discusses his experience with surgery for cortical dysplasia in West's syndrome.

INTRODUCTION

West's syndrome or "infantile spasms" is named after a physician by the name of West, who wrote a letter to Lancet in 1841 in which he described a "peculiar form of convulsion" associated with mental retardation as seen in his son. Clinically, West's syndrome consists of a triad of myoclonic spasms, hypsarrhythmic EEG and mental retardation.

PATHOLOGY

The structural abnormality in symptomatic infantile spasms may be quite small or may be extensive involving an entire hemisphere. The usual abnormality is cortical dysplasia. A dysembryoplastic neuroepithelial tumor, however, which is a mixture of malformation and tumor, can also give rise to the abnormality. Patients with hemimeganencephaly, in which one hemisphere is diffusely enlarged with disorganized cortical laminae, can also present with infantile spasms. Patients with focal tubers and with a variety of tumors and cysts also present with infantile spasms.[1,2,3,4,5]

Cortical dysplasia, however, is the commonest condition that gives rise to infantile spasms. Patients with cortical dysplasia benefit from surgical excision of the area of malformation.[6] In cortical dysplasia, there is a cortical laminar disorganization. There may be heterotopic neurons in the white matter, and there can be subpial granular cells. There also may be glioneural nodules in the subarachnoid space. The neurons may be enlarged and there can be "balloon cell" changes very similar to the balloon cell changes that one sees in tuberous sclerosis. Some of the patients with cortical dysplasia have lissencephaly, polymicrogyrea and schizencephaly.

Focal cortical dysplasia is commonest in the central insular region. The irritative zone is usually much more extensive than the visible structural lesion. The dysplastic lesions can be visible or invisible on routine MRI. Occasionally they can be seen on CT and they are visible with PET scanning. Echoplaner MRI will usually show the lesion. The area of cortical dysplasia has an intrinsic epileptogenicity. The patients with balloon cells do not do as well as those without balloon cells.

SURGERY

Frameless stereotaxy has been very useful in defining the anatomical boundaries of focal cortical dysplasia. Electrocorticography is also useful in defining the electrical abnormality in these patients. Similarly, patients with dysembryoplastic neuroepithelial tumors, other tumors, cysts and focal tubers also benefit from frameless stereotaxy in defining the lesion and helping remove the entire lesion.

When cortical dysplasia involves one hemisphere—and the other hemisphere is normal—one can do a hemispherectomy. The type of hemispherectomy that I favor consists of a decortication in which one peels off the cortex, while leaving the white matter. If one gets into the ventricle, this is repaired with surgical and fibrin glue. The calvarium (skull) over the side of the hemispherectomy is then morcellated to reduce its size so that there is very little space present between the residual hemisphere and the overlying skull.

Using this technique, we have never seen hydrocephalus, and the functional results are much better. Frequently hand movement and even finger movement is preserved.

RESULTS: THE TORONTO EXPERIENCE

Among 11 patients with symptomatic infantile spasms and cortical dysplasia who underwent surgery, none was older than eight months when their infantile spasms began. All 11 patients had focal cortical dysplasia. In four of these 11 patients a second operation had to be done to remove residual cortical dysplasia.

Five of the 11 patients are now completely free of seizures. Two patients have an occasional seizure. Two patients have only nocturnal seizures, and one patient is improved but still has an ongoing seizure problem.

DISCUSSION

Patients with poorly controlled symptomatic infantile spasms face a possible risk of Lennox-Gastaut syndrome, with intractable seizures and mental retardation. With

Table 1. Indications for Surgery in West's Syndrome

– symptomatic
– presence of a structural abnormality, such as cortical dysplasia

early effective treatment of infantile spasms in patients who have a malformative or neoplastic lesion in their brain, one can stop the seizures and allow these children to function with a relatively normal intelligence and no need for medication.

CONCLUSIONS

Young children with infantile spasms who harbor a malformative or benign neoplastic lesion in their brain can be relieved of their infantile spasms by removing the offending lesion.

REFERENCES

1. Asanumu, H., Wakai, S., Tanaka, T., and Chiba, S. Brain tumors associated with infantile spasms, *Pediatr Neurology*, 12 (1995) 361–364.
2. Branch, C.E. and Dykea, P.R. Choroid plexus papilloma and infantile spasms, *Ann Neurol*, 24 (1979) 1–8.
3. Chugani, H.T., Shield, W.D., Shewmon, D.A., Olson, D.M., Phelps, M.E., and Peacock, W.J. Infantile spasms: Pet identifiers cortical dysgenesis in cryptogenic cases for surgical treatment, *Ann Neurol*, 27 (1990) 406–407.
4. Minaki, T., Ono, J., and Yabauuchi, H. Temporal lobe astrocytoma with infantile spasms, *Ann Neurol*, 14 (1983) 695.
5. Palm, D.G. and Korinthenberg, R. West syndrome and Lennox-Gastaut syndrome in children with porencephalic cysts: Long-term follow-up after neurosurgical treatment. In: Niedermyer, E. and R. Degen (Eds.) *The Lennox-Gastaut Syndrome*, Allen R. Liss, New York (1988) pp. 419–426.
6. Shields, W.D., Shewmon, D.A., Chugani, H., and Peacock, W.J. The role of surgery in the treatment of infantile spasms. *J Epilepsy*, 3(Supl) (1990) 321–324.

8

LENNOX-GASTAUT SYNDROME

Clinical Description and Diagnosis

Ernst Niedermeyer

305 Colonial Court
Towson, MD
21204
tel: (410) 828-1396 or (410) 828-4380
fax: (410) 828-4456

The Lennox-Gastaut syndrome (LGS) is widely known as one of the most severe and prognostically unfavourable epileptic conditions. Its entity as a syndrome is based on the occurrence of certain characteristic types of seizure (particularly typical drop attacks and axial tonic seizures) and on EEG patterns such as the slow spike-wave complex and runs of rapid spikes (10–25/sec).

INTRODUCTION

"Syndromes" are clusters of *clinical* signs and symptoms, most commonly caused by various etiologies. In the diagnosis of the Lennox-Gastaut syndrome (LGS), however, the clinical semeiology is not sufficient. The clinical signs must be combined with typical EEG features. It is their integrated appearance that is the true basis for diagnosis of LGS.

Clinical description was sufficient for the establishment of West's syndrome. Dr. West's[45] letter to Lancet is considered its first valid description, and the hypsarrhythmic EEG pattern, detected by Gibbs and Gibbs,[25] served as corroborative evidence. The discovery of LGS as an epileptological entity, however, had to wait for the advent of EEG. Once Gibbs, Gibbs and Lennox[24] had shown that children with slow spike-waves have a much poorer prognosis than those with the classical 3/sec spike-wave complex, epileptologists had to find the clinical underpinning for the ominous EEG pattern, i.e. the slow spike-wave complex. This task was done by Lennox and Davis[30] and Lennox,[29] with special emphasis being placed on the ictal symptomatology which has proved to be quite peculiar and almost specific for LGS. Gastaut and his coworkers[20] provided further clinical and EEG detail in rich measure. This has justified the use of the term LGS, which was introduced at the American EEG Society meeting in 1967.[32,33]

CLINICAL SYMPTOMATOLOGY AND CONSIDERATIONS

Even though the observation of certain EEG patterns was the starting point in the discovery of the LGS, the clinical aspects should be presented first.

Age of Onset

Lennox-Gastaut syndrome starts most commonly between ages 1 and 10 years, although in some cases, it starts as early as age 6 to 12 months. Onset in the second decade is much less common. The figure of 16% given by Komai[27] appears to be quite high. Onset in adult life is possible under special circumstances,[3,4] as will be discussed later.

Prevalence

In a large epilepsy center (St. Paul, Marseille), the prevalence of LGS was found to be 5.1%. This figure rose to 10.2% in patients below age 15 and fell to 0.6% in patients above age 15.[21] There is no special gender predominance. Racial differences (LGS more often found in white rather than black population) do not reach statistical significance.[40]

Onset and Etiologies (Idiopathic vs. Symptomatic Forms)

There is a distinct dichotomy within the LGS: the existence of idiopathic and symptomatic forms. There is no demonstrable structural brain damage found in idiopathic forms. The seizures start “out of the blue sky” in a physically and mentally normal child, often beginning with major convulsions (generalized tonic-clonic) before the typical ictal semeiology of LGS manifests itself. In the symptomatic form, the LGS is engrafted upon pre-existing cerebral impairment, such as a variety of cerebral palsies and numerous other types of pathology.

Epileptologically, LGS may follow a preceding West’s syndrome or a sequence of severe neonatal seizures and subsequent West’s syndrome. In LGS, there is no justification for the term “epileptic encephalopathy”:[22] it is an epileptic syndrome with or without structure lesions.

There is reason to presume that symptomatic forms of LGS are more common than idiopathic cases. In the material of Ohtahara et al.,[41] only 21.1% (56 out of 265 cases) were found to be idiopathic.

Neurological Deficits

Neurological defects are incompatible with the idiopathic form of LGS. In symptomatic cases, all forms of cerebral palsy may become associated with LGS. Neurological deficits, hydrocephalic and microcephalic skull anomalies are found in about 20% of the cases.[33]

Mental Changes

With the usual onset prior to age 10 years, LGS is almost always associated with some degree of mental retardation. The latter ranges from mild to profound, and, as a rule of thumb, it can be said that the earlier the onset of the seizures, the more serious is the degree of mental deficiency. If the seizures start in the second decade of life, there is usually no evidence of mental retardation.

Serious behavioral changes are uncommon in LGS; if they exist, toxic medication effects have to be ruled out. In late onset LGS, however, drug-unrelated behavioral changes—even reaching psychotic degrees—may occur.

Neuroimaging

CT scan and Magnetic Resonance Imaging (MRI) are essential in the demonstration of structural impairment. Normal results support a diagnosis of idiopathic LGS.

ICTAL SYMPTOMATOLOGY

Table 1 lists: 1) seizures also occurring in other epileptic conditions, and 2) seizures occurring mainly or almost exclusively in LGS. In addition, there are certain forms of status epilepticus that are typically found in association with LGS.

Seizures Found in Other Forms of Epilepsy

Generalized Tonic-Clonic (Grand Mal). Tonic-clonic seizures play a major role only at the onset of LGS, especially in its idiopathic form. In general, however, major convulsions are uncommon. It is worthwhile to ponder the fact that the most dreaded epileptic syndromes are characterized by the predominance of minor seizure types. Is there a special element in the functional anatomy of LGS that precludes the evolution of a tonic-clonic attack?

Complex Partial (Psychomotor) Seizures. Complex partial seizures with typical temporal lobe automatisms are exceptional until after age 30, when a trend toward "secondary temporalization" indicates gradually evolving temporal lobe epilepsy with typical ictal and EEG features.[31,34,39] The dynamics of this development will be discussed later.

Myoclonus. There has been some controversy about the role of myoclonic seizures in LGS. According to Aicardi,[1] "part of the difficulty arises from the fact that there exist several types of brief seizures manifested by sudden, brief jerks". One can agree with Dravet[15] and Giovanardi Rossi et al.[26] that myoclonic seizures do occur in LGS but that their importance pales in comparison with other myoclonic epilepsies.

Focal Motor and Other Simple Partial Seizures. The occurrence of classical focal motor and other simple partial seizures (including adversive seizures) is not common in the LGS. Interestingly, Gastaut et al.[20] and Beaumanoir and Dravet[6] do not mention these seizure types.

Seizures Occurring Almost Exclusively in the LGS

Tonic Seizures. Tonic seizures in the LGS are mostly axial but may also show a proximal appendicular and global character. Either extension or flexion may prevail. These attacks are facilitated by NREM sleep and may manifest themselves during nocturnal EEG recordings very frequently. They appear as subtle and often barely detectable seizures with prominent EEG features. They may also be promoted by awakening from sleep.[6] There is no doubt that tonic seizures are the most common ictal manifestation in LGS and may be regarded as almost pathoynomonic.

Drop Attacks (Mostly Atonic). Sudden loss of muscle tone may lead to lightning-like and often traumatizing falls, necessitating the use of protective headgear. The atonia may be initiated by subtle myoclonus.

In some cases, the myoclonic component is quite powerful and may propel the patient to the ground.[17,40] Even a tonic component can be essential in drop attacks. There is no loss of consciousness with these falls. Atonic drop attacks are presumed to arise from the premotor cortex[9] which might be functionally "suspended" for the duration of the attack.[36]

Clonic Seizures. Rapidly repetitive widespread clonic movements characterize this seizure type, which may evolve after a tonic seizure. ("When they last more than a few tens of seconds, they can culminate in a tremor pattern which is constituted by a series of very rapid, small amplitude jerks affecting the whole body".[6])

Hemiclonic Seizures. Hemiclonic seizures are infrequent but clearly documented (personal data).

Atypical Absences. This is a very common manifestation with LGS. Their occurrence tends to be periodical, and there is a strong tendency toward prolongation and evolution into an atypical absence status. The investigations of Beaumanoir and coworkers[5] have shed more light on the clinical semeiology of absences and absence status in LGS. These conditions, with constantly repetitive slow spike-wave bursts, are notorious for their duration and lack of response to i.v. benzodiazepines. A length of several weeks is not uncommon.

In the more serious forms of atypical absence status, intellectual functions are markedly impaired, and a confusional state may dominate the picture. Additional tonic seizure activity and rapid clonic seizures ("vibratory seizures"; 5) are frequently found during these episodes.

EEG CHANGES IN LENNOX-GASTAUT SYNDROME

The Slow Spike-Wave Complex

The slow spike-wave complex is the classical EEG pattern of the LGS. The complexes range from 1–2.5/sec (Fig. 2). The occasional occurrence of 3–4/sec spike-waves, however, by no means precludes the diagnosis of LGS. The spike-waves usually occur as generalized-synchronous bursts, and tend to show the same frontal-midline-maximum

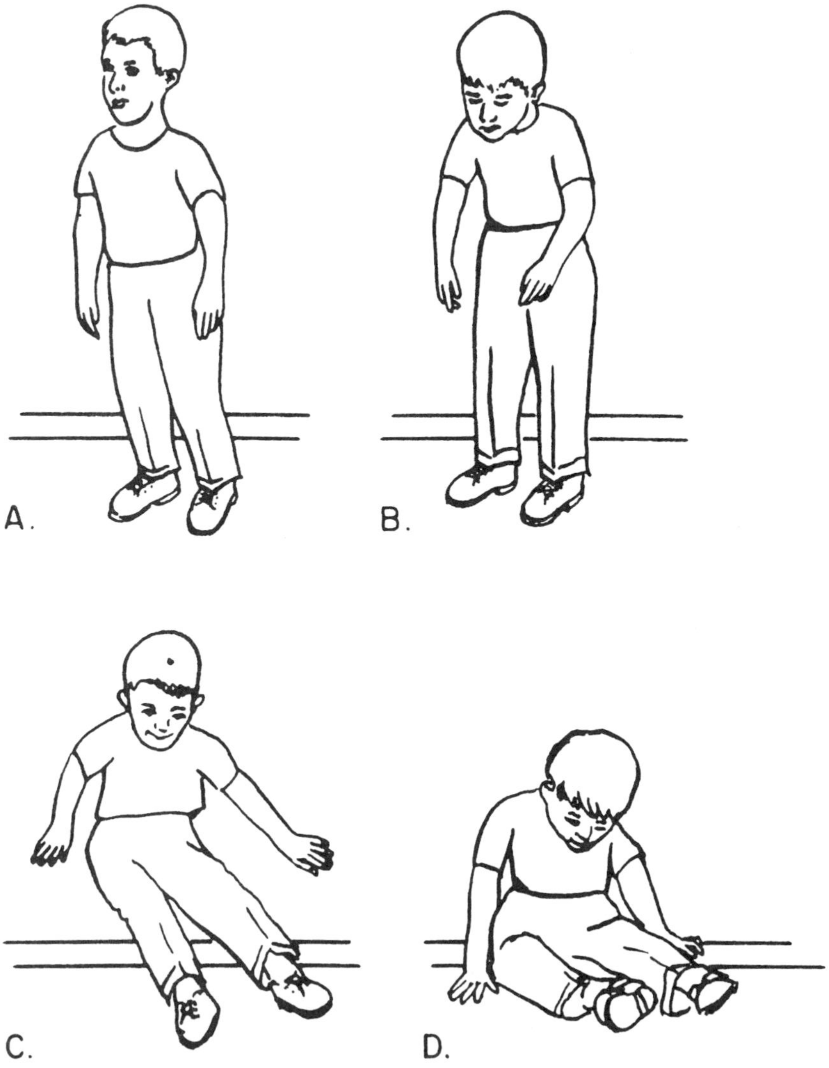

Figure 1. Atonic drop attack in Lennox-Gastaut Syndrome. Note that the boy could have fallen on his face with severely traumatizing effects. Reprinted with kind permission of Lippincott, Williams and Willkins Publishers.

Table 1. Types of Seizure Occurring in Lennox-Gastaut Syndromc

Incidence in Lennox-Gastaut Syndrome	
A) SEIZURES OCCURRING IN OTHER SEIZURE DISORDERS	
Generalized Tonic-Clonic (grand mal)	+
Complex Partial (psychomotor)	+
Myoclonus	+
Simple Elementary Partial (focal motor and other focal)	±
Adversive	±
B) SEIZURES OCCURRING ALMOST EXCLUSIVELY IN THE LENNOX-GASTAUT SYNDROME	
Tonic	+ + + +
Atonic drop (with tonic and myoclonic elements)	+ + +
Clonic (rapidly repetitive clonic movements)	+
Hemiclonic	+
Atypical absences (with slow spike-waves)	+ +

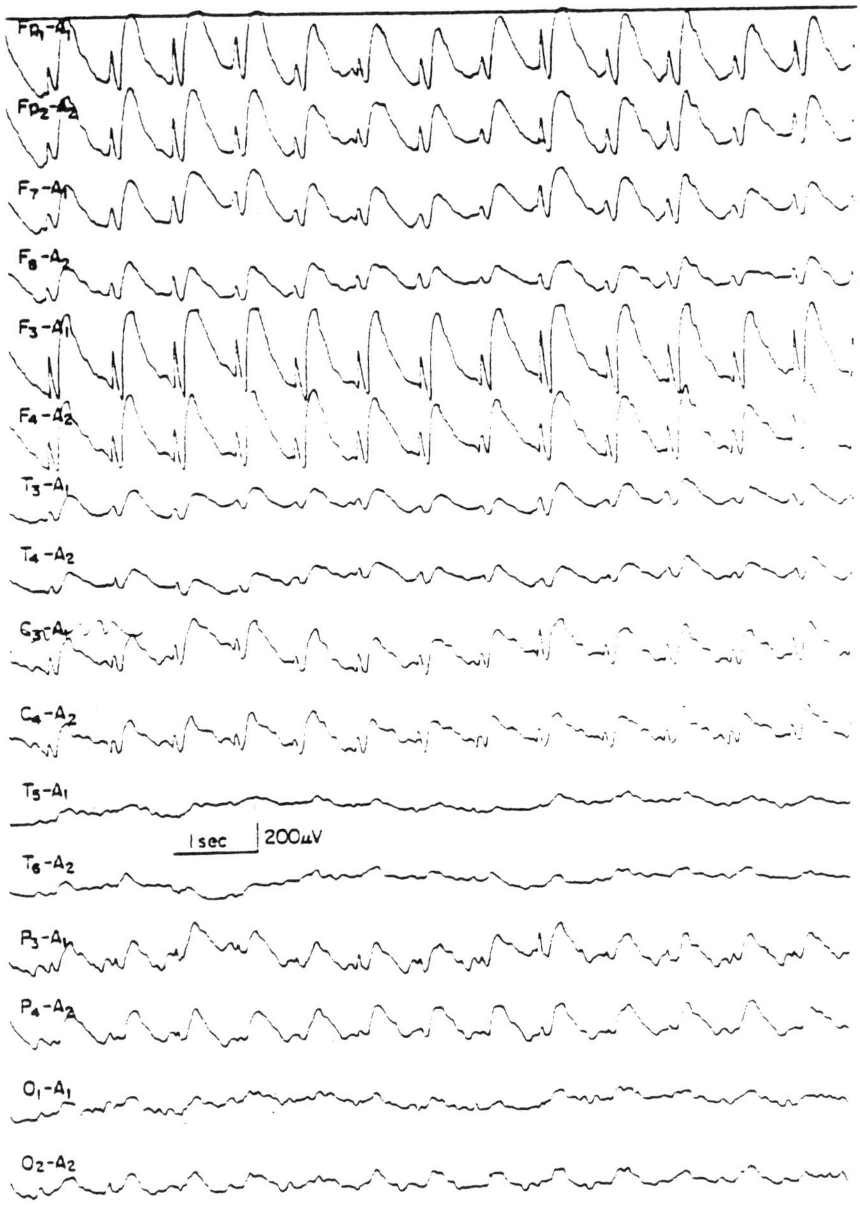

Figure 2. Slow spike-wave complexes, 1.5–2/sec, generalized-synchronous with frontal maximum, in a waking 11-year-old child with Lennox-Gastaut syndrome. There were no accompanying ictal manifestations. It must be said, however, that such a degree of clock-like rhythmicity is seldom found in patients with slow spike-waves. Reprinted with kind permission of Lippincott, Williams and Willkins Publishers.

which is the rule in primary generalized epilepsy (e.g. classical 3/sec spike-waves in "petit mal" absences). Incidentally, this fact should cast some doubt on the classification of LGS as a symptomatic epileptic seizure disorder with secondary generalization. Slow spike-waves may be found in wakefulness, drowsiness, and NREM sleep.

While the slow spike-wave complex is the classical EEG pattern of LGS, it must not be considered diagnostic, as will be discussed later.

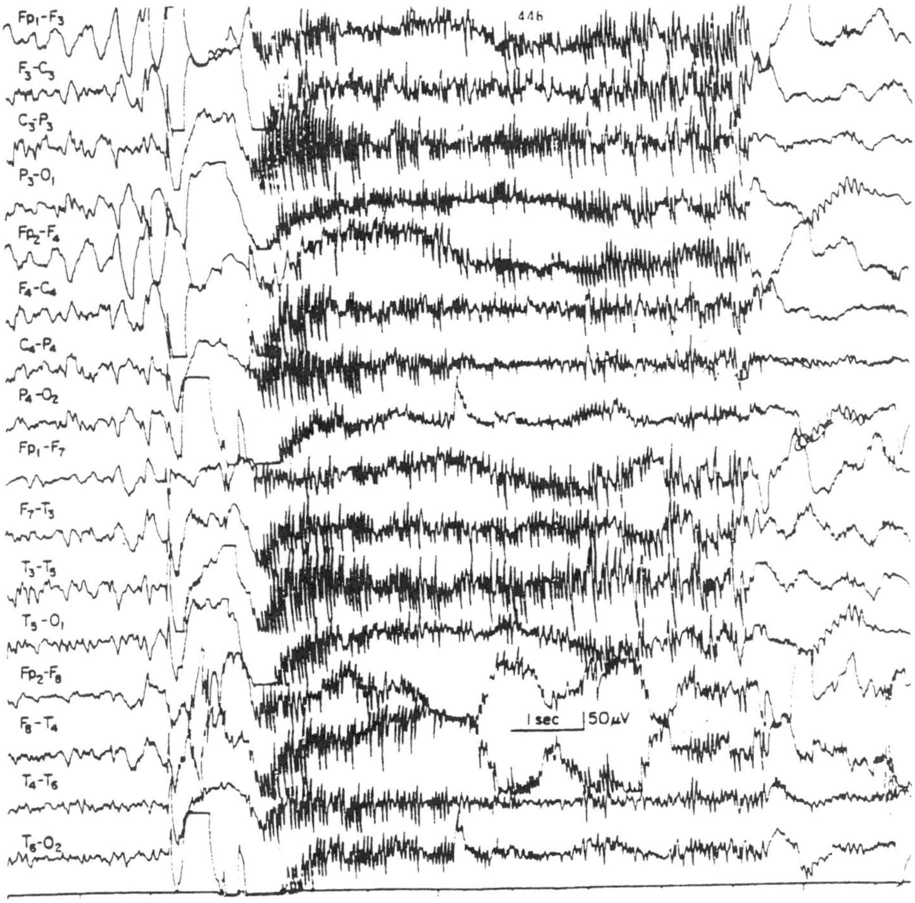

Figure 3. An example of runs of rapid spikes in a 20-year-old patient with Lennox-Gastaut syndrome, occurring in light NREM sleep and associated with mild axial tonic stretching. Reprinted with kind permission of Lippincott, Williams and Willkins Publishers.

Runs of Rapid Spikes

This pattern was originally described as "grand mal discharge",[23,25] because of similarities to the EEG pattern seen during the tonic phase of generalized tonic-clonic convulsions. It is characterized by bilateral-synchronous spiking at a rate of about 10–25/sec, with frontal maximum. It occurs mostly in sleep, often quite abundantly, either as brief subclinical bursts or in association with tonic axial seizures (Fig. 3).

Runs of rapid spikes in sleep constitute an almost specific pattern for LGS.

Other Paroxysmal Patterns

Classical spike-waves (3–4/sec) are not uncommon in LGS. Bursts of polyspikes—shorter and even faster than the runs of rapid spikes (20–60/sec)—occur mainly in sleep. Single spikes and sharp waves are quite common in depth recording of patients with LGS.[18]

Non-Paroxysmal Abnormalities

In the vast majority of the patients, the EEG shows excessive slowing, affecting the posterior basic rhythm and also diffusely distributed. In the earlier stages, a normal background of activity may be present.

ETIOLOGICAL INFLUENCES

Many etiologies produce the full-blown picture of LGS; all forms of prenatal, perinatal and postnatal brain damage. These produce LGS after preceding periods of neonatal seizures and—in frequent cases—West's syndrome). All types of cerebral palsy may be associated with LGS, but, most often, the evolving epileptic seizure disorder does not assume LGS features. This is also true for tuberous sclerosis and other neurodegenerative diseases. There is also a post-traumatic variant of LGS, with typical EEG changes but atypical clinical and ictal signs (see below).

It often appears that exactly the same type of sustained cerebral injury (including biochemical disturbances) produces cerebral palsy and an associated "ordinary" seizure disorder in child A and an identical type of cerebral palsy but with LGS in child B. This clearly demonstrates the role of genetic factors, which, however, have thus far defied full elucidation.

THE END OF THE LGS: SECONDARY TEMPORALIZATION

Around the age of 30 years, ictal and EEG changes of the LGS tend to change. This is not always the case. In a personal and rather exceptional observation, all of the ictal and EEG criteria of the LGS were still present in a 39-year-old patient. In general, however, ictal and EEG signs tend to become dominated by temporal lobe signs. Psychomotor (complex partial) automatisms become preponderant, while slow spike-waves vanish and prominent sharp waves arise from the anterior temporal regions. Most of these patients are mentally in a deteriorated state and are often institutionalized.

This development has been described as "secondary temporalization".[31,34,39] On the other hand, the typical ictal-clinical and EEG pattern (slow spike-waves) may persist—even up to the age of 64.[7]

THE DIAGNOSIS

The Core of the Diagnosis in Typical Cases

What leads to a diagnosis of LGS? The occurrence of more than one of the minor types of seizures including drop attacks (atonic seizures), plus typical EEG changes (slow spike-waves, as well as runs of rapid spikes), provide very strong evidence for LGS. Mental retardation (except for late-onset cases) further corroborates the diagnosis. Presence or absence of pre-existing brain damage (normal or abnormal neuro-imaging) is not a crucial factor. The underlying type may be *either* idiopathic or symptomatic (see Table 2).

The integration of clinical and EEG evidence is absolutely crucial.

Table 2. Diagnosis of Lennox-Gastaut Syndrome: Important Features

1) The occurrence of more than one of the minor clinical seizure types, including drop attacks (atonic seizures).
2) The appearance of characteristic patterns in the EEG, including slow spike-wave complexes (1–2.5/sec) and runs of rapid spikes (10–25/sec).
3) Mental retardation (except in late-onset) cases is corroborative.
4) Evidence of structural brain abnormalities is not required. Perhaps 20–30% of cases are idiopathic.

Problems in Differential Diagnosis

The core of the LGS ("LGS proper") is easily diagnosed, but the fringe is beset with great difficulties for differential diagnosis. In particular, the slow spike-wave complex may occur in a number of other conditions. The LGS is also subject to dynamic forces that may cause changes from one epileptic syndrome to another one. Some cases, therefore, may be difficult to diagnose. Particular problems are indicated below:

1) The clinical features are present but the EEG criteria are missing.

In general, the EEG abnormalities of the LGS are quite readily demonstrable. Hence the logical answer when they are absent would be "then forget about LGS!". A wiser response is: repeat the tracing and make sure that the record contains portions of wakefulness, drowsiness and NREM sleep. In some cases, the typical anomalies are limited to one of these states. If the clinical-ictal signs of LGS are correctly interpreted, the absence of typical EEG abnormalities may be just temporary. They will surely materialize somewhat later.

2) The EEG shows typical changes, but there are no clinical-ictal signs.

This situation is not rare. The lack of clinical-ictal signs, however, may be more apparent than real. The seizures may be so inconspicuous that they escape detection. This is particularly true for tonic-axial seizures occurring in sleep only.

3) The EEG shows typical changes, and clinical-ictal signs are present, but there are neither neurological nor mental deficits.

This is a common situation in late onset LGS (after age 10). Lack of neurological disorder and/or mental deficits does not preclude diagnosis.

4) The EEG shows typical changes, but the ictal symptomatology is different.

This constellation is found in a special form of posttraumatic epilepsy in older children, adolescents and young adults. Generalized tonic-clonic and psychomotor (complex-partial) seizures are present (rather than the variety of typical minor seizures). Neurological and mental deficits may be present, provided that the CNS trauma was particularly severe. We are hence dealing with a "strictly electrical LGS".

This variant of the LGS was described by Niedermeyer, Walker and Burton[38] and has been confirmed repeatedly. Its prognosis is clearly more favorable than that of the classic LGS.

5) The EEG shows slow spike-waves (and no runs of rapid spikes). The ictal symptomatology differs; gelastic (laughing) seizures predominate. The children show precocious puberty, and a hypothalamic hamartoma is demonstrable.

This is a rare clinical entity known since the work of Andermann and Berkovic.[2]

6) The EEG shows *unusually* slow spike-waves. There are significant neurological and mental deficits. Seizures (generalized tonic-clonic, focal) are present.

This syndrome is found in a severe encephalopathy generated by radiation necrosis—usually in children with deep-seated midline brain tumors.[35]

7) The EEG shows slow spike-waves of occipital (rather than frontal) maximum. There is a moderately mild seizure disorder, with various forms of visual seizures and other types (mostly followed by severe headache).

This situation may occur in benign occipital lobe epilepsy.[19,38] It shows that the slow spike-wave complex is not necessarily associated with severe forms of epileptic seizure disorder.

8) The EEG shows slow spike-waves and runs of rapid spikes, but their maximum is occipital. There is the typical clinical and ictal symptomatology of LGS.

This is the "occipital variant of LGS", according to Gastaut. It is a very rare finding.

9) The EEG shows typical spike-waves, especially in NREM sleep, but their maximum is midtemporal. There are major convulsions or various minor seizures. Clinically, a severe form of aphasia dominates the picture.

This is found in the Aphasia-Convulsion-Syndrome (Landau-Kleffner syndrome).

10) The EEG shows slow spike-waves of undetermined localization. The children (invariably girls) have profound mental retardation along with strange motions. Seizure types vary (major convulsions, minor seizures).

This may occur in the Rett syndrome. Most of these girls, however, show central spikes in sleep. Slow spike-waves are an exception, not the rule.

11) Throughout all NREM phases of sleep, the EEG shows continuous slow spike-wave-like activity (no true spike-waves) and their maximum is variable. There are epileptic seizures of various types (not LGS-like). Mild mental retardation is common.

This syndrome is found in ESES: Electrical Status Epilepticus of Sleep.[42,44] The sharp eyes of the electroencephalographer will note the poor development of the slow-wave component, whereas the spikes are quite prominent in this generalized-synchronous pattern, which is absent in REM sleep and in the waking state.

12) Atypical Astatic Epilepsy of Childhood.

Table 3. Differential Diagnosis of the Lennox-Gastaut Syndrome

1) POSTTRAUMATIC EPILEPSY WITH SLOW SPIKE-WAVE COMPLEXES
   Imitates Lennox-Gastaut syndrome electrically but not clinically (seizure types usually grand mal or psychomotor)
2) FRONTAL LOBE EPILEPSY WITH SECONDARY BILATERAL SYNCHRONY
   EEG and clinical differentiation very difficult or even impossible
3) ESES SYNDROME ("electrical status epilepticus of sleep")
   Children with almost continuous generalized slow spike-wave-like (i.e., no typical spike-wave complexes) activity throughout non-REM sleep, not in waking state and REM.
   No upper frontal maximum as in Lennox-Gastaut syndrome (instead, posterior or vertex maximum).
   Clinically, a seizure disorder much milder than Lennox-Gastaut syndrome.
4) APHASIA-CONVULSION SYNDROME (LANDAU-KLEFFNER SYNDROME)
   In children with aphasia and seizure disorder, which is rather mild. Slow spike-wave activity most impressive in sleep, often generalized and continuous but with maximum over midtemporal area. In most cases, a self-limited disease.
5) BENIGN OCCIPITAL LOBE EPILEPSY
   A benign epileptic condition with some relationship to migraine often associated with slow spike-waves with occipital or posterior temporal maximum, especially in the waking state. Clinically, seizures usually initiated with visual symptomatology and often followed by severe headache.
6) RETT SYNDROME
   A degenerative CNS disease (only in girls). Slow spike-wave activity common in earlier stages; the maximum of spike-wave activity varies (mainly temporal or occipital).
7) RADIATION NECROSIS ENCEPHALOPATHY
   A severe encephalopathy with epileptic seizure disorder (grand mal and various types focal seizures) may develop following radiation treatment. These changes occur mainly in deep-seated midline tumors of childhood or adolescence. Slow spike-wave complexes are particularly slow and may become very pronounced, usually generalized with frontal maximum; asymmetries are common.
8) EPILEPTIC SEIZURE DISORDER (ALONG WITH PRECOCIOUS PUBERTY) CAUSED BY HYPOTHALAMIC HAMARTOMA
9) MYOCLONIC ASTATIC EPILEPSY OF CHILDHOOD
   A controversial clinical entity; has to be considered in the differential diagnosis even though the occurrence of slow spike-wave complexes has not been clearly demonstrated.

In closing, a word should be said about Atypical Astatic Epilepsy. Hardly any other epileptic syndrome has raised so much controversy as this condition. It was first described by Doose[13] as "akinetic petit mal", later by Kruse[28] as "myoclonic—astatic petit mal". Drop attacks are common but the course is thought to be more benign as compared to LGS. Doose[14] feels that this syndrome is taxonomically more closely related to primary generalized epilepsy than to the LGS.

Table 3 presents a condensed overview of differential diagnostic possibilities (based mainly on the presence of the slow spike-wave EEG pattern).

LENNOX-GASTAUT SYNDROME AND SECONDARY BILATERAL SYNCHRONY

At an international conference held in Bologna in October 1984, Gastaut pointed out that a considerable number of presumed cases of LGS may be based upon a primary epileptogenic focus giving rise to secondarily generalized EEG patterns, such as slow spike-waves and runs of rapid spikes. Since then, the question has arisen repeatedly as to whether a given case shows the "true LGS" or just LGS as a secondary phenomenon.

Clinical and ictal signs are essentially the same in both conditions, but it was felt that the prognosis of LGS as a secondary phenomenon is somewhat less severe.

In a thoughtful study, Blume[10] has addressed this problem. Unfortunately, neither the genesis of the presumed "true" or "primary" LGS nor that of secondary bilateral synchrony is well understood. A clear cut differentiation is quite difficult if not impossible.

Let us suppose that a left frontal primary epileptogenic focus leads to secondary bilateral synchrony with generalized slow spike-waves and ictal signs of LGS. Why slow spike waves and not classical 3–4/sec spike-waves? Probably because of pre-existent genetic predisposition to LGS rather than to primary generalized epilepsy.

FURTHER PROBLEMS OF DIFFERENTIAL DIAGNOSIS: THE MYOCLONIC EPILEPSIES

The core of LGS (the "LGS proper")—is easily diagnosed, whereas the fringe districts are quite fuzzy—as shown in the preceding list of LGS-like conditions. But this is not all. While the aforementioned forms of seizure disorder show distinct ictal or electrical similarities to LGS, there are epileptic conditions in which a predominantly unfavorable course raises the question of LGS. This criterion alone, however, is quite unreliable.

Myoclonus is certainly not alien to the LGS, but *strongly* predominant myoclonus is bound to cast doubt on a presumed diagnosis of LGS. This is particularly true in the following three epileptic conditions which should be separated from LGS:

Severe Myoclonic Epilepsy in Infants

This condition[15] shows onset in first year of life in (thus far) normal infants. Initially, generalized or unilateral febrile clonic seizures are followed by axial or limb myoclonus and partial types of seizures. The EEG is initially normal, but then slowing (diffuse or lateralized), spikes and generalized spike-waves (often 3/sec) emerge. Spike-wave responses to intermittent photic stimulation are often noted. Neuroimaging contributes little. Neurological deficits (ataxia, mild spasticity) and mental decline are the rule and the general prognosis is quite poor.[16]

Myoclonic Epilepsy ("Myoclonic Status") in Non-Progressive Encephalopathies

This condition[11] is characterized by severe axial hypotonia and polymorphous abnormal movements, presenting the picture of hypotonic ataxic cerebral palsy with a dystonic-dyskinetic syndrome and profound mental retardation.[12] Occasionally, chromosomal analysis may show 15q11 deletion with the picture of Angelman syndrome.

The crucial symptomatology consists of frequently recurrent myoclonic status episodes. The myoclonias may be rhythmic and associated with large generalized spike-wave discharges. The paroxysmal discharges increase during sleep. The prognosis is doubtful despite the non-progressive nature of the disorder.

Late Infantile Neuronal Lipofuscinosis (Jansky-Bielschowsky Disease)

This progressive neurodegenerative disorder is more likely to produce generalized slow spike-waves than other diseases of this major category. Bioccipital spikes generated by single flashes are quite common in this condition.

We are leaving aside the well-known forms of hereditary essential myoclonus epilepsy that should not pose any differential diagnostic problem, especially from the EEG viewpoint.

MALLEABILITY OF THE LGS DUE TO DYNAMIC FORCES

The Influences of Age

There are, however, dynamic forces that demonstrate that the LGS may undergo variations—one of them being the "secondary temporalization" in the late stage (mentioned above). These dynamic forces are also demonstrated by the frequently observed transition from West's syndrome to LGS. Because of West's syndrome's strict age limitations, the dynamic forces convert one severe epileptic condition into another one, usually an age-adjusted one.

Deterioration of a Benign Epileptic Syndrome

Nothing demonstrates dynamic forces better than the observations of Bauer et al.[3,4] Bauer et al. observed that patients initially suffering from a generalized idiopathic (primary) epilepsy with absences, generalized tonic-clonic seizures and 3/sec spike-waves in their EEG would sometimes gradually develop an LGS-like seizure disorder. This development was usually promoted by a severe status epilepticus, but "may also present a transient intermezzo during the course of the illness".[4]

In other words, LGS may indicate the deterioration of a pre-existing benign epileptic syndrome. These observations also shed light on the deteriorating forces of a generalized tonic-clonic status epilepticus.

Dynamic forces may be also the expression of genetic predisposition. Accordingly, a child with a strong epileptic focus (in particular: a frontal lobe focus) is likely to develop slow spike-waves, runs of rapid spikes and typical seizure types of the LGS due to such a predisposition. The presence of multiple spike foci may also facilitate the development of secondary bilateral synchrony imitating the LGS.[10] These dynamic forces lend support to the concept of a "biological continuum"[8,9] which exists between the extremes of purely genetic and purely acquired epilepsies.

Therapeutic considerations often show strange caprices of the sternly severe LGS but this is beyond the topic of this presentation. What we have seen is a syndrome with a clean cut core but mysterious edges.

REFERENCES

1. Aicardi, J. *Epilepsy in Children*, Raven Press, New York (1994).
2. Andermann, F. and Berkovic, S. Secondary generalized epilepsy in patients with hypothalamic hamartoma, precocious puberty, and laughing attacks. A study of five patients. In: E. Niedermeyer and R. Degen (Eds.) *The Lennox-Gastaut Syndrome*, Liss, New York (1988) pp. 433–446.

3. Bauer, G., Aichner, F., and Saltuari, L. Epilepsies with diffuse slow spike and waves of late onset, *Eur. Neurol.*, 22 (1983) 344–350.
4. Bauer, G., Benke, T., and Bohr, K. The Lennox-Gastaut Syndrome in Adulthood. In: E. Niedermeyer and R. Degen (Eds.) *The Lennox-Gastaut Syndrome*, Liss, New York (1988) pp. 317–327.
5. Beaumanoir, A., Foletti, G., Magistris, M., and Volanschi, D. Status epilepticus in the Lennox-Gastaut Syndrome. In: E. Niedermeyer and R. Degen (Eds.) *The Lennox-Gastaut Syndrome*, Liss, New York (1988) pp. 283–299.
6. Beaumanoir, A. and Dravet C. The Lennox-Gastaut Syndrome. In: J. Roger, M. Bureau, C. Dravet, F.E. Dreifuss, A. Perret, and P. Wolf (Eds.) *Epileptic Syndromes in Infancy, Childhood and Adolescence, 2nd ed.*, Libbey, London (1992) pp. 115–132.
7. Benbadis, S.R. and Dinner, D.S. Lennox-Gastaut Syndrome in the Elderly? *Clin. Electroencephalogr.*, 25 (1994) 142–147.
8. Berkovic, S.F., Andermann, F., Andermann, E., and Gloor, P. Concepts of absence epilepsy: discrete syndromes or biological continuum? *Neurology*, 37 (1987) 993–1000.
9. Berkovic, S.F., Reutens, D.C., Andermann, E., and Andermann, F. The epilepsies: specific syndromes or a neurobiological continuum? In: P. Wolf (Ed.) *Epileptic Seizures and Syndromes*, Libbey, London (1994) pp. 25–37.
10. Blume W.Y. Lennox-Gastaut syndrome and secondary bilateral synchrony: a comparison. In: P. Wolf (Ed.) *Epileptic Seizures and Syndromes*, Libbey, London (1994) pp. 285–297.
11. Dalla Bernardina, B., Trevisan E., Bondavalli, S., Colamaria, V., Bureau, M., Roger, J., and Dravet, C. Une forme particulière d'épilepsie myoclonique chez des enfants porteurs d'encéphalopathie fixée, *Boll Lega It. Epil.*, 29–30 (1980) 183–187.
12. Dalla Bernardina, B., Fontana, E., Sgro, V., Colamaria, V., and Elia, M. Myoclonic epilepsy ("myoclonic status") in non-progressive encephalopathies. In: J. Roger, M. Bureau, C. Dravet, F.E. Dreifuss, A. Perret, and P. Wolf (Eds.) *Epileptic Syndromes in Infancy, Childhood and Adolescence, 2nd ed.*, Libbey, London (1992) pp. 89–96.
13. Doose, H. Das akinetische Petit mal. II. Verlaufsformen und Beziehungen zu den Blitz-Nick-Salaam-Kraempfen und den Absencen. *Arch. Psychiat. Nervenkr.*, 206 (1964) 637–654.
14. Doose, H. Myoclonic astatic epilepsy of early childhood. In: J. Roger, M. Bureau, C. Dravet, F.E. Dreifuss, A. Perret, and P. Wolf (Eds.) *Epileptic Syndromes in Infancy, Childhood and Adolescence, 2nd ed.*, Libbey, London (1992) pp. 103–114.
15. Dravet, C. Les épilepsies graves de l'enfant, *Vie Méd.*, 8 (1978) 543–548.
16. Dravet, C., Bureau, M., Guerrini, R., Giraud, N., and Roger, J. Severe myoclonic epilepsy in infants. In: J. Roger, M. Bureau, C. Dravet, F.E. Dreifuss, A. Perret, and P. Wolf (Eds.) *Epileptic Syndromes in Infancy, Childhood and Adolescence, 2nd ed.*, Libbey, London (1992) pp. 75–88.
17. Egli, M., Mothersill, I., O'Kane, M., and O'Kane, F. The axial spasm—the predominant type of drop seizur in patients with secondary generalized epilepsy, *Epilepsia*, 26 (1985) 401–415.
18. Fisher, R.S. and Niedermeyer, E. Depth EEG studies in the Lennox-Gastaut Syndrome, *Clin. Electroencephalogr.*, 18 (1987) 191–200.
19. Gastaut, H. A new type of epilepsy: benign partial epilepsy of childhood with occipital spike-waves, *Clin. Electroencephalogr.*, 13 (1982) 13–22.
20. Gastaut, H., Roger, J., Soulayrol, R., Tassinari, C.A., Regis H., Dravet, C., Bernard, R., Pinsad, N., and Saint-Jean, M. Childhood epileptic encephalopathy with diffuse slow spike-waves (otherwise known as "petit mal variant") or Lennox syndrome, *Epilepsia*, 7 (1966) 139–179.
21. Gastaut, H., Gastaut, J.L., Gonçalves e Silva, E., and Fernandez Sanchez, G.R. Relative frequency of different types of epilepsy: A study employing the classification of the International League against Epilepsy, *Epilepsia*, 16 (1975) 457–461.
22. Genton, P. and Dravet C. Lennox-Gastaut Syndrome and Other Childhood Epileptic Encephalopathies. In: J. Engel, Jr. and T.A. pedley (Eds.) *Epilepsy: A Comprehensive Textbook*, Vol. 3, Lippincott-Raven, Philadelphia, (1998), pp. 2355–2366.
23. Gibbs, F.A., Davis, H., and Lennox, W.G. The electro-encephalogram in epilepsy and conditions with impaired consciousness, *Arch. Neurol. Psychiat.*, 34 (1935) 1133–1148.
24. Gibbs, F.A. and Gibbs, E.L. *Atlas of Electroencephalography. Vol 2*, Addison-Wesley, Cambridge, MA (1952).
25. Gibbs, F.A., Gibbs, E.L., and Lennox, W.G. The influence of the blood sugar level on the wave and spike formation in petit mal epilepsy, *Arch. Neurol. Psychiat.*, 41 (1939) 1111–1116.
26. Giovanardi Rossi, P., Gobbi, G., Melideo, G., Parmeggiani, A., Tullini, A., and Santucci, M. Myoclonic manifestations in the Lennox-Gastaut syndrome and other childhood epilepsies. In: E. Niedermeyer and R. Degen (Eds.) *The Lennox-Gastaut Syndrome*, Liss, New York (1988) pp. 137–158.

27. Komai, S. Lennox-Gastaut's syndrome. Prognosis of the secondary generalized epilepsies, *Epilepsia*, 18 (1977) 131 (abstract).
28. Kruse, R. *Das Myoklonisch-Astatische Petit Mal*, Springer, Berlin (1968).
29. Lennox, W.G. *Epilepsy and Related Disorders*, Little, Brown and Co., Boston (1960).
30. Lennox, W.G. and Davis J.P. Clinical Correlates of the Slow Spike-Wave Electroencephalogram, *Pediatrics*, 5 (1950) 626–644.
31. Niedermeyer, E. Considerations of the centrencephalic (generalized) type of epilepsy, *Delaw. Med. J.*, 38 (1966) 341–348.
32. Niedermeyer, E. The Lennox-Gastaut syndrome: A severe type of childhood epilepsy, *Electroencephalogr. Clin. Neurophysiol.*, 24(3) (1968) 283 (abstract).
33. Niedermeyer, E. The Lennox-Gastaut syndrome: A severe type of childhood epilepsy, *Deutsch Z. Nervenheilk*, 195 (1969) 263–282.
34. Niedermeyer, E. *The Generalized Epilepsies*, Thomas, Springfield (1972).
35. Niedermeyer, E. The Electroencephalogram in the Differential Diagnosis of the Lennox-Gastaut Syndrome. In: E. Niedermeyer and R. Degen (Eds.) *The Lennox-Gastaut Syndrome*, Liss, New York (1988) pp. 177–220.
36. Niedermeyer, E. Frontal Lobe Epilepsy: The Next Frontier. *Clin. Electroencephalogr.*, 29 (1998) 163–169.
37. Niedermeyer, E., Riggio, S., and Santiago, M. Benign occipital lobe epilepsy, *J. Epilepsy*, 1 (1988) 3–11.
38. Niedermeyer, E., Walker, A.E., and Burton, D. The slow spike-wave complex as a correlate of frontal or fronto-temporal post-traumatic epilepsy, *Eur. Neurol.*, 3 (1970) 330–346.
39. Nogueira de Melo, A. and Niedermeyer, E. Considerations of Secondary Temporalization, *Clin. Electroencephalogr.*, 22 (1991) 161–171.
40. Nolte, R., Wolff, M., and Krägeloh-Mann I. The atonic (astatic) drop attacks and their differential diagnosis. In: E. Niedermeyer and R. Degen (Eds.) *The Lennox-Gastaut Syndrome*, Liss, New York (1988) pp. 95–108.
41. Ohtahara, S., Ohtsuka, Y., Yoshinaga, H., Iyoda, K., Amano, R., Yamatogi, Y., and Oka, E. Lennox-Gastaut Syndrome: Etiological Considerations. In: E. Niedermeyer and R. Degen (Eds.) *The Lennox-Gastaut Syndrome*, Liss, New York (1988) pp. 47–63.
42. Patry, G., Lyagoubi, S., and Tassinari, C.A. Subclinical "electrical status epilepticus" induced by sleep in children, *Arch. Neurol.*, 24 (1971) 242–252.
43. Santiago, M. and Niedermeyer, E. Racial factors and epileptic seizure disorders, *J. Epilepsy*, 1 (1988) 31–33.
44. Tassinari, C.A., Daniele, O., Dravet, C., Bureau, M., Dalla Bernardina, B., Michelucci, R., Picornell-Darder, I., Vigevano, F., and Roger, J. Sleep polygraphic studies in some epileptic encephalopathies from infancy to adolescence. In: R. Degen and E. Niedermeyer (Eds.) *Epilepsy, Sleep and Sleep Deprivation*, Elsevier, Amsterdam (1984) pp. 175–189.
45. West, W.J. On a peculiar form of infantile convulsion, *Lancet*, 1 (1841) 724.

9

DRUG THERAPY IN LENNOX-GASTAUT SYNDROME

Kevin Farrell

Neurology
British Columbia's Children's Hospital
4480 Oak Street
Room A303
Vancouver, B.C. V6H 3V4
tel: (604) 875-2121
fax: (604) 875-2285

Like West's syndrome, Lennox-Gastaut syndrome resists most of the standard anticonvulsants. Dr. Farrell discusses the first- and second-line drugs which may be used—plus the ketogenic diet.

INTRODUCTION

A logical approach to the pharmacological management of Lennox-Gastaut syndrome (LGS) is complicated by several factors. There have been few well-designed studies of antiepileptic drugs in LGS. In addition, the criteria used for the diagnosis of LGS vary between studies. The multitude of different approaches to drug treatment reflects the evidence-based vacuum in which the treatment of LGS exists. This chapter will discuss some of these issues and will describe the authors approach to drug treatment of this condition.

DRUG STUDIES: PROBLEMS WITH THE EXISTING DATA

Lennox-Gastaut syndrome is characterized by the presence of multiple seizure types, and specific interictal and ictal EEG abnormalities.[4,40] Progressive cognitive dysfunction occurs in most patients, but is usually not apparent in the early stages. Tonic seizures, occurring particularly during sleep and associated with paroxysmal fast 10 Hz rhythms, are considered to be an integral feature of this disorder.[4,47]

In contrast, many reports include under the rubric of LGS syndrome any patient with symptomatic generalized epilepsy and seizures that cause the patient to fall. The problem with such an approach is that there are other epileptic syndromes characterized by drop attacks in which the prognosis is much better. Patients with atypical partial benign epilepsy will often present between 2 and 6 years of age with clusters of myoclonic and atonic seizures lasting between 2 and 4 weeks. Nocturnal partial seizures are prominent, particularly in the early stages of this type of epilepsy, but tonic seizures do not occur. In most of these patients, the seizures remit before 10 years of age and any mental regression is minimal.[2] Similarly, children with myoclonic astatic epilepsy are characterized by myoclonic, astatic and tonic-clonic seizures.[16] Tonic seizures, when they do occur, do so only late in the course of this disorder. The prognosis of these patients is also much better than the prognosis in those with LGS and their response to treatment differs. Thus, the lack of a widely accepted operational definition for LGS has complicated the evaluation of drug treatment.

Most reports on antiepileptic drug treatment of LGS involve open studies, and the patient populations are poorly described. The relative absence of well-designed studies relates to several factors. Classification of the seizure types on the basis of parental history is often inaccurate.[11] The subtlety of the seizures and their frequent occurrence during sleep makes measurement of seizure frequency difficult. The situation is further complicated by the fact that a diagnosis of LGS can often only be made several months or years after the onset of the seizures. Most patients present with partial or tonic-clonic seizures in early infancy. The characteristic tonic, myoclonic and astatic seizures often do not occur until some months later. In addition, the characteristic interictal slow spike-wave complexes often do not occur until 1–2 years after the onset of the seizures.[9] There is also a reluctance by neurologists to recruit children with frequent seizures into controlled studies. Finally, although seizures that become intractable usually do so during childhood, and are often generalized, phase 2 and 3 drug studies are almost always restricted to adults with partial epilepsy. The importance of a placebo-control study in the measurement of efficacy of antiepileptic drugs in LGS is highlighted by a study of cinromide in this disorder. Although treatment with cinromide was associated with a reduction in seizure frequency from the baseline period, a similar effect was noted with the placebo.[30]

GENERAL PRINCIPLES FOR PHARMACOLOGICAL THERAPY

Monotherapy is difficult to achieve in the treatment of LGS because of the poor response to antiepileptic drugs. This results in many patients being treated with multiple medications, a strategy which is usually associated with cognitive and behavioural side-effects. Avoidance of sedation is important because of the higher incidence of seizures during drowsiness.[42] It is often forgotten that the introduction of a new antiepileptic drug implies failure of the previous treatment, and should normally be accompanied by withdrawal of an ineffective drug.

Tolerance to the anti-epileptic effects of a drug appears to be a particular problem in patients with LGS. Tolerance has been described most frequently with benzodiazepines.[24] Animal studies have demonstrated that tolerance to antiepileptic effect occurs with most drugs. A history indicating that the child responded well to a variety of medications, but that the effect was short-lived, is heard commonly from parents of children with LGS. This suggests that tolerance may also depend partly on the type of epilepsy.

Withdrawal of the drug for a period of several weeks may be associated with recovery of seizure control when the drug is reintroduced, but this rarely results in prolonged seizure control.

FIRST-LINE DRUGS

In most LGS patients, the initial seizure type has been partial or generalized tonic-clonic. The seizure types characteristic of LGS do not appear until several months later. Similarly, the EEG may not demonstrate the classical interictal EEG changes for 1–2 years.[9] The term "first-line" drug is used here to describe a drug that the author would consider as the initial treatment, when a diagnosis of LGS is considered. Valproic acid, the benzodiazepines, lamotrigine and topiramate fall into this category.

Valproic Acid

Valproic acid has been considered by some to be the drug of choice in the treatment of LGS.[1,20] Valproic acid appears to be particularly effective in the control of myoclonic, atypical absence and atonic seizures.[32,41] It may be more effective in cryptogenic (idiopathic) LGS.[41]

Valproic acid is associated with many interactions, which complicate the use of this drug. Phenobarbital, phenytoin and carbamazepine induce the cytochrome P-450 metabolism of valproic acid. Valproic acid inhibits the metabolism of phenobarbital and lamotrigine. Valproic acid also displaces phenytoin from protein binding sites, increasing the free fraction of phenytoin.

The major drawback to the use of valproic acid is the risk of serious hepatotoxicity, which is greater in the age group of children with LGS.[17] The use of carnitine supplementation (100 mg/kg/day) has been proposed as a method of reducing the risk of hepatotoxicity, but data to support this assertion are somewhat limited. Valproic acid may also cause pancreatitis.[61,62] Also, while valproic acid is associated with fewer neurologic side effects than the benzodiazepines, it *may* be associated with stupor and coma.[17,49] Anorexia and gastrointestinal upset are the most frequent adverse effects in young children, particularly in those receiving valproic acid syrup. Thrombocytopenia, abnormalities of platelet aggregation and hypofibrinogenemia have also been reported in some patients using valproic acid.[52]

Benzodiazepines

The 1,4-benzodiazepines (clonazepam, nitrazepam, diazepam) and a 1,5-benzodiazepine (clobazam) have been reported to be particularly useful in the management of patients with LGS.[25,26,53] The 1,4- and 1,5-benzodiazepines appear to have different pharmacologic effects, and one drug may be effective when the other has failed.[39]

Clobazam is associated with much less sedation than the 1,4-benzodiazepines[12] and is also simple to use. In the author's experience, clobazam is more effective than 1,4-benzodiazepines in the management of tonic seizures, but not in the management of myoclonic seizures. The major drawback to the use of clobazam is the development of tolerance, which occurs in approximately 40% of patients.[21] When tolerance occurs, a dosage increase may improve seizure control. An alternative approach is to withdraw the drug for a period of time.

Clonazepam and nitrazepam are also helpful in the management of patients with LGS. They appear to be particularly effective in the management of myoclonic seizures and may be more effective than clobazam in the treatment of that seizure type. Clonazepam and nitrazepam are associated with significant neurotoxicity, particularly drowsiness, drooling, hyperactivity and ataxia.[39,51] Initiating therapy at a low dose and increasing the dose gradually may limit the neurotoxic effects of these drugs. There are no significant drug interactions.

Oral diazepam has also been used in the management of patients with minor motor seizures. It should be noted that both intravenous diazepam and lorazepam may induce tonic seizures in patients with LGS.[6,15,56]

Lamotrigine

Lamotrigine is a phenyltriazine which may have a particular role to play in the management of LGS. Patients with LGS have been reported to respond particularly well to lamotrigine, and up to one-third of patients may become seizure free.[18] In a placebo-controlled study in patients with Lennox-Gastaut syndrome, those receiving lamotrigine had a significant reduction in the frequency of all major seizures.[38] In one open study of LGS, 10 of 11 patients had a greater than 50% improvement in seizure control with lamotrigine and 1 patient became seizure free.[59] In another study, three of 10 patients with LGS became seizure free on lamotrigine and a further 3 had greater than 50% improvement.[50] In a prospective open study, 3 of 15 children with LGS associated with tonic seizures became seizure free and a further 8 had a greater than 50% improvement in seizure control.[22]

Lamotrigine is rapidly absorbed, and the peak concentration is reached within 2–3 hours. The oral bioavailability is almost complete.[65] The plasma protein binding is approximately 50%.[10] Lamotrigine is extensively metabolized in the liver by glucuronidation and exhibits linear kinetics.

Drug interactions with lamotrigine are common. The metabolism of lamotrigine is enhanced by enzyme-inducing drugs and inhibited by co-medication with valproic acid.[64] These interactions are clinically significant, and in order to achieve comparable blood levels, the dose of lamotrigine must be doubled in the presence of enzyme-inducing drugs and halved in the presence of valproic acid.[43] Neurotoxicity has been reported in patients receiving lamotrigine and carbamazepine at blood levels not usually associated with toxicity. The data regarding the influence of lamotrigine on carbamazepine-epoxide levels are conflicting.[43] The occurrence of clinical signs of neurotoxicity when lamotrigine is used in combination with oxcarbazepine, which is not metabolized to an epoxide, suggests that the interaction may be pharmacodynamic rather than pharmacokinetic. The metabolism of lamotrigine is enhanced in the presence of acetaminophen and the elimination half-life is decreased by approximately 20%.[13]

Skin rash, anorexia, drowsiness, headache and vomiting are the most frequently reported adverse effects associated with lamotrigine therapy. In placebo-controlled studies, rash occurred in 10% of adults receiving lamotrigine and 5% of those receiving placebo.[46] Lamotrigine was discontinued because of rash in 22 of 323 children between 2–16 years of age in open studies.[66] In approximately 4,700 patients in clinical trials, a higher incidence of skin rashes was noted in those started at a higher lamotrigine dose and in those receiving valproic acid co-medication. A lower incidence was observed in patients receiving enzyme-inducing drugs.[46] The association between rash and high initial drug concentration has been observed also with phenytoin and carbamazepine.[8]

The occurrence of rash when a patient is receiving lamotrigine may not necessarily preclude the further use of this drug. In our study of 56 patients with generalized epilepsy, we were eventually able to restart the medication without recurrence of rash in all 5 children. In one patient, the rash recurred on the first rechallenge, but did not recur after valproic acid had been withdrawn. There have been two previous reports of successful reintroduction of lamotrigine following the recurrence of rash. Lamotrigine was restarted without recurrence of rash in 6 of 8 patients studied by Tavernor et al.[57] Reintroduction of lamotrigine in children with rash was also successful in all 4 patients not receiving valproic acid who were rechallenged by Schlumberger et al.,[50] but was not successful in the patient receiving valproic acid.

Topiramate

Topiramate is a structurally novel compound that has a broad spectrum of antiepileptic activity. It is almost completely absorbed and has low protein binding (9–17%). There is limited hepatic metabolism, and most of the drug is eliminated unchanged via the kidney.[34] The plasma clearance of topiramate, however, is enhanced in patients receiving drugs that induce hepatic microsomal metabolism.[27]

In an open label study of topiramate in children with Lennox-Gastaut, a ≥50% reduction in seizure frequency was observed in tonic-atonic seizures in 6 of 7 children, and in atypical absence and generalized tonic-clonic seizures in 4 of 5 children.[27] In a double-blind randomized placebo-controlled study of topiramate in Lennox-Gastaut syndrome, a significant reduction in drop attacks and generalized tonic-clonic seizures was observed in patients on topiramate.[48]

Anorexia is a common adverse effect, and weight loss may be a problem in some children receiving topiramate. Mental dulling can also be a problem, particularly when the drug is used at higher dosages as often occurs in children with intractable seizures. Renal stones have been reported in 1.5% of patients receiving topiramate.[55]

SECOND-LINE DRUGS

The second-line drugs are those which are either less effective or which are associated with significant side-effects. These include acetazolamide, corticosteroids, ethosuximide, felbamate, and vigabatrin. The author would include also under this heading the ketogenic diet, on the basis that food is being used as a drug.

Acetazolamide

Acetazolamide is a sulfonamide which is a potent inhibitor of carbonic anhydrase. Acetazolamide has been demonstrated to be effective in the treatment of myoclonic seizures[36] and generalized tonic-clonic seizures. Tolerance appears to be a common problem associated with the use of acetazolamide. Anorexia and nausea are also common problems. Renal stone formation may occur.

Ethosuximide

Ethosuximide is a succinimide which has been used for many years in the treatment of absence seizures. Ethosuximide may also be helpful in the treatment of atypical absence seizures.[54] A combination of valproic acid and ethosuximide has been observed to be effective in poorly controlled patients with atypical absence or myoclonic absence seizures.

Adrenocorticotrophin

Corticotrophin (ACTH) treatment has sometimes been associated with good seizure control, particularly when used soon after the onset of cryptogenic (idiopathic) LGS.[63] Seizure control may not be achieved until up to 6 weeks after the onset of treatment.[31,63] Doses of 30–40 international units have been recommended.[41,63] Corticotrophin may have a particular role in the management of non-convulsive status and during periods of frequent seizures.[1,7,19]

Ketogenic Diet

The ketogenic diet may be effective in the treatment of LGS and symptomatic generalized epilepsy.[7,33] The ketogenic diet has been used for many years in the treatment of patients with intractable epilepsy. Fasting has been known since biblical times to have an influence on seizure control. The ketogenic diet was developed in an attempt to mimic fasting. Marked ketosis is achieved by limiting the number of calories the patient receives and by ensuring that between 75–80% of calories are derived from fat. The mechanism of action is unknown. Interest in the diet has been renewed since the publication in 1992 of an open study of 58 children.[33] Most of these children had multiple seizure types, were receiving multiple anti-epileptic drugs, and were considered among the most intractable patients in the clinic. Seizure control was improved in two-thirds of the patients and increased alertness and improved behaviour were observed in a significant number of patients. It is not clear whether the improvement in neurologic function relates to the seizure control, the diet or the withdrawal of medication.

In order to achieve ketosis, a prolonged fast of up to 48–72 hours may be necessary. The availability of an experienced dietician appears to be critical to the success of the diet. Acetazolamide should be discontinued prior to the diet because of the danger of severe metabolic acidosis when these treatments are used in combination.[14] The ketogenic diet appears to potentiate the drowsiness associated with benzodiazepine and phenobarbital therapy. In order to reduce the neurotoxicity associated with these drugs when used in combination with the ketogenic diet, it is the author's practice to reduce their dosage at the onset of the diet, and to withdraw these drugs over the next 2–3 weeks. Side-effects of the ketogenic diet include renal stones, hyperuricemia and acidosis. (For further discussion of the ketogenic diet, see Vining, this volume.)

Felbamate

Felbamate is a dicarbamate which has a broad spectrum of pharmacologic activity in animal models. It is almost completely absorbed and is 20–25% bound to plasma protein.[58] The elimination half-life ranges from 13–23 hours.

In a placebo controlled study in 73 patients with LGS between 4 and 36 years of age, there was a significant difference between felbamate and placebo in seizure counts by the parents.[23] Atonic seizures were reduced by 34% with felbamate and 9% with placebo. Felbamate appeared to be more effective at a dose of 45 mg/kg/day than at a dose of 15 mg/kg/day.

Anorexia, vomiting and sleepiness are the most common side effects associated with felbamate. The occurrence of aplastic anemia and liver failure in patients receiving felbamate, however, significantly limits the usefulness of this drug.

Vigabatrin

Vigabatrin was designed to increase brain GABA levels by the inhibition of GABA transaminase (GABA-T), the enzyme involved in the metabolism of this neurotransmitter. Vigabatrin replaces GABA as the substrate for GABA-T and acts as a "suicide" substrate by binding covalently to the active site.[29] Cerebrospinal fluid levels of GABA rise in a dose-related manner in patients treated with vigabatrin.[5]

Vigabatrin is rapidly absorbed, reaching peak concentrations within 1–2 hours.[44] Vigabatrin is not protein bound, and the major route of elimination is renal excretion. Pharmacokinetic studies in children do not show any effect of maturation on the disposition of the pharmacologically active enantiomer.[44] The bioavailability of vigabatrin may be lower in children. Vigabatrin is associated with a 20% fall in plasma phenytoin concentrations. The mechanism underlying this interaction is poorly understood. It has been suggested that a change in compliance might be a factor in the decrease in plasma phenytoin concentration.[45]

Improvement in seizure control in children with LGS has been reported in two uncontrolled studies of vigabatrin in children.[37,35] Patients with non-progressive myoclonic epilepsy appeared to respond less well, and the seizure frequency increased in some patients.

Agitation has been the most commonly reported side-effect of vigabatrin in children. Agitation occurs most often at the onset of therapy and is more common in children with mental retardation.[37] Marked hypotonia may occur in patients with infantile spasms receiving vigabatrin. In addition, myoclonic status may occur occasionally in these patients during treatment with vigabatrin. Visual field defects have been detected in some adults and children receiving vigabatrin. These may relate to inhibition of GABA-transaminase in the retina and altered cone function. Formal perimetry prior to treatment and at 6–12 month intervals has been recommended in children receiving vigabatrin who have a cognitive age of more than 9 years.[3] Visual evoked potentials and electroretinography are of no value in the detection of this abnormality.[3] Perimetry is not

Table 1. Medical Treatment of Lennox-Gastaut Syndrome

First-Line Drugs
Valproic Acid
The Benzodiazepines (clobazam, clonazepam, nitrazepam)
Lamotrigine
Topiramate
Second-Line Drugs
Acetozolamide
Ethosuximide
Adrenocorticotropin (ACTH)
Felbamate
Vigabatrin
Ketogenic Diet

feasible in children with Lennox-Gastaut syndrome, and the decision to use vigabatrin has to consider the potential benefits and risk. In that respect, most adults with vigabatrin-attributed visual field defects have been detected by formal perimetry techniques and only a small number have spontaneously complained of symptoms.

The dosages of vigabatrin used in pediatric studies have varied widely. All patients reported by Luna et al.[37] to have a 90–100% seizure reduction were treated with vigabatrin at a dose of 40–80 mg/kg/day. Based on these observations, a starting dose of 40 mg/kg/day has been recommended, with dosage increases up to 80 mg/kg/day if necessary.[28]

REFERENCES

1. Aicardi, J. *Epilepsy in Children*, 2nd Edition, Raven Press, New York (1994) International Review of Child Neurology Series.
2. Aicardi, J. and Chevrie, J.J. Atypical benign epilepsy of childhood, *Dev. Med. Child Neurol.*, 24 (1982) 281–292.
3. Appleton, R.E. Guideline may help in prescribing vigabatrin, *Br. Med. J.*, 317 (1998) 1322.
4. Beaumanoir, A. and Dravet, C. The Lennox-Gastaut syndrome. In: J. Roger, M. Bureau, C. Dravet, F.E. Dreifuss, A. Perret, and P. Wolf (Eds.) *Epileptic Syndromes in Infancy, Childhood and Adolescence*, 2nd Edition, John Libbey Eurotext, London, England (1992) pp. 115–134.
5. Ben-Menachem, E. Pharmacokinetic effects of vigabatrin on cerebrospinal fluid amino acids in humans, *Epilepsia*, 30(Suppl. 3) (1989) S12–S14.
6. Bittencourt, P.R.M. and Richens, A. Anticonvulsant-induced status epilepticus in Lennox-Gastaut syndrome, *Epilepsia*, 22 (1981) 129–134.
7. Brett, E.M. The Lennox-Gastaut syndrome: therapeutic aspects. In: E. Niedermeyer and R. Degen (Eds.) *The Lennox-Gastaut Syndrome*, Alan R Liss Inc., New York (1988) pp. 329–339.
8. Chadwick, D., Shaw, M.D., Foy, P., Rawlins, M.D., and Turnbull, D.M. Serum anticonvulsant concentrations and the risk of drug induced skin eruptions, *J. Neurol. Neurosurg. Psychiatr.*, 47 (1984) 642–644.
9. Chevrie, J.J. and Aicardi, J. Childhood epileptic encephalopathy with slow spike-wave. A statistical study of 80 cases, *Epilepsia*, 13 (1972) 259–271.
10. Cohen, A.F., Land, G.S., Breimer, D.D., Yuen, W.C., Winton, C., and Peck, A.W. Lamotrigine, a new anticonvulsant: Pharmacokinetics in normal humans, *Clin. Pharmacol. Ther.*, 4 5) (1987) 535–541.
11. Connolly, M.B., Farrell, K., Karim, Y., Smith, S., and Wong, P.K.H. Video-EEG in paediatric outpatients, *Epilepsia*, 35 (1994) 477–481.
12. Cull, C.A. and Trimble, M.R. Anticonvulsant benzodiazepines and performance, *Roy. Soc. Med. Int. Cong. Symp. Ser.*, 74 (1985) 121–128.
13. Depot, M., Powell, J.R., Messenheimer, J.A. Jr., Cloutier, G., and Dalton, M.J. Kinetic effects of multiple oral doses of acetaminophen on a single oral dose of lamotrigine, *Clin. Pharmacol. Ther.*, 48(4) (1990) 346–355.
14. De Vivo, D.C. How to use other drugs (steroids) and the ketogenic diet. In: P.L. Morselli, C.E. Pippenger, and J.K. Penry (Eds.) *Antiepileptic Drug Therapy in Pediatrics*, Raven Press, New York (1983) pp. 137–144.
15. Dimario, F.J. and Clancy, R.R. Paradoxical precipitation of tonic seizures by lorazepam in a child with atypical absence seizures, *Pediatr. Neurol.*, 4 (1988) 249–251.
16. Doose, H. Myoclonic-astatic epilepsy of early childhood. In: J. Roger, M. Bureau, C. Dravet, F.E. Dreifuss, A. Perret, and P. Wolf (Eds.) *Epileptic Syndromes in Infancy, Childhood and Adolescence*, 2nd edition, John Libbey Eurotext, London, England (1992) pp. 103–114.
17. Dreifuss, F. Valproate: Toxicity. In: R.H. Levy, F.E Dreifuss, R.H. Mattson, B.S. Meldrum, and J.K. Penry (Eds.) *Antiepileptic Drugs*, Raven Press, New York (1989) pp. 643–651.
18. Dulac, O. The use of lamotrigine in children. In: S. Johnson and F.N. Johnson *Rev. Contemp. Pharmacother.*, Marius Press (1994) pp. 133–139.
19. Dulac, O. and N'Guyen, T. The Lennox-Gastaut syndrome, *Epilepsia*, 34(Suppl.7) (1993) S7–S17.
20. Eeg-Olofsson, O. Treatment of epileptic disorders of children. In: P. Robb (Ed.) *Epilepsy Updated: Causes and Treatment*, Yearbook Medical Publishers, Chicago (1980) pp. 199–213.

21. Farrell, K. Benzodiazepines in the treatment of children with epilepsy, *Epilepsia*, 27 (1986) S45–S51.
22. Farrell, K., Connolly, M.B., Munn, R., Peng, S., and MacWilliam, L.M. Prospective, open-label, add-on study of lamotrigine in 56 children with intractable generalized epilepsy, *Pediatric Neurology*, 16 (1997) 201–205.
23. Felbamate Study Group in Lennox-Gastaut. Efficacy of felbamate in childhood epileptic encephalopathy (Lennox-Gastaut syndrome), *N. Engl. J. Med.*, 328 (1993) 29–33.
24. Frey, H-H. Experimental evidence for the development of tolerance to anticonvulsant drug effects. In: H-H. Frey, W. Fröscher, W.P. Koella, and H. Meinardi (Eds.) *Tolerance to Beneficial and Adverse Effects of Antiepileptic Drugs*, Raven Press, New York (1986) pp. 7–14.
25. Gastaut, H. and Lowe, M.D. Antiepileptic properties of clobazam, a 1,5 benzodiazepine, in man, *Epilepsia*, 20 (1979) 437–446.
26. Gastaut, H., Roger, J., Soulayrol, R., Tassinari, C.A., Regis, H., and Dravet, C. Childhood epileptic encephalopathy with diffuse spike-waves (otherwise known as "petit mal variant") or Lennox syndrome, *Epilepsia*, 7 (1966) 139–179.
27. Glauser, T.A. Preliminary observations on topiramate in pediatric epilepsies, *Epilepsia*, 38(Suppl 1) (1997) S37–S41.
28. Gram, L., Sabers, A., and Dulac, O. Treatment of Pediatric Epilepsies with g-Vinyl GABA (Vigabatrin), *Epilepsia*, (1992) S26–S29.
29. Grant, S. and Heel, R.C. Vigabatrin: A review of its pharmacodynamic and pharmacokinetic properties, and therapeutic potential in epilepsy and disorders of motor control, *Drugs*, 41(6) (1991) 889–926.
30. Group for the evaluation of cinromide in Lennox-Gastaut syndrome. Double-blind placebo-controlled study of cinromide in patients with Lennox-Gastaut syndrome, *Epilepsia*, 30 (1989) 422–429.
31. Hasaerts, D. and Dulac, O. Traitement par hydrocortisone des epilepsies generalisees secondaire de grand enfant, *Arch. Fr. Pediatr.*, 46 (1989) 635–639.
32. Jeavons, P.M., Clark, J.E., and Maheshwari, M.C. Treatment of the generalized epilepsies of childhood and adolescence with sodium valproate (Epilim), *Dev. Med. Child Neurol.*, 19 (1977) 9–25.
33. Kinsman, S.L., Vining, E.P.G., Quaskey, S.A., Mellits, D., and Freeman, J.M. Efficacy of the ketogenic diet for intractable seizure disorders: review of 58 cases, *Epilepsia*, 33 (1992) 1132–1136.
34. Kramer, L.D. and Reife, R.A. Topiramate. In: J.E. Jr. and T.A. Pedley (Eds.) *Epilepsy: a comprehensive textbook*, Lippincott-Raven Publishers, Philadelphia (1997) pp. 1593–1598.
35. Livingston, J.H., Beaumont, D., Arzimanoglou, A., and Aicardi, J. Vigabatrin in the treatment of epilepsy in children, *Br. J. Clin. Pharmacol*, 27 (1989) 109S–112S.
36. Lombroso, C.T. and Forsythe, I. A long-term follow-up of acetazolamide in the treatment of epilepsy, *Epilepsia*, 1 (1960) 493–500.
37. Luna, D., Dulac, O., Pajot, N., and Beaumont, D. Vigabatrin in the treatment of childhood epilepsies: A single-blind placebo-controlled study, *Epilepsia*, 30(4) (1989) 430–437.
38. Motte, J., Trevathan, E., Arvidsson, J.F., Barrera, M.N., Mullens, E.L., and Manasco, P. Lamotrigine for generalized seizures associated with Lennox-Gastaut syndrome, *N. Engl. J. Med.*, 337 (1997) 1807–1812.
39. Munn, R.I. and Farrell, K. Open study of Clobazam in refractory epilepsy, *Pediatr. Neurol.*, 9 (1993) 465–469.
40. Niedermeyer, E. The Lennox-Gastaut syndrome: Clinical Description and Diagnosis. In: W.M. Burnham, P.L. Carlen, and P.A. Hwang (Eds.) *Intractable Seizures: Diagnosis and Therapy*, Kluwer Academic/Plenum Publishers, New York (2000).
41. O'Donohoe, N.V. *Epilepsies of Childhood*, 2nd Ed., Butterworths, London (1985).
42. Papini, M., Pasquinelli, A., Armellini, M., and Orlandi, D. Alertness and incidence of seizures in patients with Lennox-Gastaut syndrome, *Epilepsia*, 25 (1984) 161–167.
43. Pisani, F. Pharmacokinetics of lamotrigine. In: E.H. Reynolds (Ed.) *Lamotrigine—a new advance in the treatment of epilepsy*, Royal Society of Medicine Services, London (1993).
44. Rey, E., Pons, G., and Olive, G. Vigabatrin, *Clin. Pharmacokinet.*, 12(4) (1992) 267–278.
45. Richens, A. Pharmacology and clinical pharmacology of vigabatrin, *J. Child Neurol.*, 5 (1991) 2S7–2S10.
46. Richens, A. Safety of lamotrigine, *Epilepsia*, 35(Suppl. 5) (1994) S37–S40.
47. Roger, J., Dravet, C., and Bureau, M. The Lennox-Gastaut syndrome, *Cleve. Clin. J. Med.*, 56 (1989) S172–S180.
48. Sachdeo, R.C., Glauser, T.A., Ritter, F., Reife, R., Lim, P., and Pledger, G. A double-blind, randomized trial of topiramate in Lennox-Gastaut syndrome, *Neurology*, 52 (1999) 1882–1887.
49. Sackallares, J.C., Lee, S.I., and Dreifuss, F.E. Stupor following administration of valproic acid to patients receiving other antiepileptic drugs, *Epilepsia*, 20 (1979) 697–703.

50. Schlumberger, E., Chavez, F., Palacios, L., Rey, E., Pajot, N., and Dulac, O. Lamotrigine in treatment of 120 children with epilepsy, *Epilepsia*, 35 (1994) 359–367.
51. Schmidt, D. How to use benzodiazepines. In: P.L. Morselli, C.E. Pippenger, and J.K. Penry (Eds.) *Antiepileptic drug therapy in paediatrics*, Raven Press, New York (1983) pp. 271–278.
52. Schmidt, D. Adverse effects of valproate, *Epilepsia*, 25(Suppl. 1) (1984) S44–S49.
53. Schmidt, D. Benzodiazepines—an update. In: T.A. Pedley and B.S. Meldrum (Eds.) *Recent Advances in Epilepsy*, Churchill Livingstone, Edinburgh (1985) pp. 125–135.
54. Sherwin, A.L. Ethosuximide: Clinical use. In: R.H. Levy, R.H. Mattson, and B.S. Meldrum (Eds.) *Antiepileptic Drugs*, 4th Ed., Raven Press, New York (1995) pp. 667–673.
55. Shorvon, S.D. Safety of topiramate: adverse events and relationships to dosing, *Epilepsia*, 37(Suppl 2) (1996) S18–S22.
56. Tassinari, C.A., Dravet, C., Roger, J., Cano, J.P., and Gastaut, H. Tonic status epilepticus precipitated by intravenous benzodiazepines in five patients with Lennox-Gastaut syndrome, *Epilepsia*, 13 (1972) 421–435.
57. Tavernor, S.J., Wong, I.C., Newton, R., and Brown, S.W. Rechallenge with lamotrigine after initial rash, *Seizure*, 4 (1995) 67–71.
58. Theodore, W.H., Jensen, P.K., and Kwan, R.M.F. Felbamate: clinical use. In: R.J. Levy, R.H. Mattson, and B.S. Meldrum (Eds.) *Antiepileptic Drugs*, 4th Ed., Raven Press, New York (1995) pp. 817–822.
59. Timmings, P.L. and Richens, A. Lamotrigine as an add-on drug in the management of Lennox-Gastaut syndrome, *Eur. Neurol.*, 32 (1992) 305–307.
60. Vining, E.P. The Ketogenic Diet. In: W.M. Burnham, P.L. Carlen, and P.A. Hwang (Eds.) *Intractable Seizures: Diagnosis and Therapy*, Kluwer Academic/Plenum Publishers, New York (2000).
61. Williams, L.H.P., Reynolds, R.P., and Emery, J.L. Pancreatitis during sodium valproate treatment. *Arch. Dis. Child*, 58 (1983) 543–544.
62. Wyllie, E., Wyllie, R., Cruse, R.P., Erenberg, G., and Rothner, A.D. Pancreatitis associated with valproic acid therapy, *Am. J. Dis. Child*, 138 (1984) 912–914.
63. Yamatogi, Y., Ohtsuka, Y., Ishida, T., Ichiba, N., Ishida, S., Miyake, S., Oka, E., and Ohtahara, S. Treatment of the Lennox syndrome with ACTH: a clinical and electroencephalographic study, *Brain Dev.*, 1 (1979) 267–276.
64. Yuen, A.W.C., Land, G., Weatherley, B.C., and Peck, A.W. Sodium valproate acutely inhibits lamotrigine metabolism, *Br. J. Clin. Pharmacol.*, 33 (1992) 511–513.
65. Yuen, A.W.C. and Peck, A.W. Lamotrigine pharmacokinetics: oral and i.v. infusion in man, *Br. J. Clin. Pharmacol.*, 26 (1988) 242.
66. Yuen, A.W.C. and Rafter, J.E.W. Lamotrigine (Lamictal) as add-on therapy in pediatric patients with treatment-resistant epilepsy: an overview, *Epilepsia*, 33(Suppl. 3) (1992) 82–83.

10

SURGERY IN LENNOX-GASTAUT SYNDROME

Corpus Callosum Division for Children

John R. Gates

Minnesota Epilepsy Group
310 North Smith Avenue #300
St. Paul, MN 55102
tel: (651) 220-5290
fax: (651) 220-5248

Atonic or "drop" attacks—often seen in Lennox-Gastaut syndrome—are some of the most feared childhood seizures, since they may lead to head injuries. Corpus callosum division will often control atonic seizures, and other types of generalized seizure.

INTRODUCTION

Corpus callosotomy was first performed by Van Wagenem and Herren in 1939 in Rochester, New York as a treatment for severe epilepsy in humans. The authors stated, "As a rule, consciousness is not lost when the spread of the epileptic wave is not great, or when it is limited to one cerebral cortex."[34] As the procedure has evolved in the subsequent half century, its indications, surgical techniques, results and consequences, especially as relates to children, are still the object of continuous clinical and investigational research. This chapter reviews the more prominent contributions to the fields of basic and clinical research relating to corpus callosotomy as a treatment for medically refractory epilepsy.

BACKGROUND: ANATOMY AND PHYSIOLOGY OF THE CORPUS CALLOSUM

There is considerable variation in the anatomy of the corpus callosum as one ascends the evolutionary scale. In mammals, for example, the corpus callosum develops in proportion to neocortical volume, and, therefore, reaches its maximum expression

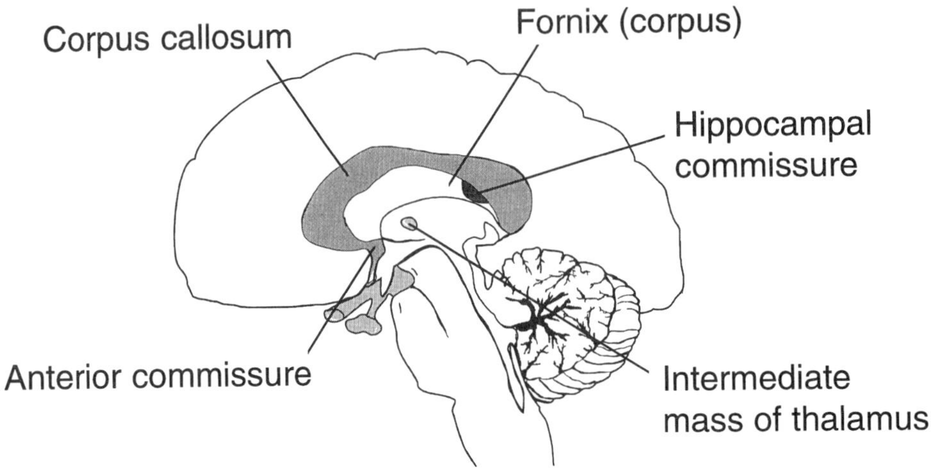

Figure 1.

in adult humans, where it measures approximately 10 cm. in length from the anterior commissure to the posterior commissure.[38] Figure 1 demonstrates the different anatomical components of the corpus callosum and their relationship to the other interhemispheric fiber tracts. It is estimated that the human corpus callosum contains 180 million axons.[2]

A fundamental concept in the physiology of the callosum is that the cerebral hemispheres may be connected either homotopically or heterotopically, as elaborated by Pandya in the Rhesus monkey.[25] In the Rhesus, for example, the primary auditory cortex sends projections mainly to the opposite auditory cortex, whereas the primary somatosensory cortex sends connections not only to the opposite somatosensory cortex, but also to secondary and supplementary sensory cortex. Consequently, auditory cortex may be considered more homotopic in Rhesus, and somatosensory cortex more heterotopic, as a consequence of these connections.

It is now known that different sets of neurons may project ipsilaterally as well. The connections between the cerebral hemispheres are organized in a rostral-caudal manner, with the frontal lobes occupying the rostral portion of the callosum and the other lobes following the anterior posterior anatomy. Grafstein in 1959[17] concluded that the homotopic connections, the callosal axons, terminated at roughly the same depth and relationship to the cerebral cortex as they arose in the opposite hemisphere. The nature of these connections appears to be both stimulatory and inhibitory, but the data do fully support the anatomical/physiological hypothesis that the corpus callosum plays the principle role in the propagation of electrical discharges from one hemisphere to the other.

Innocenti in 1986[19] introduced the concept of "exuberance" in relationship to the corpus callosum. It was meant to describe the widespread distribution of callosal projection neurons in the visual cortex of the neonatal cat. The widespread (heterotopic) distributions of such callosal neurons are not only seen in different mammalians, but are also subjected to substantial changes in the mature brain of the primate. Kass,[20] studying a different species of new world monkey, the prosimian galagos, and the old world macaque, demonstrated a significant reduction of density and distribution of callosal connections to some cortical fields, suggesting that alteration in the connections is part of a pattern related to the evolution of more complex processing systems with more cortical fields. Thus, it could be argued that a pathological preservation of ontogenic exuberance could theoretically work as a facilitator, by virtue of the heterotopic connections, the rapid development of secondarily generalized seizures. This is potentially prevented by callosal section.[15]

BACKGROUND: PROBLEMS IN EXTRAPOLATING FROM ANIMAL STUDIES TO HUMANS

The second Dartmouth International Conference on Epilepsy and the Corpus Callosum reported two important limiting factors in corpus callosum research:[36] 1) The simple extrapolation of findings in animal models to humans involves a significant risk, due to the considerable differences in callosal anatomy and physiology; 2) Although the corpus callosum constitutes the largest commissural pathway, it is not the only structure which functions as a commissure. The other structures include the hippocampal commissure, the anterior commissure, the massa intermedia, the habencular commissure, the posterior commissure, the supramammalary commissure, and the collicular commissure—plus the brain stem, cerebellum, and diencephalic structures.

Several studies emphasize the danger of extrapolating across species. Musgrave and Gloor in 1980[23] studied feline generalized penicillin epilepsy and assessed the role of the corpus callosum in the bilateral interhemispheric synchrony of spike and slow-wave discharges. Four different groups of felines were subjected to: 1) complete section of the corpus callosum and intercommissure; 2) division of the massa intermedia alone; 3) partial section of the massa intermedia; or 4) disconnection of the callosum, plus a slab of cortex, from all its subcortical inputs. The results of these experiments clearly demonstrated a persistence of generalized discharges in all but the first group, i.e., the complete callosal-sectioned group. Therefore, it was concluded that, in the feline generalized epilepsy model, the corpus callosum is the main, if not the exclusive, pathway ensuring bilateral synchrony of the epileptic discharge.

The Wistar rat, on the other hand,[35] spontaneously presents with generalized nonconvulsive seizures correlated to a 7–10 Hz generalized synchronous spike-and-wave discharge on the EEG. Vergnes et al.[35] demonstrated the presence of at least a few bilateral spike-and-wave discharges after callosum division. The persistence of generalized discharges suggested that, although the callosum plays a major role, it may not be the only structure involved.

The Singhalese baboon, Papio papio, has been extensively studied by Wada and colleagues.[37] The animals respond to intermittent photic stimulation at 25 Hz with a photo-myoclonic and eventually photo-convulsive discharge. Some of the animals had spontaneous generalized tonic-clonic seizures. These subjects, when subjected to kindling from the amygdala or frontal cortex, eventually develop bilateral synchronous discharges and convulsions. After collosal section, the synchronization of these paroxysmal discharges and seizures induced by photic stimulation was diminished.

The Rhesus monkey, on the other hand, presents us with a different observation. When these animals have the anterior 2/3 of the callosum severed, they develop partial motor seizures five times faster than non-bisected monkeys, and the seizures spread twice as fast as in nonoperated animals. In this species, the corpus callosum has an inhibitory role.[37]

Consequently, in three species, the cat, rat, and baboon, the corpus callosum has a facilitory role in the interhemispherical spread of epileptic discharges, whereas in the Rhesus monkey it appears to have an inhibitory role. Obviously, extrapolations from the animal models to the human must be made carefully.

EFFICACY: CLINICAL STUDIES IN CHILDREN

Van Wagenem and Herren[34] in 1940 subjected 10 patients to corpus callosotomy, 3 of them performed in 2 stages, and published their observations on this procedure in the Archives of Neurology. They apparently proceeded totally unaware of the excellent work by Erickson[9] on the spread of the epileptic discharge, also published in 1940 in the Archives of Neurology and Psychiatry. Instead, they based their decision to proceed on their own clinical observations of improvement in seizures in patients with damage to the corpus callosum due to natural disease processes. Subsequent clinical studies on adults included the work of Bogen and Vogel in 1962[5] and Bogen in 1965[3] and 1969,[4] who by this time was performing a multiple commisurotomy procedure which included the corpus callosum and several other structures as well.

It was in 1970 that Luessenhop[21] and colleagues became the first investigators to address the issue of surgical disconnection of the cerebral hemispheres for intractable seizures in children in infancy. Their series included four patients with cerebral hemiatrophy. Three of the 4 patients had a complete callosal division with a section of the anterior commissure at one of the fornices. The final patient had a complete corpus callosum and anterior commissure section with sparing of the fornix. Three of the 4 patients (1 with generalized tonic-clonic seizures and 2 with atonic-tonic seizures) had significant improvement. The electroencephalographic data from the series suggested that the best outcomes were obtained in children with clear hemispheric lateralization. Patients with independent bilateral spiking apparently did more poorly. Since this introduction of the surgical procedure, various modifications in the technique have occurred, resulting in a dramatically decreased morbidity and mortality.[6,14,16,22,24,26,28,29,31–33,39–41]

Various investigators have demonstrated the effectiveness of commissurotomy as a palliative procedure in individuals with uncontrolled atonic-tonic, and generalized tonic-clonic seizures.[21,14,16,28,39–40,2,11,18] Harbaugh, Wilson, Reeves, and Gazzaniga in 1983[6] concluded that atonic seizures responded favorably to corpus callosum section. Gates et al.[14] in 1987 similarly reported a highly statistically significant reduction of post-operative tonic or atonic seizures in their series of 24 adult patients specifically selected for that seizure type. The 17 patients with concomitant tonic-clonic episodes also had a highly statistically significant reduction in those events.

In a review specifically looking at children, Gates et al. in 1990[13] found only 21 cases in the peer-reviewed literature. More recently, however, a significant number of patients have been added to this group in studies by Spencer and Spencer, 1989,[30] Black, 1992,[1] Raffel, 1992[27] and Cendes, 1993.[7]

Gates and De Paola[12] summarize the literature to date on children. Work from our own group (Courtney and Gates) will be expanded (Tables 1 and 2).

METHODS: THE MINNESOTA STUDY

Twenty-two patients, age 3–9 years (mean: 5.8 years), had partial, complete, or two-staged complete corpus callosotomies (CC) using Wilson's technique.[39,40] The patients who were considered for corpus callosotomy fulfilled the following criteria (Tables 1 and 2):

1. Tonic, atonic, or tonic-clonic seizures were the patients' primary seizure type, although other seizure types (myoclonic, atypical absence, simple partial, complex partial seizures) were also present.
2. The patients' primary seizures were uncontrolled for at least 1–1/2 years (mean 3.9 years), despite appropriate trials of multiple antiepileptic drugs with documented therapeutic serum concentrations.
3. Seizures were of sufficient frequency or character as to result in serious injuries from seizure-related falls or to interfere significantly with activities of daily living.
4. All patients were documented to have multifocal secondary generalized electroencephalographic correlates for their ictal events.

PATIENT EVALUATION

Preoperative evaluation consisted of 3–14 days of continuous long-term monitoring (video EEG) utilizing surface electrodes with and without sphenoidal electrodes in order to record the patients' ictal events. Seizure frequencies pre- and post-surgery were obtained from the patients and their families, who kept meticulous seizure records. Each patient's seizures were categorized as tonic, atonic, tonic-clonic, myoclonic, atypical absence, simple partial or complex partial utilizing the International Classification of Seizures following an analysis of videotaped clinical events and family descriptions.

All patients experienced several of their primary seizures (i.e., either tonic, atonic, tonic-clonic, or a combination of the three types) which were captured on video EEG.

Electroencephalographically, the clinical events were associated with either an electrodecremental epileptiform fast (EEF) or slow spike-wave (SSW) pattern (Figs. 2, 3). On a few occasions, moderate to high-voltage theta slowing was seen, followed by either SSW or low-voltage fast activity at the onset of the patient's clinical events. When this particular EEG pattern was seen, it was classified as a SSW ictal pattern.

Satisfactory outcome was defined as 50% reduction in the primary seizure type, though response to all seizure types was assessed. Follow-up was 0.5–6.5 years (mean: 3.4 years).

Eight patients had single-stage complete corpus callosotomy (Tables 1 and 2). The remaining 14 patients underwent a partial callosal section, which was completed if there was no significant reduction in the patient's seizures. Four of the 22 patients had this two-stage corpus callosotomy procedure (Tables 1 and 2).

RESULTS

Of the 22 patients that underwent corpus callosotomy as described in the Methods Section, 50% achieved a satisfactory outcome (Table 1). In terms of seizure type: 1) 53% of patients with tonic seizures as their primary seizure type had a successful outcome; 2) 50% of patients evidencing atonic seizures as their primary seizure type demonstrated a satisfactory outcome; and 3) 50% of patients with either an EEF or SSW ictal electroencephalographic pattern demonstrated a satisfactory outcome. Patients displaying

Table 1. Satisfactory Outcome (>80% Reduction in Primary Seizure Type)

Minnesota Epilepsy Group Corpus Callosum Results: Seizure Freq (Per Month) Pre/Postoperatively in Patients Less than 10 Years of Age

Patient/ Sex	Age at Surgery (years)	Extent of Division	Tonic	Tonic-Clonic	Atonic	Simple Partial	Complex Partial	Myoclonic	Absence	Ictal Pattern Prior to Surgery	# AEDs
1/F	4	P	*750/120	–/–	–/–	–/–	–/–	120/90	120/90	2	3/3
2/F	5	P, C	–/–/–	16/–/–	*150/–/–	–/–/–	0.05/30/one after surgery	–/–/–	900/30/–	2	1/1/1
3/F	8	P	*945/30	–/–	–/–	–/–	–/–	–/–	–/–	1	3/2
4/F	5	C	–/–	0.5/–	*180/28	–/–	–/–	–/–	6,000/270	2	2/3
5/M	5	P^S	*1,500/–	–/–	–/–	–/–	60/–	–/–	–/–	2	2/2
6/M	5	C	*510/–	32/45	–/–	–/–	–/–	–/–	TNTC/12	2	2/2
7/F	5	P	*480/–	–/–	–/–	–/–	–/–	–/–	150/–	1	3/–
8/M	6	C	*300/32	–/–	–/–	–/–	–/–	300/–	600/9	1	2/3
9/M	8	C	*3,000/–	Freq status/15	300/8	–/–	–/–	300/–	–/–	1	4/2
10/M	9	P	*900/58	–/6	–/–	–/–	–/–	–/–	–/–	2	2/2
11/M	4	C	*165/30	–/–	–/–	–/–	–/–	–/–	–/–	1	2/2

*: Primary seizure type.
H: Hemi-tonic.
P^S: Partial callosotomy in patient S/P left frontal lobectomy 11/91.
C: Complete callosum section.
P: Anterior 2/3 to 3/4 of callosum section.
1: Electrodecremental epileptiform fast ictal EEG pattern.
2: Slow spike-wave ictal EEG pattern.
TNTC: Too numerous to count.

Table 2. Unsatisfactory Outcome (<80% Reduction in Primary Seizure Type)

Minnesota Epilepsy Group Corpus Callosum Results: Seizure Freq (Per Month) Pre/Postoperatively in Patients Less than 10 Years of Age

Pt/ Sex	Age at Surgery (years)	Extent of Division	Tonic	Tonic-Clonic	Atonic	Simple Partial	Complex Partial	Myoclonic	Absence	Ictal pattern prior to Surgery	# AEDs
1/F	9	C	*110/90	–/–	90/–	–/–	–/–	–/120	–/–	2	2/1
2/F	7	P	*600/180	150/–	240/–	–/–	–/–	120/–	600/–	1	3/2
3/M	5/5	P, C	*7,500/3,000/18,750	–/–/–	–/–/–	–/–/–	–/–/–	–/–/–	390/–/375	1	2/2/3
4/F	7/7	P, C	–/12[H]/–	*100/50/60[H]	–/–/–	–/–/–	–/–/–		TNTC/TNTC/TNTC	2	3/4/3
5/M	5	C	*92/33	40/14	–/–	–/–	40/–	–/–	240/–	2	3/4/3
6/M	4	P	*450/282	2 in lifetime/–	–/–	–/–	–/–	60 clusters/ –	120/–	1	2/2
7/F	6	P	*8/16	–/–	–/–	–/–	0.17/4	–/–	–/–	2	1/3
8/M	3/3	P, C	4/90/–	–/–/–	*90/60/300	–/–/–	90/16/8	–/–/–	–/–/–	1, 2	1/2/3
9/F	9	P	–/–	–/–	*30/900	–/–	–/–	300/–	–/270	1	2/2
10/F	3	P	*300/180	–/–	–/–	–/–	–/120	–/–	–/–	1	2/3
11/F	5	C	*45/38	0.5/–	–/–	–/68	67/–	2,250/–	–/–	2	2/3

*: Primary seizure type.
H: Hemi-tonic or hemi-tonic-clonic.
C: Complete callosum section.
P: Anterior 2/3 to 3/4 of callosum section.
1: Electrodecremental epileptiform fast ictal EEG pattern.
2: Slow spike-wave ictal EEG pattern.
TNTC: Too numerous to count.

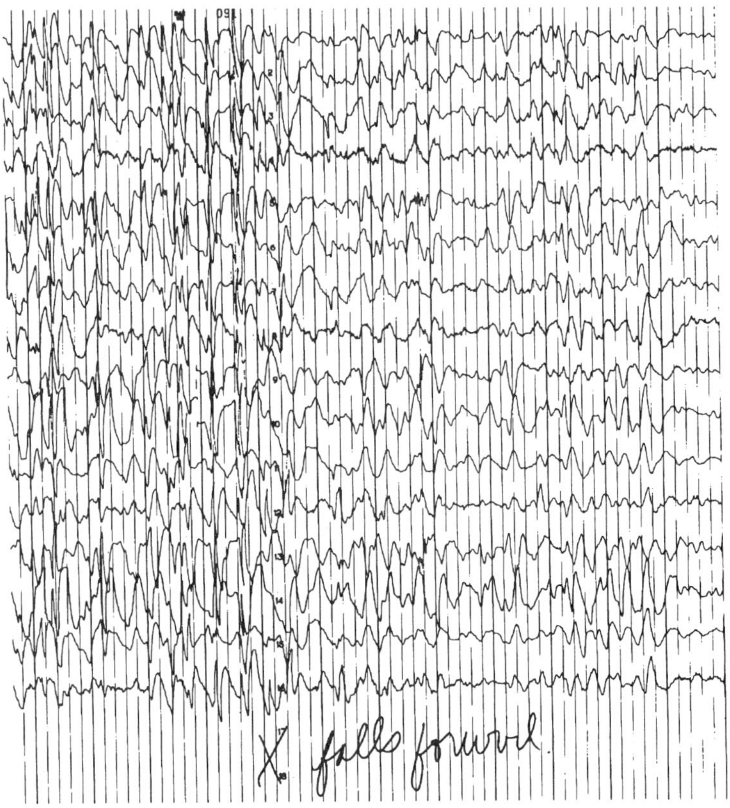

Figure 2.

primarily tonic seizures associated with an EEF ictal EEG pattern did not demonstrate a significantly better outcome (56%) compared to patients without these features (50%). There was no statistical difference seen with regard to age at seizure onset, patient's sex, or partial vs. single- or two-staged complete corpus callosotomy.

These results in patients younger than 10 years of age are to be contrasted with the patients greater than age 10 with a primary seizure type of tonic seizures and an ictal EEF pattern. In 37 patients, ages 10–49 with a mean of 21.8 years, Courtney et al. observed that 75% of patients who had tonic seizures achieved a satisfactory outcome. Those with both tonic seizures and EEF seizure onset had a 92% chance of a satisfactory outcome, as compared to 33% of patients not demonstrating both features ($p < .001$).[8]

NEUROPSYCHOLOGICAL OUTCOME AFTER CALLOSAL SECTION

The literature is limited. Ferguson[20] concluded that callosotomy produced no alterations in basic personality and had beneficial effects on behavior as a result of the case-specific changes in seizure characteristics and frequency. The gains outweighed

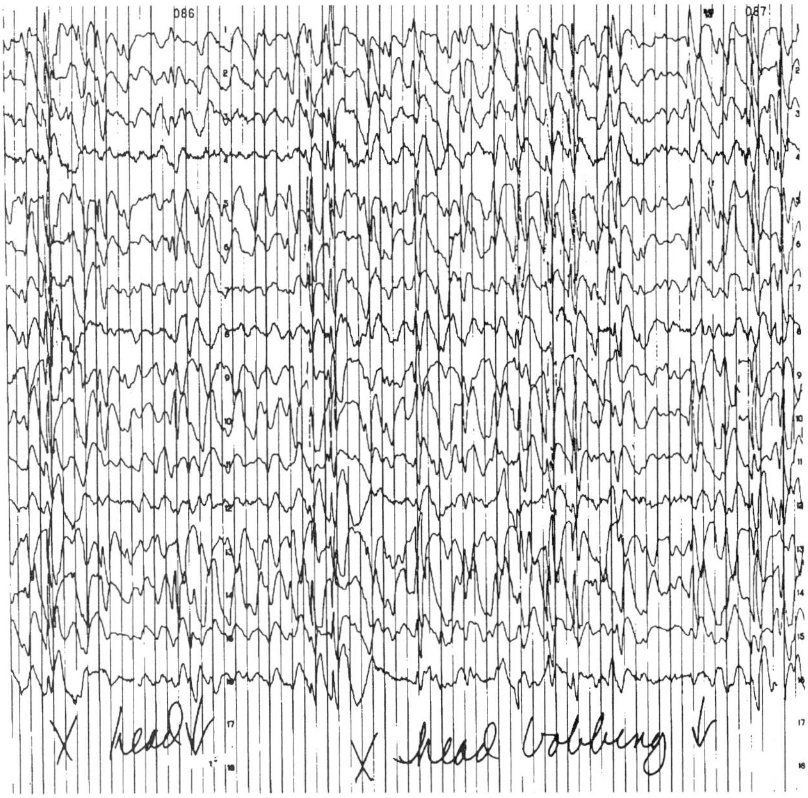

Figure 3.

the negative consequences of callosal section. Improvements were observed in the areas of independence, scope of daily activities, reduction of difficult pre-operative behavior.[10] Similar observations have been made from the Minnesota Epilepsy Group series in 1990.[13]

SUMMARY AND CONCLUSIONS

1. The corpus callosum has its most highly developed expression in the human. Extrapolation from animal models to the human must be carefully approached.
2. Even for children under 10, it appears that callosotomy improves tonic and atonic seizures. Satisfactory outcomes (defined as a greater than 80% reduction of the main seizure type) are reported in 50–100% of these patients. Under the age of 10, however, the epileptiform fast pattern (EEF), as opposed to the slow spike and wave, does not predict the superior outcome seen in patients older than 10.
3. Generalized tonic-clonic seizures, as the main indication for callosal section, demonstrate lower rates of successful outcome.

4. Surgical indication for other seizure types is unclear at this point. Complex partial seizures, as well as myoclonic seizures when associated with tonic or atonic seizures, appear to improve, but their role as a main indication has not been proven.
5. A substantial volume of literature now supports the indication of callosotomy for patients who meet the definition of Lennox-Gastaut Syndrome.
6. The role of anterior vs. complete callosotomy has not been clarified for patients with tonic or atonic seizures as a primary indication. Most will substantially benefit from the anterior 2/3 section and some of the failures of the 2/3 section will improve after completion.
7. The more recent studies report significantly fewer complications after callosotomy, particularly from the neuropsychological perspective. This is most likely due to improvements in technique, with less cortical damage from retraction in this technically difficult population who may have adhesions from the subclinical subarachnoid hemorrhages caused by their multiple falls and injuries.
8. Corpus callosotomy is a palliative procedure. It does not render the patients seizure-free. Utilization of the Engel Criteria Outcome, therefore, is probably not appropriate. We clearly need to develop a new definition of satisfactory seizure outcome for these patients.
9. A primitive, but effective, measure of the success of the procedure is the dramatic reduction in the frequency with which these children are seen in emergency rooms for sutures or other injuries related to the falls that used to devastate them prior to the surgical intervention.

REFERENCES

1. Black, P.M., Holmes, G., and Lombroso, C. Corpus callosum section for intractable epilepsy in children, *Pediatr Neurosurg*, 18 (1992) 298–304.
2. Blume, W.T. Corpus callosum section for seizure control: Rationale and review of experimental and clinical data, *Cleve Clin Quart*, 51 (1984) 319–332.
3. Bogen, J.E., Fisher, E.D., and Vogel, P.J. Cerebral Commissurotomy: a second case report. *J Am Med Assoc*, 194(2) (1965) 1328–1329.
4. Bogen, J.E., Sperry, R.W., and Vogel, P.J. Addendum: commissureal section and the propagation of seizures. In: H.H. Jasper, A.A. Ward, and A. Pope (Eds.) *Basic Mechanisms of Epilepsy*, Little Brown, Boston (1969) pp. 439.
5. Bogen, J.E. and Vogel, P.J. Cerebral commissurotomy in man: preliminary case report, *Bull LA Neurol Soc*, 27 (1962) 169–172.
6. Bouvier, G., Mercier, C., St. Hilaire, J.M., Giard, N., Labrecque, R., and Beique, R.A. Anterior callosotomy and chronic depth electrode recording in the surgical management of some intractable seizures, *Appl Neurophysiol*, 46 (1983) 52–56.
7. Cendes, F., Ragazzo, P.C., da Costa, V., and Martins L.F. Corpus callosotomy in treatment of medically resistant epilepsy: preliminary results in a pediatric population, *Epilepsia*, 34(5) (1993) 910–917.
8. Courtney, W., Gates, J., Ritter, F., Frost, M.D., Spiegel, R.H., Moriarty, G.L., Penovich, P.E., and Hempel, A. Prediction of Seizure Outcome Following Corpus Callosotomy in Patients Ten Years or Older, *Epilepsia*, 34(S6) (1993) 43.
9. Erickson, T.E. Spread of the epileptic discharge: an experimental study of the after discharge induced by electrical stimulation of the cerebral cortex, *Arch Neurol Psychiatry*, 43 (1940) 429.
10. Ferguson, S.M., Rayport, M., Blumer, D.P., Fenwick, P.B.C., and Taylor, D.C. Postoperative psychiatric changes. In: J. Engel Jr. (Ed.) *Surgical Treatment of the Epilepsies*, 2nd Edn. Raven Press Ltd, New York (1993) pp. 649–662.
11. Gates, J.R. Candidacy for corpus callosum section. In: H. Luders (Ed.) *Epilepsy Surgery*, Raven Press, New York (1991).

12. Gates, J.R. and DePaola, L. Chapter 56: Corpus Callosum Section. In: *The Treatment of Epilepsy*, Blackwell Science, Ltd., 1996, pp. 722–739.
13. Gates, J.R., Ritter, R.J., Ragazzo, P.C., and Reeves, A.G. Corpus callosum section in children: seizure response, *Epilepsy*, 3(S) (1990) 271–278.
14. Gates, J.R., Rosenfeld, W.E., Maxwell, R.E., and Lyons, R.E. Response of multiple seizure types to corpus callosum sectioning, *Epilepsia*, 28 (1987) 28–34.
15. Gates, J.R., Wada, J.A., Reeves, A.G., Lassonde, M., Pappo, I., Spencer, S.S., Anderman, F., Risse, G., Ritter, F.J., and Purves, S. Reevaluation of corpus callosotomy. In: J. Engel Jr (Ed.) *Surgical Treatment of The Epilepsies*, 2nd Ed. Raven Press Ltd, New York (1993) pp. 637–648.
16. Geoffroy, G., Lassonde, M., Delisle, F., and Decarie, M. Corpus callosotomy for control of intractable epilepsy in children, *Neurol*, 33 (1983) 891–897.
17. Grafstein, B. Organization of callosal connections in suprasylvian gyrus of cats, *J Neurophysiol*, 22 (1959) 504–515.
18. Harbaugh, R.E., Wilson, D.H., Reeves, A.G., and Gazzaniga, M.S. Forebrain commissurotomy for epilepsy. Review of 20 consecutive cases, *Acta Neurochir* (Wien) 68 (1983) 263–275.
19. Innocenti, G.M. What is so special about callosal connections? In: F. Lepore, M. Ptito, and H.H. Jasper (Eds.) *Two hemispheres—One Brain: Functions of the Corpus Callosum*, Alan R. Liss, New York (1986) pp. 75–81.
20. Kass, H.J. Organization of callosal projections in primates. At the 2nd Dartmouth International Conference on Epilepsy and the Corpus Callosum, at the Dartmouth College, Hanover, New Hampshire, 1991.
21. Luessenhop, A.J., de la Cruz T.C., and Fenichel, G.M. Surgical disconnection of the cerebral hemispheres for intractable seizures. Results in infancy and childhood, *JAMA*, 213 (1970) 1630–1636.
22. Marino, R Jr. Neurosurgical aspects of epilepsy in adults. In: J.R. Youmans (Ed.) *Neurological Surgery*, third edition, W.B. Saunders, Philadelphia (1990) pp. 4228–4306.
23. Musgrave, J. and Goor, P. The role of the corpus callosum in bilateral interhemispheric synchrony of spike and wave discharge in feline generalized penicillin epilepsy, *Epilepsia*, 21 (1980) 369–278.
24. Oguni, H., Oliver, A., Anderman, F., and Comair, J. Anterior callosotomy in the treatment of medically intractable epilepsies: A study of 43 patients with a mean follow-up of 39 months, *Annals of Neurology*, 30 (1991) 357–364.
25. Pandya, D.N. and Seltzer, B. The topography of commissural fibers. In: F. Lepore, M. Ptito, and H.H. Jasper (Eds.) *Two hemispheres—One Brain: Functions of the Corpus Callosum*, Alan R. Liss, New York (1986) pp. 47–73.
26. Papo, I., Quattrini, A., Provinciali, L., Rychlicki, F., Paggi, A., Del Pesce, M., and Ortenz. Callosotomy for the management of intractable non-focal epilepsy: A preliminary personal assessment, *Acta Neurochir*, 96 (1989) 46–53.
27. Raffel, C., Kongelbeck, S.R., and Snead, O.C. Corpus callosotomy for intractable epilepsy in children, *Pediatr Neurosurg*, 18 (1992) 305–309.
28. Rayport, M., Ferguson, S.M., and Corrie, S.W. Outcome and indications of corpus callosotomy section for intractable seizure control, *Appl Neurophysiol*, 46 (1983) 47–51.
29. Spencer, S.S. Corpus callosum section and other disconnection procedures for medically intractable epilepsy, *Epilepsia*, 29(S2) (1988) 85–99.
30. Spencer, D.D. and Spencer, S.S. Corpus callosotomy in the treatment of medically intractable secondarily generalized seizures in children, *Cleveland Clin Med J*, 56(S 1) (1989) S69–S78.
31. Spencer, S.S., Spencer, D.D., Glaser, G.H., Williamson, P.D., and Mattson, R.H. More intense focal seizure types after callosal sectioning: The role of inhibition, *Ann Neurol*, 16 (1984) 686–693.
32. Spencer, S.S., Spencer, D.D., Sass, K., Westerveld, M., Katz, A., and Mattson, R. Anterior, total, and two-staged corpus callosum section: differential and incremental seizure responses, *Epilepsia*, 34(3) (1993) 561–567.
33. Spencer, S.S., Spencer, D.D., and Williamson, P.D. Corpus callosotomy for epilepsy: Seizure effects, *Neurol*, 38 (1988) 19–24.
34. Van Wagenem, W.P. and Herren, R.Y. Surgical division of commissural pathways in the corpus callosum—relation to spread of an epileptic attack, *Arch Neurol Psychiatry*, 44 (1940) 740–759.
35. Vergnes, M., Marescaux, C., Lannes, B., Depaulis, A., Micheletti, G., and Warter, J.M. Interhemispheric desynchronization of spontaneous spike and wave discharges by corpus callosum transection in rats with petit mal-like epilepsy, *Epilepsy Res*, 4 (1989) 8–13.
36. Wada, J. Midline structures for transhemispheric ictal and interictal transmission. At the 2nd Dartmouth International Conference on Epilepsy and the Corpus Callosum. Hanover, New Hampshire, 1991.

37. Wada, J.A. and Komai, S. Effect of anterior two-thirds callosal bisection upon bisymmetrical and bisynchronous generalized convulsions kindled from amygdala in epileptic baboon, Papio, Papio. In: A.G. Reeves (Ed.) *Epilepsy and the Corpus Callosum*, Plenum Press, New York and London (1985) pp. 75–97.
38. Williams, P.L. and Warwick, R. The corpus callosum. In: P.L. Williams and R. Warwick (Eds.) *Functional Neuroanatomy of the Man*, WB Saunders Company, Philadelphia, pp. 965–967.
39. Wilson, D.H., Reeves, A., and Gazzaniga, M. Division of the corpus callosum for incontrollable epilepsy, *Neurol*, 28 (1978) 649–653.
40. Wilson, D.H., Reeves, A., Gazzaniga, M., and Culver, C. Cerebral commissurotomy for the control of intractable epilepsy, *Neurol*, 27 (1977) 708–715.
41. Wyllie, E. Corpus callosotomy for intractable generalized epilepsy, *J Pediatr*, 8 (1988) 255–261.

11

EARLY VERSUS LATE SURGERY FOR INTRACTABLE SEIZURES

Jerome Engel, Jr.

Departments of Neurology and Neurobiology and
the Brain Research Institute
UCLA School of Medicine,
Reed Neurological Research Center
710 Westwood Plaza
Los Angeles, CA 90095-1769
tel: (310) 825-5745
fax: (310) 206-8461

Certain drug-resistant epilepsies are highly responsive to surgery. Dr. Engel argues that there is no need to pursue exhaustive drug trials with these patients. If the first-line anticonvulsants fail, they should be referred directly to an epilepsy center and considered for surgical treatment.

INTRODUCTION

It has been estimated that there are approximately 100,000 patients with medically refractory epilepsy in the United States. These patients are potential candidates for surgical treatment, with an additional 5,000 to 10,000 added to this pool each year. The data, however, indicate that only 1,500 surgical procedures for epilepsy were performed in 1991.[9]

There has been increasing interest in surgical treatment for epilepsy over the past 5 to 10 years, as evidenced by the number of recent books published on this subject.[1,3,6,12,17,18,19] Many new epilepsy surgery facilities are being developed, and surgical intervention has now become an option for a larger number of patients. In order to take advantage of this alternative mode of therapy, referring physicians must know how to identify potential surgical candidates and when to refer them to epilepsy surgery centers. This requires an understanding of the term "medical intractability".

With the recent addition of several new antiepileptic drugs to our pharmacological armamentarium, and the promise of more in the near future, assessment of medical

Table 1. Surgically Remediable Syndromes and the Indications for Early Surgical Intervention

Mesial Temporal Lobe Epilepsy (Anteromesial Temporal Lobectomy)
Intractable disabling complex partial seizures
Unilateral hippocampal atrophy or temporal lobe hypometabolism
Focal sphenoidal ictal EEG onset
Lesional Localization-Related Epilepsy (Neocortical Resection)
Intractable disabling partial seizures
Discrete structural lesion in a surgically approachable area
Interictal and ictal epileptiform EEG abnormalities in the area of the lesion
Catastrophic Unilateral or Secondary Generalized Epilepsy of Infants and Small Children (Hemispherectomy or Multilobar Resection)
Intractable unilateral or generalized seizures
Developmental delay
Diffuse structural lesion or hypometabolism limited to one hemisphere
Interictal and ictal epileptiform EEG abnormalities limited to the same hemisphere (or generalized epileptiform abnormalities with attenuation in the region of the lesion for infantile spasms)
Hemiparesis with a useless hand when hemispherectomy is considered
Secondary Generalized Epilepsy in Older Children and Adults (Corpus Callosotomy)
Intractable drop attacks as the most disabling seizure type
Not a candidate for resective surgery

intractability can become a lengthy process. Is it really necessary to try every antiepileptic drug, alone and in combination, before concluding that a patient is a surgical candidate? A major conclusion deriving from a recent international conference on the surgical treatment of epilepsy, held in Palm Desert, California in 1992,[6,7] was that surgical treatment should not necessarily be considered a therapy of last resort. Rather, a number of surgically remediable epileptic syndromes have been identified which have a relatively well-established pathophysiology, are associated with an invariably poor response to pharmacotherapy but a predictably excellent response to surgical intervention[8,9] (Table 1). Patients with these syndromes should be referred for surgery early in the course of their disorder, after failure of the two, or perhaps three, most effective antiepileptic drugs for their condition.

EARLY SURGERY FOR SURGICALLY REMEDIABLE SYNDROMES

Mesial Temporal Lobe Epilepsy

The prototype of a surgically remediable syndrome is mesial temporal lobe epilepsy, a condition of medically refractory temporal lobe seizures associated with hippocampal sclerosis. It usually begins in the first decade of life and can have disastrous psychosocial consequences.[5,21] The characteristic features of this syndrome are shown in Table 2.

The presurgical evaluation of mesial temporal lobe epilepsy can usually be accomplished noninvasively with long-term EEG and video monitoring, high resolution magnetic resonance imaging (MRI), neuropsychological testing, and perhaps functional imaging such as positron emission tomography (PET) or single photon emission

Table 2. The Syndrome of Mesial Temporal Lobe Epilepsy (MTLE)

History
- Increased incidence of complicated febrile convulsions
- Increased incidence of a family history of epilepsy
- Onset in latter half of first decade of life
- Auras common and occur in isolation
- Secondarily generalized seizures occur infrequently
- Seizures often remit for several years until adolescence or early adulthood
- Seizures often become medically intractable
- Interictal behavioral disturbances can occur (most commonly depression)

Clinical Seizure
- *Aura* is usually present—most common is epigastric rising, often other autonomic or psychic symptoms, with emotion (e.g., fear), can be olfactory or gustatory sensation (several seconds).
- *Complex partial seizure*—often begins with arrest and stare, oroalimentary automatisms and complex automatisms common. Posturing of one upper extremity may occur contralateral to the ictal discharge (one to two minutes).
- *Postictal phase*—usually includes disorientation, recent memory deficit, amnesia for the event, and dysphasia if seizures begin in the language dominant hemisphere (several minutes)

Neurological Examination
- Usually normal
- May have recent memory deficit

EEG
- Unilateral or bilateral independent anterior temporal spikes, maximum amplitude in basal electrodes.
- May be intermittent or continuous rhythmic slowing in one mesial temporal area.
- Extracranial ictal activity appears only with complex partial symptoms, usually initial or delayed focal onset pattern of 5–7/sec rhythmic activity, maximum amplitude in one basal temporal derivation.
- Depth electrode ictal onset most often high amplitude rhythmic spikes or sharp waves, less commonly low voltage fast or suppression.
- Propagation to contralateral side is slow (>5 sec, but may be minutes), or does not occur at all.

Focal Functional Deficits
- Usually temporal lobe hypometabolism on interictal FDG-PET, often involves ipsilateral thalamus and basal ganglia.
- Usually temporal lobe hypoperfusion on interictal SPECT and characteristic pattern of hyper- and hypoperfusion on ictal SPECT.
- Usually material specific memory disturbances on neuropsychological testing and amnesia with contralateral intracarotid sodium amobarbital injection.
- Mesial temporal EEG slowing and attenuation of normal rhythms can be seen with scalp/sphenoidal electrodes, but more common with depth electrodes; exacerbated by iv pentothal test.

Structural Imaging
- May have small hippocampus on one side (on MRI).
- May have small temporal lobe on one side.
- May have enlarged temporal horn on one side.

Pathophysiology
- Hippocampal sclerosis (>30% cell loss with specific patterns).
- Sprouting of dentate granule cell mossy fibers.
- Selective loss of certain hilar neurons (somatostatin and NPY-containing cells).
- Hamartomas and heterotopias may occur as "dual pathology".
- Microdysgenesis common.
- Seizures may originate in sclerotic hippocampus but much larger area appears to be included in the epileptogenic region.

Features That Place Diagnosis in Doubt
- History of severe head trauma, encephalitis or other specific causal events.
- Focal motor or specific sensory symptoms at seizure onset or postictally.
- Interictal focal neurological deficits.
- Marked cognitive impairment on neuropsychological testing.
- Bilaterally synchronous, generalized or extratemporal focal EEG spikes.
- Diffuse or extratemporal focal EEG slowing.
- Cerebral lesion other than hippocampal sclerosis on MRI

From Engel.[7]

computed tomography (SPECT). Seventy to 80% of patients so identified can expect to become seizure free following a relatively limited anteromedial temporal lobe resection, and the remainder will experience a marked reduction in seizure frequency.

The prognosis with respect to epileptic seizures, and perhaps more importantly, the degree of psychosocial rehabilitation, however, depends upon the age at surgery and the duration of epilepsy prior to operation.[14] Because early surgical intervention provides the greatest opportunity to return the patient to a completely normal life style, prolonged trials with second-line drugs and drug combinations can be counterproductive. Once this syndrome has been identified, and it has been ascertained that the primary pharmacologic agents, carbamazepine, phenytoin, and possibly vigabatrin, are unable to control the habitual seizures, patients should be referred directly to an epilepsy center and considered for surgical treatment.

Partial Epilepsy Due to a Discrete Structural Lesion

A second surgically remediable epileptic syndrome is partial epilepsy due to a discrete structural lesion that can be easily resected without causing additional neurologic deficit.[2] Such lesions include small neoplasms, hamartomas, vascular malformations, traumatic scars and congenital disturbances, including migration defects such as focal cortical dysplasia. Most of these lesions can now be readily demonstrated with high resolution MRI. A more detailed presurgical evaluation is necessary, however, to ascertain that the observed lesion is, in fact, responsible for the habitual epileptic events. In some cases, the structural disturbance is multifocal and the epileptogenic lesion is missed. In others, the observed lesion is a fortuitous finding unrelated to the epileptic condition. Consequently, long-term EEG and video monitoring is usually required to localize ictal and interictal epileptiform discharges to the area of the lesion, and removal of epileptogenic cortex adjacent to the lesion provides the best chance of a seizure-free outcome.[10] Results of limited surgical resection in this patient population are as good as those for anteromedial temporal resection in patients with mesial temporal lobe epilepsy. Consequently, there is no need to pursue lengthy drug trials in these patients once the first-line antiepileptic medications fail.

Unilateral and Secondary Generalized Epilepsy Due to a Diffuse Hemispheric Lesion

Other surgically remediable epilepsies include a variety of unilaterally diffuse hemispheric disturbances in infants and small children that produce catastrophic seizures and profound developmental delay (e.g., hemimegencephaly, Sturge-Weber syndrome and Rasmussen's encephalitis).[16] There is also a subpopulation of patients who present with signs and symptoms of infantile spasms who have extensive unilateral cortical dysplasia, usually confined to the posterior part of one hemisphere, that can be identified with PET,[4] and, in many cases now, MRI.[15] Because seizures are very frequent in these patients, the ineffectiveness of the most appropriate antiepileptic drugs can usually be ascertained relatively quickly. Whereas these children have a dismal prognosis with medical therapy, 70% to 80% become seizure free following early hemispherectomy or a large multilobar hemispheric resection and, with timely intervention, developmental delay can be largely reversed. Early surgical intervention is essential in this situation, not only because of the deleterious effect that the seizures appear to have on behavioral development, but also because the plasticity of the immature brain reduces the degree of neurological impairment resulting from removal of large cortical areas.

Secondary Generalized Epilepsy with Drop Attacks as the Most Disabling Seizure Type

A final class of surgically remediable syndromes are those associated with "drop attacks" as the most disabling seizure type. In most patients with drop attacks (for instance those with the Lennox-Gastaut syndrome), there are also other types of epileptic seizures which contribute to the disability, as well as mental retardation or other neurologic deficits. Drop attacks are among the most medically refractory seizure types, often cause repeated serious injuries, and place a severe limitation on activities of daily living. Corpus callosotomy can be very effective in eliminating drop attacks. This is only a palliative procedure, however, because it usually does not affect the other seizure types, nor does it reduce the associated cognitive and neurological impairments.[11] Nevertheless, there is no need to consider extensive trials of antiepileptic drugs to control drop attacks that are severely disabling in patients who could be candidates for corpus callosotomy.

SURGICAL CONSIDERATIONS FOR OTHER INTRACTABLE EPILEPSIES

Patients with medically refractory partial seizures associated with symptomatic forms of epilepsy, who have no demonstrable lesions on MRI and no localizing functional deficit on PET or SPECT, are much less likely to benefit from surgical treatment than patients with surgically remediable syndromes.[20] Furthermore, when the boundaries of the surgical resection are defined only on the basis of electrophysiological recordings, this usually requires chronic intracranial recordings, using either depth or subdural grid electrodes. Such studies are expensive and associated with a small, but not insignificant, morbidity.[13] Consequently, decisions to proceed to surgery in this patient population are only justified after thorough attempts at pharmacotherapy have been tried and have failed. In addition, older children and adults with extensive structural lesions, evidence of bilateral independent ictal onsets, or other discordant findings on presurgical evaluation, and who also would require intracranial recording to determine whether surgical treatment is advisable and what should be removed, have a poorer prognosis than patients with surgically remediable syndromes. For these patients as well, extended trails of most of the available antiepileptic drugs might be appropriate before proceeding to invasive testing.

For patients with symptomatic partial epilepsy who might be considered candidates for surgical treatment, but who do not have an obvious surgically remediable syndrome, decisions concerning how far to pursue pharmacotherapy before referral to an epilepsy surgery center should be based, in part, on psychosocial considerations. For instance, patients who are not cognitively impaired, but who are experiencing increasing difficulties at school or work as a result of their seizures, should be referred relatively quickly. Examples might include a child who had been at the top of the class, but is now receiving failing grades, either because of frequent seizures or drug side effects, or a young adult who is well-trained and capable but is in jeopardy of losing a job because of seizures or their consequences. Not only do the ictal events themselves interfere with quality of life, but they can be associated with progressive memory loss, or behavior problems such as depression, which are also an indication to consider surgical treatment sooner rather than later. When quality of life is not severely affected by epileptic seizures, either because they are relatively mild or because activities of daily living are more impaired as a result of other neurological disturbances or mental retardation, new drugs and drug combinations can be pursued for longer periods of time before considering surgical referral, if at all.

ACKNOWLEDGMENTS

Original research reported by the author was supported in part by Grants NS-02808, NS-15654, NS-33310, and GM-24839, from the National Institutes of Health, and Contract DE-AC03-76-SF00012 from the Department of Energy.

REFERENCES

1. Apuzzo, M.L.J. *Neurosurgical Aspects of Epilepsy*, American Association of Neurological Surgeons, Park Ridge, Illinois (1991).
2. Cascino, G.D., Boon, P.A.J.M., and Fish, D.R. Surgically remediable lesional syndromes. In: J. Engel, Jr. (Ed.) *Surgical Treatment of the Epilepsies, 2nd Edition*, Raven Press, New York, (1993) pp. 77–86.
3. Chauvel, P., Delgado-Escueta, A.V., Bancaud, J., and Halgren, E. (Eds.) *Frontal Lobe Seizures and Epilepsies*, Raven Press, New York (1992), p. 750.
4. Chugani, H.T., Shields, W.D., Shewmon, D.A., Olson, D.M., Phelps, M.E., and Peacock, W.J. Infantile spasms: I. PET identifies focal cortical dysgenesis in cryptogenic cases for surgical treatment, *Ann. Neurol.*, 27 (1990) 406–413.
5. Engel, J. Jr. Recent Advances in Surgical Treatment of Temporal Lobe Epilepsy, *Acta Neurol. Scand.*, 86 (Suppl 5) (1992) 71–80.
6. Engel, J. Jr. (Ed.) *Surgical Treatment of the Epilepsies, 2nd Edition*, Raven Press, New York (1993a) p. 786.
7. Engel, J. Jr. Update on surgical treatment of the epilepsies, *Neurology.* 43 (1993b) 1612–1617.
8. Engel, J. Jr. Current concepts: surgery for seizures. *N. Engl. J. Med.*, 334 (1996) 647–652.
9. Engel, J. Jr. and Shewmon, DA. Overview: Who Should Be Considered a Surgical Candidate? In: J. Engel Jr. (Ed.) *Surgical Treatment of the Epilepsies, 2nd Edition*, Raven Press, New York (1993) pp. 23–34.
10. Fried, I. and Cascino, G.D. Lesional surgery. In: J. Engel Jr. (Ed.) *Surgical Treatment of the Epilepsies, 2nd Edition*, Raven Press, New York (1993) pp. 501–509.
11. Gates, J.R., Wada, J.A., Reeves, A., Lassonde, M., Papo, I., Spencer, S.S., Andermann, F., Risse, G., Ritter, F.J., and Purves, S.J. Re-evaluation of corpus callosotomy. In: J. Engel Jr. (Ed.) *Surgical Treatment of the Epilepsies, 2nd Edition*, Raven Press, New York (1993) pp. 637–648 .
12. Lüders, H.O. (Ed.) *Epilepsy Surgery*, Raven Press, New York (1992) p. 854.
13. Pilcher, W.H., Roberts, D.W., Flanigin, H.F., Crandall, P.H., Wieser, H.G., Ojemann, G.A., and Peacock, W.J. Complications of epilepsy surgery. In: J. Engel Jr. (Ed.) *Surgical Treatment of the Epilepsies, 2nd Edition*, Raven Press, New York (1993) pp. 565–581.
14. Rausch, R. and Crandall, P.H. Psychological status related to surgical control of temporal lobe seizures, *Epilepsia*, 23 (1982) 191–202.
15. Sankar, R., Curran, J.G., Kevill, J.W., Rintahaka, P.J., Shewmon, D.A., and Vinters, H.V. Microscopic cortical dysplasia in infantile spasms: Evolution of white matter abnormalities, *A.J.N.R.*, 16 (1995) 1265–1272.
16. Shields, W.D., Duchowny, M.S., and Holmes, G.L. Surgically remediable syndromes of infancy and early childhood. In: J. Engel Jr. (Ed.) *Surgical Treatment of the Epilepsies, 2nd Edition*, Raven Press, New York (1993) pp. 35–48.
17. Silbergeld, D.L. and Ojemann, G.A. Neurosurgery Clinics of North America, *Epilepsy Surgery*, 4 (1993) p. 356.

18. Spencer, D.D. and Spencer, S.S. (Eds.) *Epilepsy Surgery*, Blackwell, Cambridge (1991).
19. Theodore, W. (Ed.) Surgical Treatment of Epilepsy, *Epilepsy Res.*, (Suppl 5) (1992).
20. Van Ness, P. Surgical outcome for neocortical (extratemporal) focal epilepsy. In: H.O. Lüders (Ed.) *Epilepsy Surgery*. Raven Press, New York (1992) pp. 613–624.
21. Wieser, H.G., Engel, J. Jr., Williamson, P.D., Babb, T.L., and Gloor, P. Surgically remediable temporal lobe syndromes. In: J. Engel Jr. (Ed.) *Surgical Treatment of the Epilepsies, 2nd Edition*. Raven Press, New York (1993) pp. 49–63.

12

WHY ARE COMPLEX PARTIAL SEIZURES INTRACTABLE?

W. McIntyre Burnham

Bloorview Epilepsy Research Program
Department of Pharmacology
University of Toronto
1 King's College Circle
Toronto, Ontario, CANADA M5S 1A8
tel: (416) 978-0779
fax: (416) 971-2433

Dr. Burnham argues that data from animal studies—and especially kindling studies—can provide insights into why complex seizures are so common and drug resistant.

INTRODUCTION

Complex partial seizures are the most common seizures in adults,[3] and some of the most drug resistant.[4] They most commonly arise in the temporal lobes,[3] and involve the limbic structures.

Why do the limbic structures so often give rise to epileptic discharge, and why does that discharge resist drug control? Animal studies—and particularly studies involving the kindling model—provide some insights into this question.

THE KINDLING MODEL

The kindling model is an animal model of partial epilepsy in which stimulating electrodes are implanted chronically into the forebrain—often into a limbic structure—and low-intensity current is used to trigger focal seizure activity. With repetition, the focal seizures begin to propagate into more and more extra-focal sites, until eventually secondarily generalized motor seizures occur. Thus, limbic kindling provides a model of partial limbic seizures, secondarily generalized.[5,7]

SEIZURE THRESHOLDS IN LIMBIC STRUCTURES: KINDLING STUDIES

An important feature of the kindling model—which distinguishes it from other partial seizure models—is that seizure thresholds can easily be measured through the chronically implanted electrodes. This has been done in a large number of studies, providing a sizable amount of information on seizure thresholds in different parts of the brain. Table 1 summarizes the data from studies in which thresholds for electrical stimulation were measured in male, Long-Evans rats using the following standard kindling parameters: one second of 60 Hz, balanced biphasic (positive- and negative-going) 1 ms square-wave pulses. All values are peak-to-peak.

As Table 1 indicates, thresholds are different in different parts of the brain. The lowest thresholds are seen in the limbic structures. The threshold in the amygdala, for instance, is often around 100 μA. The hippocampal threshold is also around 100 μA and may actually be lower, especially in the dorsal hippocampus. Curiously, septal thresholds are higher, being around 300–350 μA. Still, limbic thresholds are generally low, and the amygdaloid and hippocampal thresholds are some of the lowest in the brain.

Thresholds in juxtallo (limbic) cortex (4–5 layered) tend to be somewhat higher. The threshold in the piriform cortex—closely associated to the amygdala—is around 130 μA, whereas the threshold in the cingulate is 175–200 μA. Data on the other limbic cortices are lacking, but the impression is that thresholds in limbic cortex are higher than thresholds in the amygdala or hippocampus.

Thresholds in the neocortex are distinctly higher, often around 300 μA, and the threshold in the brain-stem reticular core is extremely high. In the brain-stem reticular core—with 10 seconds of stimulation—threshold is 700 μA to 1,000 μA.

Thus, the amygdala and hippocampus have the lowest thresholds in the brain for electrical seizures, whereas the brain stem has some of the highest. Assuming that these data can be generalized to spontaneous seizures in humans, they would explain why seizures seldom arise in the brain stem[8,14] whereas the amygdala and hippocampus frequently initiate seizures and are often targets for seizure surgery.[9]

THRESHOLD DROP IN LIMBIC STRUCTURES: KINDLING STUDIES

Thresholds are not only low in limbic structures, they tend to drop lower with repeated activation.

Data in this field must be viewed with some caution, since Loscher et al.[10] have recently demonstrated a spontaneous drop in amydaloid threshold during the first month after electrode implantation. This spontaneous drop—which is unrelated to stimulation—disappears by two months.[10]

Data from studies in which stimulated threshold drop was measured two or more months after implantation are summarized in the second column of Table 1. The data are limited, but they seem to indicate that the largest drops occur in the amygdala (60%), with moderate reductions being seen in the pyriform and cingulate cortex (40%). Smaller reductions occur in the neocortex and brain-stem reticular formation (20–30%). Curiously, thresholds in the hippocampus show only a very small reduction (around 25%). This may relate to the fact that they are so low to begin with.

Table 1. Mean Afterdischarge Threshold (μA) in Long-Evans Rats

	Initial Threshold (mean)	% Drop after 40–60 stimulations
AMYGDALA	111.0–158.3	59.0%[11]
HIPPOCAMPUS		
Dorsal (region not specified)	91.7–179.0	25.0%[11]
Dorsal fascia dentata	62.0	—[13]
Dorsal CA1	40.6	—[13]
Ventral fascia dentata	109.0	—[13]
Ventral CA1	76.0	—[13]
SEPTAL AREA	328.0–340.0	—[13]
PYRIFORM CORTEX	128.0–130.4	43.1%[12]
CINGULATE CORTEX	176.6–193.8	39.0%[12]
NEOCORTEX		
Area 2	293.8–325.0	27.9%[12]
Area 6	265.0–328.1	31.4%[12]
Area 17	306.0–326.0	15.0%[12]
Area 18a	331.2–343.8	19.3%[12]
MRF (10 seconds)		
Forelimb Clonus	772	—[6]
Forelimb Tonic Extension	940.0–1034.0	—[6]

It has been suggested that complex partial seizures tend to become more frequent over time. If seizure thresholds in the human limbic system drop with repetition—as they do in the kindled rat—an increase in spontaneous seizure frequency might well be expected.

SEIZURE THRESHOLDS AND ANTICONVULSANT MEDICATIONS

The low—and falling—thresholds seen in limbic structures provide some insight into why complex partial seizures are so common. The resistance of limbic structures to anticonvulsant medications may offer insight into why they are so drug resistant.

The first data related to limbic drug resistance came from a study by Albright and Burnham.[2] Albright and Burnham kindled rats from either the amygdala or the motor cortex after 30 generalized (stage 5) seizures – when threshold was stable. Stimulation intensity was set at threshold plus 20% and anticonvulsants were administered. It was found that anticonvulsants easily suppressed cortical focal seizures in this paradigm, but that they suppressed amygdala focal seizures only at high, often toxic, doses. In other words, anticonvulsants easily elevated cortical thresholds by 20%, but were much less able to elevate limbic thresholds by 20%. Albright[1] subsequently did actual measurements of the effects of anticonvulsant on cortical and amygdala thresholds. At the ED_{50} for generalized convulsions, phenytoin and carbamazepine raised cortical thresholds by 45% and 60%, respectively, but had no significant effect on amygdala thresholds. Significant elevations of amygdala thresholds were only seen at high, toxic drug doses. A similar study by Chiu and Burnham[6] found that brain-stem triggered seizures—like cortical focal seizures—are highly responsive to drugs.

These data suggest that anticonvulsants are effective at raising thresholds for cortical and brain-stem seizures, but only weakly effective at raising thresholds for limbic seizures. If these data can be generalized to humans, they may explain why the anticonvulsant drugs are effective against tonic-clonic and simple partial attacks, but often ineffective against complex partial seizures.[3]

CONCLUSIONS

Interpretation of these findings depends on the assumption that data on thresholds for electrical stimulation in rats can be generalized to thresholds for spontaneous seizures in humans. Assuming that they can, the kindling data provide important insights. They suggest that complex partial seizures are common because limbic seizure thresholds are low—and drop lower—and that they are drug-resistant because anticonvulsants are poor at raising thresholds in the limbic system.

These data suggest that it will be important to find drugs that *can* elevate limbic thresholds. These drugs may well be effective against complex partial seizures in humans.

REFERENCES

1. Albright, P.S. Effects of carbamazepine, clonazepam, and phenytoin on seizure threshold in amygdala and cortex, *Experimental Neurology*, 79 (1983) 11–17.
2. Albright, P.S. and Burnham, W.M. Development of a new pharmacological seizure model: effects of anticonvulsants on cortical- and amygdala-kindled seizures in the rat, *Epilepsia*, 21 (1980) 681–689.
3. Blume, W.T. Partial seizures: clinical description and diagnosis. In: W.M. Burnham, P.L. Carlen, and P.A. Hwang (Eds.) *Intractable Seizures: Diagnosis and Therapy*, 2000.
4. Bruni, J. Drug therapy for complex partial seizures. In: W.M. Burnham, P.L. Carlen, and P.A. Hwang (Eds.) *Intractable Seizures: Diagnosis and Therapy*, 2000.
5. Cain, D.P. (1992) Kindling and the amygdala. In: *The Amygdala: Neurobiological Aspects of Emotion, Memory, and Mental Dysfunction*, Wiley-Liss, Inc., pp. 539–560
6. Chiu, P. and Burnham, W.M. The effect of anticonvulsant drugs on convulsions triggered by direct stimulation of the brainstem, *Neuropharmacology* 21 (1982) 355–359.
7. Corcoran, M.E. (1988) Characteristics and mechanisms of kindling. In: P. Kalivas and C. Barnes (eds.) *Sensitization of the Nervous System* The Telford Press, Caldwell, N.J. pp. 81–116.
8. Fromm, G.H. The brain-stem and seizures: summary and synthesis. In: G.H. Fromm, C.L. Faingold, R.A. Browning, and W.M. Burnham (Eds.) *Neurology and Neurobiology, Volume 27: Epilepsy and the Reticular Formation—The Role of the Reticular Core in Convulsive Seizures*, Alan R. Liss, Inc., New York (1987) pp. 203–218.
9. Girvin, J.P. Surgery for complex partial seizures. In: W.M. Burnham, P.L. Carlen, and P.A. Hwang (Eds.) *Intractable Seizures: Diagnosis and Therapy*, 2000.
10. Loscher, W., Wahnschaffe, U., Honack, D., and Rundfelt, C. Does prolonged implantation of depth electrodes predispose the brain to kindling? *Brain Research*, 697 (1995) 197–204.
11. Racine, R.J. Modification of seizure activity by electrical stimulation: I. After-discharge threshold, *Electroencephalography and Clinical Neurophysiology*, 32 (1972) 269–279.
12. Racine, R.J. Modification of seizure activity by electrical stimulation: cortical areas, *Electroencephalography and Clinical Neurophysiology*, 38 (1975) 1–12.
13. Racine, R., Rose, P.A., and Burnham, W.M. Afterdischarge thresholds and kindling rates in dorsal and ventral hippocampus and dentate gyrus, *Canadian Journal of Neurological Sciences*, 4–4 (1977) 273–278.
14. Williams, D. The thalamus and epilepsy, *Brain*, 88 (1965) 539–556.

13

PATHOPHYSIOLOGY OF INFANTILE SPASMS

Harry T. Chugani

Division of Pediatric Neurology and the PET Center
Children's Hospital of Michigan, Wayne State University
3901 Beaubien Blvd
Detroit, MI 48201
tel: (313) 993-2867
fax: (313) 993-3845

Infantile spasms—seen in West's Syndrome—are often associated with cortical abnormalities. The spasms themselves, however, appear to be generated subcortically. Dr. Chugani reviews the clinical data related to the pathophysiology of infantile spasms and proposes a hypothesis which involves both cortical and subcortical mechanisms.

INTRODUCTION

Recent observations have contributed significantly to our understanding of the pathophysiology of infantile spasms, and have allowed a number of hypotheses to be generated. Any consideration of the pathophysiology of infantile spasms must take into account the following features common to the population of subjects with infantile spasms:

1) The spasms are generalized seizures which consist of brief muscle extension, flexion, or mixed flexion-extension jerks involving the neck, trunk, or extremities.[47,49] Lateralizing or localizing clinical features may be present during the spasms.
2) Infantile spasms can be associated with a wide variety of neurological conditions, including metabolic, dysplastic or dysgenetic abnormalities, and various prenatal, perinatal, and postnatal insults.[51] In the *symptomatic* category, a specific neurological condition can be identified, whereas in the *cryptogenic* subgroup an underlying condition is suspected based on clinical features, EEG or other tests, but a specific diagnosis cannot be made.

3) There is a small group (<5%) of infants with spasms, the *idiopathic* subgroup, which is believed not to be associated with any underlying condition. Neurodevelopment is normal at presentation, and there is a lack of focal features on clinical examination, seizure semiology or EEG. Reappearance of hypsarrhythmia between consecutive spasms in a cluster is said to be an important feature in this subgroup.[34] There is rapid response of the spasms to medical treatment and an ultimately benign course is expected. This rare situation may represent a case in which, during development, there is a chemical or physiological disturbance resulting in spasms, but which can be totally reversed with medical treatment.
4) With rare exceptions,[38] the spasms typically begin between age 3 and 8 months and disappear by age 2 to 3 years. In other words, there appears to be a developmental window during which various brain abnormalities may result in infantile spasms as one manifestation of the underlying disorder.
5) Although the vast majority of brain lesions that have been identified in patients with infantile spasms involve the cerebral cortex, their locations are not specific to any cortical region.
6) With some exceptions, the spasms do not appear to respond to conventional anticonvulsants, but rather, to therapeutic agents not typically used in epilepsy, such as corticosteroids. If the spasms respond to such treatment, the response is often abrupt and dramatic rather than graded, as is seen in other forms of epilepsy.

In this chapter, the basic mechanisms that have been proposed as integral to the pathophysiology of infantile spasms are reviewed. This will be followed by the formulation of a scheme of events which we believe may be important in the generation of infantile spasms. The scheme proposed will take into account many of the observations reviewed in this chapter.

NEUROTRANSMITTERS AND BASIC MECHANISMS

There have been many studies performed on patients with infantile spasms aimed at determining whether there are chemical disturbances that might provide clues as to the pathophysiology. Although the excellent review on this subject by Pranzatelli [1994][70] concluded that our knowledge of the pathophysiology of spasms is still quite rudimentary, it is still useful to summarize some of the major findings.

Following the demonstration by Coleman[23] that the serotonin precursor 5-hydroxytryptophan *induced* infantile spasms when administered to patients with Down syndrome, Silverstein and Johnston[76] were somewhat surprised to find *lower* cerebrospinal fluid (CSF) concentrations of the serotonin metabolite 5-hydoxyindole acetic acid (5-HIAA) in infants with spasms, as compared to age-matched controls. Silverstein and Johnston,[76] therefore, proposed that a state of supersensitivity of serotonin receptors in infants with spasms may exist and may result in diminished presynaptic serotonin turnover and release from serotonin-containing neurons, which are concentrated in the raphe region of the pons. Several other studies have also suggested abnormal metabolism of tryptophan in patients with infantile spasms. For example, Yamamoto et al.[82]

reported lower kynurenic acid and higher 3-hydroxykynurenine concentrations in the CSF of patients with infantile spasms, as compared to controls. These disturbances in monoamine metabolism may be important because of the convulsive effect of 3-hydroxykynurenine and the excitatory amino acid antagonism of kynurenic acid. Abnormal metabolism of tryptophan is also seen in phenylketonuria, in which there is an inability to hydroxylate phenylalanine to tyrosine. Levels of serotonin and 5-HIAA are decreased in urine in infants with phenylketonuria, and, clinically, seizures in the form of infantile spasms are typical in infancy.[57]

Hrachovy et al. (1981)[43] have suggested, based on long term video-EEG monitoring and analysis of sleep patterns in a large number of patients with infantile spasms (see Chapter 5), that increased activity of adrenergic and/or serotonergic systems or decreased activity of cholinergic systems of the brain stem may account for some of the clinical and EEG features. Based on their observations, they treated 12 patients with the antiserotonergic and antiadrenergic drug tetrabenazine.[41] None of the infants, however, showed improvement in control of spasms, EEG features, or developmental status. Further trials using methysergide (a serotonin receptor blocker) and α-methylparatyrosine (a competitive inhibitor of tyrosine hydroxylase, the rate-limiting enzyme in catecholamine synthesis) also failed to show benefit in the majority of treated infants.[40] Since these neurotransmitter systems, particularly the serotonergic system, are very complex with multiple receptor subtypes, it is possible that more specific antagonists, which target specific receptor subtypes, would be required to be of benefit.

Levels of the neurotransmitter GABA have been reported to be low in the CSF of patients with infantile spasms.[45,53] Disturbances of the GABA system may be related to the observation that medications which increase GABA levels or promote GABA-mediated neurotransmission are often beneficial in patients with infantile spasms. Examples of such agents include vitamin B_6 used in high doses,[5,67] the benzodiazepine nitrazepam[31] and vigabatrin, an irreversible inhibitor of GABA transaminase.[52,10]

Low CSF concentrations of ACTH have been reported in infantile spasms, with normal concentrations of corticotropin-releasing hormone and b-endorphin.[62,1,3] Some investigators believe that abnormally increased corticotropin-releasing hormone (CRH) synthesis and activity due to injury or stress may cause hyperexcitability of specific neurons resulting in infantile spasms.[2] It is difficult, however, to presume that such a wide variety of prenatal, perinatal and postnatal insults, ranging from metabolic diseases to traumatic brain injury, can have an increase of CRH as a common feature. Furthermore, spasms often have an insidious onset not associated with stress. Indeed, clusters of spasms typically occur at a time of calm and maternal comfort when stress is minimal.

Recently, Izumi et al.[46] have found lower levels of gangliotetraose-series gangliosides in the cerebrospinal fluid of patients with infantile spasms, as compared to disease controls. Since gangliosides may play a role in neuronal maturation, these investigators suggested that a disturbance of neuronal maturation may contribute to the pathogenesis of infantile spasms.

Serologic HLA typing studies in patients with infantile spasms have shown an increased representation of DRw52, as compared to controls, suggesting that the pathophysiology of infantile spasms may include immunological mechanisms yet to be defined.[41] Other studies on the genetic predisposition to infantile spasms have found that after the exclusion of familial diseases known to be associated with spasms, the recurrence risk in families was less than 1%.[33]

NEUROIMAGING

Clearly, the most important advances in recent years towards understanding the pathophysiology of infantile spasms have come from various forms of neuroimaging. The application of magnetic resonance imaging (MRI) and positron emission tomography (PET) with 2-deoxy-2[^{18}F]fluoro-D-glucose (FDG) has demonstrated the frequent occurrence of cerebral malformations in the cryptogenic infantile spasms group and allowed their spasms to be reclassified as symptomatic. In a study of 46 patients with West syndrome, computed tomography (CT) was normal in 38 patients and showed diffuse atrophy in 8. MRI showed, however, delayed myelination in one patient and focal lesions in four.[77]

Even more sensitive than MRI in identifying focal abnormalities in patients with infantile spasms is FDG-PET. Interictally, these focal abnormalities appear as cortical areas of decreased glucose metabolism, but hypermetabolism is seen ictally or when a persistent spike or spike-and-wave focus is present on the EEG during the scan.[21,19] Neuropathological studies have shown that the anatomical correlate of focal PET abnormalities not apparent on the MRI is typically cortical dysplasia.[78] A recent analysis of 140 patients with infantile spasms showed that, with the benefit of FDG-PET, the total number of symptomatic cases rose from 42 (30%) to 134 (95.7%) in this series which was heavily biased towards cryptogenic cases.[12]

Serial FDG-PET studies in a small number of infants with spasms have shown that focal cortical hypometabolism sometimes may be transient. Conversely, some infants with initial normal scans may show focal abnormalities on subsequent scans. Although the suggestion has been made that, in some instances, focal PET abnormalities may be truly functional and not represent organic lesions,[54,55] neuropathological confirmation of this notion is lacking. Other possible explanations include developmental factors[14,15] and differences in neurophysiological activity between scans during the tracer uptake period.[16]

Studies with [^{99m}Tc]HMPAO or ^{133}Xenon single photon emission computed tomography (SPECT) confirm that functional imaging is far more sensitive than MRI at showing focal cortical abnormalities in patients with infantile spasms.[11,59] In the study of Miyazaki et al.,[59] 7 of 10 infants had localized cortical hypoperfusion on SPECT, and only 3 had MRI lesions. This study found that the temporal lobes were affected in all 7 infants and suggested that the temporal lobes may be important in the pathophysiology of the spasms. In a separate study,[60] these same investigators found various degrees of brain stem atrophy on MRI, associated with abnormal auditory brain stem evoked responses, in 6 of 10 patients, with infantile spasms, supporting the now generally accepted notion that cortical brain stem interactions are important in the generation of infantile spasms.

Further studies with PET have shown that the lenticular nuclei and brain stem of patients with infantile spasms are often metabolically prominent whether or not there is an associated cortical abnormality (Fig. 1).[19] Although cortical and brain stem abnormalities might have been expected based on available neurophysiological data, the finding of lenticular nuclei hypermetabolism was unexpected, and suggested recruitment of lenticular nuclei in the secondary generalization of a cortical focus. Symmetric hypermetabolism of lenticular nuclei may also account for the semiology of infantile spasms, which appear relatively symmetric even when focal cortical lesions are present. Removal of a focal lesion, if present, can lead to both the cessation of spasms[6,58,66,22,75] and the normalization of subcortical metabolic patterns (unpublished observations). These findings

Figure 1. Positron emission tomography (PET) images of glucose metabolism from an infant with spasms showing increased glucose metabolism in the left fronto-temporal cortex (short arrows), brain stem (long arrow), and lenticular nuclei (arrowheads).

are consistent with previous suggestions that an abnormal cortical-subcortical interaction is important in the pathophysiology of infantile spasms.[65,32,83,25]

ELECTROPHYSIOLOGY

The electrophysiological features of infantile spasms are somewhat unique. For example, the *interictal* EEG feature is hypsarrhythmia, which consists of poorly-synchronized, high-amplitude spikes and slow waves. Hypsarrhythmia changes with behavioral states, and is most prominent in sleep stages 2 and 3, usually disappearing during rapid eye movement (REM) sleep.[81] The characteristic *ictal* EEG pattern of spasms in a recent study of 955 spasms was a positive-vertex slow wave.[36] Other EEG patterns also associated with the spasms were spindle-like fast activity (corresponding to a clinical stare) and decremental activity (thought to represent a postictal event). When the spasms themselves are asymmetric, a cerebral lesion is almost always present.

The observation of a cluster of spasms immediately preceded by a partial seizure was first made by Dalla Bernardina et al.,[26] and subsequently confirmed by other investigators.[83,30,8] In such cases, it appears as if the partial seizure actually triggers a cluster of spasms. Further observations have found that 85% of infants with spasms and partial seizures have an asymmetric interictal EEG background, and that the presence of partial seizures is associated with a poor prognosis for seizure control and psychomotor development.[69] It is important to identify this group of infants early, so that they can be evaluated for cortical resection,[21,20] but this can be difficult in the presence of hypsarrhythmia on the EEG.

Donat and Lo[29] found that 38% of 77 consecutive infantile spasms patients had focal or lateralized features on video-EEG monitoring and always indicated the presence of a lesion on CT or MRI. Most infants with symmetric hypsarrhythmia and spasms also had structural changes, but these tended to be bilateral and diffuse. Importantly, patients with normal developmental outcome almost always had symmetric findings.

Video-EEG studies on patients with infantile spasms have documented a marked decrease in rapid eye movements (REM) sleep, as compared to normal infants.[35,43] Infants with spasms were also found to have less total sleep time when compared to normals.[43] Infants whose spasms and hypsarrhythmic pattern on EEG improved with ACTH or prednisone showed a reversal of the REM sleep abnormality. Based on these findings, it was proposed that the site of pathogenesis in infantile spasms may be at the pontine level in close proximity to centers that regulate sleep cycles,[43] and that structural abnormalities may be present in the brain stems of such infants. Although the neuropathologic study of Morimatsu et al.[61] in patients who had earlier manifested infantile spasms found a number of pontine abnormalities, other neuropathology studies have not reported a high incidence of pontine or brain stem abnormalities.[56,48]

The origin of hypsarrhythmia has also been a subject of some debate. The observation that following total corpus callosotomy in two children with spasms, the hypsarrhythmia became unilateral suggests that a transcallosal cortico-cortical pathway may be required for the generalization of hypsarrhythmia.[68]

Overwhelmingly, the data point to a primary role of the cerebral cortex in the generation of infantile spasms, with secondary involvement of brain stem and possibly basal ganglia. The abnormality in the cortex can be structural, whether diffuse (e.g., widespread dysgenesis or damage) or focal (e.g., porencephaly, focal dysgenesis). In other cases, the cortical abnormality can be functional, as in some inborn errors of metabolism.

PATHOPHYSIOLOGY OF INFANTILE SPASMS: AN HYPOTHESIS

It is clear from this review that the pathophysiology of infantile spasms has yet to be worked out. Based on published observations, including our own, however, we have proposed a neuronal circuitry which we believe may be important in the pathogenesis of infantile spasms. The essential components of this circuitry are summarized in Fig. 2.

We believe that the primary offending lesion is a focal or diffuse cortical abnormality, which, at a critical stage of maturation, causes abnormal functional interactions

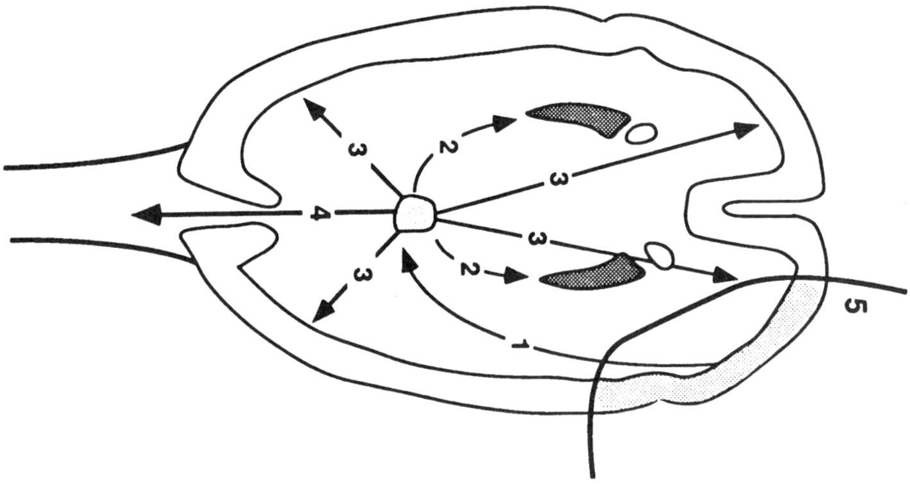

Figure 2. A schematic diagram illustrating secondary generalization from a cortical focus through subcortical recruitment, leading to infantile spasms. (1) An abnormal cortical region triggers the brain stem raphe area through epileptic discharges; (2) the raphe area, in turn, stimulates the striatum through serotonergic raphe-striatal projections, which are under tonic control by corticosteroids; (3) raphe-cortical and cortical-cortical interactions result in the hypsarrhythmia pattern on the EEG; (4) direct or indirect propagation to the spinal cord and lenticular nuclei result in infantile spasms; (5) surgical ablation of the primary cortical focus prevents this circuit from being activated.

with brain stem raphe nuclei. Serotonin-synthesizing neurons in the raphe region project widely throughout the brain.[64] Raphe-cortical, as well as cortical-cortical, projections could possibly mediate the hypsarrhythmic changes on EEG. Moreover, the prominent raphe-striatal pathway in humans involves strong connections between raphe neurons and putamen bilaterally, which could cause the lenticular and brain stem metabolic activation found in PET scanning studies. Descending spinal pathways from the raphe or from the closely associated reticular formation are also likely to be involved in the generation of infantile spasms in a symmetric fashion. With maturation, conditions no longer favor this type of secondary generalization, and there is an evolution to other types of epilepsy, such as partial epilepsy or the seizures of the Lennox-Gastaut syndrome. Interestingly, PET studies in Lennox-Gastaut syndrome have also found focal or multifocal cortical abnormalities supporting a different type of secondary generalization without the recruitment of basal ganglia.[13,44]

There are many pieces of evidence to support the circuitry which we have proposed:

1) Even if a structural cortical abnormality is present in the neonate, infantile spasms do not appear until physiological and metabolic maturation of the cortical region which harbors the lesion, i.e., 3–4 months in posterior cortical regions and 6–10 months in frontal cortex.[14,17] This supports a primary role of cerebral cortex in the generation of infantile spasms.
2) Since the raphe-striatal projection is serotonergic,[24] increased activity of this projection would be consistent with the finding of decreased 5-HIAA in CSF of infants with spasms.[76] An iatrogenic increase in brain serotonin activity can also be accomplished by exogenous administration of 5-hydroxytryptophan, resulting in infantile spasms.[23] There are multiple types of serotonin receptors in the brain, of which 5-HT_{1D} receptors are the predominant type in the terminals of the raphe-striatal projections.[79,39,80] Since there are no specific antagonists to 5-HT_{1D} receptors available for clinical use, it is not surprising that Hrachovy et al.,[41,42] who also postulated increased serotonergic activity, found less specific serotonin antagonists to be ineffective in suppressing infantile spasms.
3) The fact that glucocorticoids modulate raphe projections may explain the mechanism of action of steroid therapy in suppressing infantile spasms. Glucocorticoid receptors occur in significant numbers in rat striatum,[28] and both ACTH immunoreactive fibers[72] and glucocorticoid binding sites[71] have been detected in the raphe region. Functionally, serotonin receptors in the raphe-hippocampal projection, which have a number of features in common with the raphe-striatal pathway, are clearly regulated by glucocorticoids.[27,4]
4) Dysfunction in the area around the raphe nuclei, which are known to be involved in sleep organization, may account for the sleep disturbances in children with infantile spasms.[35,43] In addition, the spasms, which so often occur soon after arousal of the infant from sleep[49], may be the consequence of a cortical epileptic focus interacting with the raphe area which is undergoing transitional changes during arousal.
5) Interruption of the cortical-subcortical interaction through surgical resection of an offending focal cortical lesion leads to cessation of the spasms and relative normalization of the EEG.[22,73,74,75]
6) There is a well-established relationship between abnormal serotonin metabolism and myoclonic disorders, which infantile spasms clearly resemble.[9,50,63]

It should be pointed out that secondary generalization through cortical-subcortical interaction (Fig. 2) is neither novel nor unique to infantile spasms, but is known to occur in other types of epilepsy in humans[18] as well as in some animal models of epilepsy.[37,7] We believe that the scheme outlined in Fig. 2 is testable, and provides an initial framework which can be eventually refined. The development of an animal model of infantile spasms will be crucial in this regard.

REFERENCES

1. Airaksinen, E., Mustonen, K., and Tuomisto, L. ACTH therapy in infantile spasms: Effects on cerebrospinal amino acids, HVA, 5-hydroxyindoleacetic acid, and histamine, *Cleve Clin. J. Med.*, 56 (1989) S-269.
2. Baram, T.Z. Pathophysiology of massive infantile spasms: Perspective on the putative role of brain adrenal axis, *Ann. Neurol.*, 33 (1993) 231–236.
3. Baram, T.Z., Mitchell, W.G., Snead III, O.C., Horton, E.J., and Saito, M. Brain-adrenal axis hormones are altered in the CSF of infants with massive infantile spasms, *Neurol.*, 42 (1992) 1171–1175.
4. Biegon, A., Rainbow, T.C., and McEwen, B.S. Corticosterone modulation of neurotransmitter receptors in rat hippocampus: A quantitative autoradiographic study, *Brain Res.*, 332 (1985) 309–314.
5. Blennow, G. and Starck, L. High dose B6 treatment in infantile spasms, *Neuropediatrics*, 17 (1986) 7–10.
6. Branch, C.E. and Dyken, P.R. Choroid plexus papilloma and infantile spasms, *Ann. Neurol.*, 5 (1979) 302–304.
7. Browning, R.A. Role of the brain-stem reticular formation in tonic-clonic seizures: lesion and pharmacological studies, *Fed. Proc.*, 44 (1985) 2425–2431.
8. Carrazana, E.J., Lombroso, C.T., Mikati, M., Helmers, S., and Holmes, G.L. Facilitation of infantile spasms by partial seizures, *Epilepsia*, 34 (1993) 97–109.
9. Chadwick, D., Harris, R., Jenner, P., Reynolds, E.H., and Marsden, C.D. Manipulation of brain serotonin in the treatment of myoclonus, *Lancet*, ii (1975) 434–443.
10. Chiron, C., Dulac, O., Beaumont, D., Palacios, L., Pajot, N., and Mumford, J. Therapeutic trial of vigabatrin in refractory infantile spasms, *J. Child Neurol.*, 6 (Suppl. 2) (1991) S52–S59.
11. Chiron, C., Dulac, O., Bulteau, C., Nuttin, C., Depas, G., Raynaud, C., and Syrota, A. Study of regional cerebral blood flow in West syndrome, *Epilepsia*, 34 (1993) 707–715.
12. Chugani, H.T. and Conti, J.R. Etiological classification of infantile spasms in 140 cases: Role of positron emission tomography, *J. Child Neurol.*, 11(1) (1996) 44–48.
13. Chugani, H.T., Mazziotta, J.C., Engel, J. Jr., and Phelps, M.E. Lennox-Gastaut syndrome: metabolic subtypes determined by 18FDG positron emission tomography, *Ann. Neurol.*, 21 (1987) 4–13.
14. Chugani, H.T. and Phelps, M.E. Maturational changes in cerebral function in infants determined by 18FDG positron emission tomography, *Science*, 231 (1986) 840–843.
15. Chugani, H.T., Phelps, M.E., and Mazziotta, J.C. Positron emission tomography study of human brain functional development, *Ann. Neurol.*, 22 (1987) 487–497.
16. Chugani, H.T., Phelps, M.E., and Mazziotta, J.C. Positron emission tomography study of human brain functional development, *Ann. Neurol.* 22 (1987) 487–497.
17. Chugani, H.T., Rintahaka, P.J., and Shewmon, D.A. Ictal patterns of cerebral glucose utilization in children with epilepsy, *Epilepsia*, 35 (1994) 813–822.
18. Chugani, H.T., Shewmon, D.A., Khanna, S., and Phelps, M.E. Interictal and postictal focal hypermetabolism on positron emission tomography, *Pediatr. Neurol.*, 9 (1993) 10–15.
19. Chugani, H.T., Shewmon, D.A., Sankar, R., Chen, B.C., and Phelps, M.E. Infantile spasms: II. lenticular nuclei and brain stem activation on positron emission tomography, *Ann. Neurol.*, 31 (1992) 212–219.
20. Chugani, H.T., Shewmon, D.A., Shields, W.D., Sankar, R., Comair, Y., Vinters, H.V., and Peacock, W.J. Surgery for intractable infantile spasms: neuroimaging perspectives, *Epilepsia*, 34 (1993) 764–771.
21. Chugani, H.T., Shields, W.D., Shewmon, D.A., Olson, D.M., Phelps, M.E., and Peacock, W.J. Infantile spasms: I. PET identifies focal cortical dysgenesis in cryptogenic cases for surgical treatment, *Ann. Neurol.*, 27 (1990) 406–413.
22. Chugani, H.T., Shields, W.D., Shewmon, D.A., Olson, D.M., Phelps, M.E., and Peacock, W.J. Infantile spasms: I. PET identifies focal cortical dysgenesis in cryptogenic cases for surgical treatment, *Ann. Neurol.*, 27 (1990) 406–413.

23. Coleman, M. Infantile spasms associated with 5-hydroxytryptophan administration in patients with Down's syndrome, *Neurology*, 21 (1971) 911–919.
24. Cross, A.J. and Slater, P. High affinity serotonin binding sites in human brain: a comparison of cerebral cortex and basal ganglia, *J. Neural Transm.*, 76 (1989) 211–219.
25. Cusmai, R., Dulac, O., and Diebler, C. Lesions focales dans les spasmes infantiles, *Neurophysiol. Clin.*, 18 (1988) 235–241.
26. Dalla Bernardina, B., Colamaria, V., and Capoville, G. Epileptic syndromes and cerebral malformations in infancy: multicentric study, *Boll Lega Ital. Epil.*, 45/46 (1984) 65–67.
27. Defiore, C.H. and Turner, B.B. [3H] Corticosterone binding in the caudate-putamen, *Brain Res.*, 278 (1983) 93–101.
28. De Kloet, E.R., Sybesma, H., and Reul, H.M. Selective control by corticosterone of serotonin1 receptor capacity in raphe-hippocampal system, *Neuroendocrinology*, 42 (1986) 513–521.
29. Donat, J.F. and Lo, W.D. Asymmetric hypsarrhythmia and infantile spasms in West syndrome, *J. Child Neurol.*, 9 (1994) 290–296.
30. Donat, J.F. and Wright, F.S. Simultaneous infantile spasms and partial seizures, *J. Child Neurol.*, 6 (1991a) 246–250.
31. Dreifuss, F.E. and Santilli, N. Valproic acid hepatic fatalities: Analysis of US cases, *Neurology*, 36 (Suppl. 1) (1986) 175.
32. Dulac, O., Chiron, C., Jambaque, I., Plouin, P., and Raynaud, C. Infantile spasms, *Progress in Clin. Neurosci.*, 2 (1987) 97–109.
33. Dulac, O., Feingold, J., Plouin, P., Chiron, C., Pajot, N., and Ponsot, G. Genetic predisposition to West syndrome, *Epilepsia*, 34 (1993b) 732–737.
34. Dulac, O. and Plouin, P. Cryptogenic/Idiopathic West Syndrome. In: O. Dulac, H.T. Chugani and B.D. Bernardina (Eds.) *Infantile Spasms and West Syndrome*, WB Saunders, London (1994) pp. 232–243.
35. Fukuyama, Y., Shionaga, A., and Iida, Y. Polygraphic study during whole night sleep in infantile spasms, *Eur. Neurol.*, 18 (1979) 302–311.
36. Fusco, L. and Vigevano, F. Ictal clinical electroencephalographic findings of spasms in West syndrome, *Epilepsia*, 34 (1993) 671–678.
37. Gale, K. Mechanisms of seizure control mediated by gamma-aminobutyric acid: role of the substantia nigra, *Fed Proc.*, 44 (1985) 2414–2424.
38. Gobbi, G., Bruno, L., Pini, A., Rossi, P.G., and Tassinari, C.A. Periodic Spasms: An unclassified type of epileptic seizure in childhood, *Develop Med. and Child Neurol.*, 29 (1987) 766–775.
39. Herrick-Davis, K., Maisonneuve, I.M., and Titeler, M. Postsynaptic localization and up-regulation of serotonin 5-HT1D receptors in rat brain, *Brain Res.*, 483 (1989) 155–157.
40. Hrachovy, R.A. and Frost, J.D. Jr. Infantile spasms: A disorder of the developing nervous system. In: P. Kellaway and J.L. Noebels (Eds.) *Problems and Concepts in Developmental Neurophysiology*, Johns Hopkins University Press, Baltimore (1989) pp. 131–147.
41. Hrachovy, R.A., Frost, J.D., Glase, D.G., and Rose, D. Treatment of infantile spasms with methysergide and alpha-methylparatyrosine, *Epilepsia*, 30 (1989) 607–610.
42. Hrachovy, R.A., Frost, J.D. Jr., and Glaze, D.G. Treatment of infantile spasms with tetrabenazine, *Epilepsia*, 29 (1988) 561–563.
43. Hrachovy, R.A., Frost, J.D. Jr., and Kellaway, P. Sleep characteristics in infantile spasms, *Neurology*, 31 (1981) 688–694.
44. Iinuma, K., Yanai, K., Yanagisa, T., Fueki, N., Tada, K., Ito, M., Matsuzawa, T., and Ido, T. Cerebral glucose metabolism in five patients with Lennox-Gastaut syndrome, *Pediatr. Neurol.*, 3 (1987) 12–18.
45. Ito, M., Mikawa, H., and Taniguchi, T. Cerebrospinal fluid GABA levels in children with infantile spasms, *Neurology*, 34 (1984) 235–238.
46. Izumi, T., Ogawa, T., Koizumi, H., and Fukuyama, Y. Low levels of CSF gangliotetraose-series gangliosides in West syndrome: implications of brain maturation disturbance, *Pediatr. Neurol.*, 9 (1993) 293–296.
47. Jeavons, P.M. and Bower, B.D. *Infantile spasms: A review of the literature and a study of 112 cases, Clinics in Develop. Med. No.15*, Spastic Society and Heinemann, London (1964).
48. Jellinger, K. Neuropathological aspects of infantile spasms, *Brain Dev.*, 9 (1987) 349–357.
49. Kellaway, P., Hrachovy, R.A., Frost, J.D., and Zion, T. Precise characterization and quantification of infantile spasms, *Ann. Neurol.*, 6 (1979) 214–218 .
50. Klawans, H., Goetz, C., and Weiner, W.J. 5-hydroxytryptophan induced myoclonus in guinea pigs and the possible role of serotonin in infantile spasms, *Neurology*, 23 (1973) 1234–1240.
51. Lacy, J.R. and Penry, J.K. Infantile Spasms, Raven Press, New York (1976).

52. Livingston, J.H., Beaumont, D., Arzimanoglou, A., and Aicardi, J. Vigabatrin in the treatment of epilepsy in children, *Br. J. Clin. Pharmacol.*, 27 (1989) S109–S112.
53. Loscher, W. and Siemes, H. Cerebrospinal fluid g-aminobutyric acid levels in children with different types of epilepsy: Effect of anticonvulsant treatment, *Epilepsia*, 26 (1985) 314–319.
54. Maeda, N., Watanabe, K., Negoro, T., Aso, K., Haga, Y., Kito, M., Ohki, T., Ito, K., and Kato, T. Transient focal cortical hypometabolism in idiopathic West syndrome, *Pediatr. Neurol.*, 9 (1993) 430–434.
55. Maeda, N., Watanabe, K., Negoro, T., Aso, K., Ohki, T., Ito, K., and Kato, T. Evolutional changes of cortical hypometabolism in West's syndrome, *Lancet*, 343 (1994) 1620–1623.
56. Meencke, H.J. and Gerhard, C. Morphological aspects of aetiology and the course of infantile spasms (West-Syndrome), *Neuropediatr.*, 16 (1985) 59–66.
57. Menkes, J. *Textbook of Child Neurology*, 4th Ed. Lea & Febiger, Philadelphia. London (1990) pp. 31–35.
58. Mimaki, T., Ono, J., and Yabuuchi, H. Temporal lobe astrocytoma with infantile spasms, *Ann. Neurol.*, 14 (1983) 695–696.
59. Miyazaki, M., Hashimoto, T., Fujii, E., Tayama, M., and Kuroda, Y. Infantile spasms: localized cerebral lesions on SPECT, *Epilepsia*, 35 (1994) 988–992.
60. Miyazaki, M., Hashimoto, T., Tayama, M., and Kuroda, Y. Brainstem involvement in infantile spasms: a study employing brainstem evoked potentials and magnetic resonance imaging, *Neuropediatrics*, 24 (1993) 126–130.
61. Morimatsu, Y., Murofushi, R., Handa, T., Shinoara, T., and Shiraki, H. Pathological studies of several physically and mentally handicapped- with special reference to four cases of infantile spasms, *Adv. Neurol. Sci.*, 16 (1972) 465–470.
62. Nalin, A., Facchinetti, F., Galli, V., Petraglia, F., Storchi, R., and Genazzani, A.R. Reduced ACTH content in cerebrospinal fluid of children affected by crytogenic infantile spasms with hypsarrhythmia, *Epilepsia*, 26 (1985) 446–449.
63. Nausieda, P.A., Carvey, P.M., and Braun, A. Long-term suppression of central serotonergic activity by corticosteroids: A possible model of steroid-responsive myoclonic disorders, *Neurology*, 32 (1982) 772–775.
64. Nieuwenhuys, R. *Chemoarchitecture of the brain*, Springer-Verlag, New York (1985).
65. Ogata, H., Mitsudome, A., Yokota, K., and Pachibana, H. Three cases of infantile spasms with brain cystic lesions, *Brain & Dev.*, 7 (1985) 184.
66. Palm, D.G., Brandt, M., and Korinthenberg, R. West syndrome and Lennox-Gastaut syndrome in children with porencephalic cysts: long-term follow-up after neurosurgical treatment. In: E. Niedermeyer and R. Degen (Eds.) *The Lennox-Gastaut syndrome*, Alan R. Liss, New York (1988) pp. 419–426.
67. Pietz, J., Benninger, C., Schäfer, H., Sontheimer, D., Mittermaier, G., and Rating, D. Treatment of infantile spasms with high-dosage vitamin B6, *Epilepsia*, 34 (1993) 757–763.
68. Pinard, J.M., Delalande, O., Plouin, P., and Dulac, O. Callosotomy in West syndrome suggests a cortical origin of hypsarrhythmia, *Epilepsia.* 34 (1993) 780–787.
69. Plouin, P., Dulac, O., Jalin, C., and Chiron, C. Twenty-four-hour ambulatory EEG monitoring in infantile spasms, *Epilepsia*, 34 (1993) 686–691.
70. Pranzatelli, M.R. Putative neurotransmitter abnormalities in infantile spasms: Cerebrospinal fluid neurochemistry and drug effects, *J. Child Neurol.*, 9 (1994) 119–129.
71. Sarrieau, A., Dussaillant, M., Sapolsky, R.M., Aitken, D.H., Olivier, A., Lal, S., Rostene, W.H., Quirion, R., and Meaney, M.J. Glucocorticoid binding sites in human temporal cortex, *Brain Res.*, 442 (1988) 157–160.
72. Schwartzberg, D.G. and Nakane, P.K. ACTH-related peptide containing neurons within the medulla oblongata of the rat, *Brain Res.*, 276 (1983) 351–356.
73. Shewmon, D.A., Shields, W.D., Chugani, H.T., and Peacock, W.J. Contrasts between pediatric and adult epilepsy surgery: Rationale and strategy for focal resection, *J. Epilepsy*, 3 (suppl. 1) (1991) 141–155.
74. Shewmon, D.A., Shields, W.D., Olson, D.M., Peacock, W.J., and Chugani, H.T. Multifocal independent epileptogenicity in children treated by focal cortical resection, *Epilepsia*, 30 (1989) 660.
75. Shields, W.D., Shewmon, D.A., Chugani, H.T., and Peacock, W.J. The role of surgery in the treatment of infantile spasms, *J. Epilepsy*, 3 (Suppl.) (1990) 321–324.
76. Silverstein, F. and Johnston, M.V. Cerebrospinal fluid monoamine metabolites in patients with infantile spasms, *Neurology*, 34 (1984) 102–105.
77. Van Bogaert, P., Chiron, C., Adamsbaum, C., Robain, O., Diebler, C., and Dulac, O. Value of magnetic resonance imaging in West syndrome of unknown etiology, *Epilepsia*, 34 (1993) 701–706.
78. Vinters, H.V., De Rosa, M.J., and Farrell, M.A. Neuropathologic study of resected cerebral tissue from patients with infantile spasms, *Epilepsia*, 34 (1993b) 772–779.

79. Waeber, C., Dietl, M.M., Hoyer, D., Probst, A., and Palacios, J.M. Visualization of a novel serotonergic recognition site (5-HT1D) in the human brain by autoradiography, *Neurosci. Lett.*, 88 (1988) 11–16.
80. Waeber, C., Hoyer, D., and Palacois, J.M. GR 43175: A preferential 5-HT1D agent in monkey and human brains as shown by autoradiography, *Synapse*, 4 (1989) 168–170.
81. Watanabe, K., Negoro, T., Aso, K., and Matsumoto, A. Reappraisal of interictal electroencephalograms in infantile spasms, *Epilepsia*, 34 (1993) 679–685.
82. Yamamoto, H., Shindo, I., Egawa, B., and Horiguchi, K. Kynurenic acid is decreased in cerebrospinal fluid of patients with infantile spasms, *Pediatr. Neurol.*, 10 (1994) 9–12.
83. Yamamoto, N., Watanabe, K., Negoro, T., Furune, S., Takahashi, I., Nomura, K., and Matsumoto, A. Partial seizures evolving to infantile spasms, *Epilepsia*, 29 (1988) 34–40.

14

FUNCTIONAL PROPERTIES OF HUMAN NEOCORTICAL NEURONS

A Window on the Physiopathogenesis of Seizures in Patients with Cortical Dysplasia

Massimo Avoli, Donatella Mattia, and Andre Olivier

Cell Biology of Excitable Tissue Research Group
Montreal Neurological Institute and Department of Neurology and Neurosurgery
McGill University
3801 University Street
Montreal, QC, Canada
H3A 2B4
tel: (514) 398-1955
fax: (514) 398-4497

Cortical dysplasia is found in many cases of intractable epilepsy. Dr. Avoli reports that dysplastic cortex—removed from human brains during seizure surgery—shows abnormal excitability **in vitro.**

INTRODUCTION

Knowledge of the fundamental mechanisms underlying epileptiform discharge has been greatly advanced by the use of *in vitro* preparations, such as cultured and acutely dissociated neurons or brain slices (see [17]). In particular, the *in vitro* brain-slice preparation has made it possible to study at cellular, pharmacological and molecular levels neurons and glial cells that retain their original connections. The relevance of the mechanisms discovered in animal models for understanding the physiopathogenesis of human epileptic syndromes, however, depends on the demonstration that such mechanisms do occur in the human condition as well.

During the last decade, several groups of researchers have used *in vitro* brain slices of human cortical tissue, obtained in the course of neurosurgical intervention (for review see [2,7]). In the present chapter we shall briefly review some of the findings that have been gathered in brain slices from patients operated on at the Montreal Neurological Institute and Hospital of McGill University. These patients include cases of temporal lobe epilepsy and those with focal cortical dysplasia.

METHODOLOGICAL AND TECHNICAL ASPECTS

Human brain tissue is obtained during neurosurgical procedures for the cure of different neurological conditions. Since the biopsies come from patients suffering from a variety of neurological diseases, it is imperative for the neurophysiologist to obtain as much information as possible about the clinical history, the results of brain imaging studies and the electrophysiological diagnostic procedures.

For characterization purposes, the samples of human neocortex that we have analyzed *in vitro* can be grouped into four main categories. First, "epileptogenic" tissue samples that are obtained from areas of the neocortex removed from epileptic patients and that display active "spiking" during intraoperative electrocorticography. Second, "juxta-epileptogenic" samples, i.e., neocortical tissue that is excised from epileptic patients, but does not show any sign of epileptiform spiking during electrocorticography. Both types of neocortical samples are mainly obtained from patients that have a diagnosis of temporal lobe epilepsy. In these patients, the site of origin of epileptiform activity most often resides in limbic structures, which implies that both the "epileptogenic" and the "juxta-epileptogenic" neocortical tissue may only be reflecting hyperexcitability that is projected from subcortical sites such as amygdala or hippocampus. The third category of neocortical tissue consists of brain samples that are removed from epileptic patients with a diagnosis of focal cortical dysplasia (FCD). The fourth category of neocortical tissue is obtained from non-epileptic patients, who have been operated on for pathologies other than epilepsy. Although these slices may be used as a control, yet one must maintain some reservations about the tissue normality.

Slices are obtained by using a McIlwain tissue chopper or a vibratome, and they have a thickness that may vary between 500 and 700 μm. The ionic composition of the standard medium used for maintaining human tissue in vitro is similar to that employed for studies of rat or guinea pig cortical structures (see [2]).

PROPERTIES OF HUMAN NEOCORTICAL CELLS

Membrane Properties and Intrinsic Excitability

Intracellular analysis of neurons in slices of the human temporal or frontal neocortex has shown the existence of several subthreshold responses that deviate from the expected passive behavior of the neuron membrane during intracellular injection of hyperpolarizing and depolarizing current pulses. Some of the ionic mechanisms underlying these subthreshold responses have been analyzed in previous reports from our laboratory.[3,5] They include voltage- and time-dependent responses in the subthreshold range that confer on these cells a very limited passive behavior. Similar types of rectification have been reported in a variety of central neurons. Human neocortical neurons generate rhythmic, regular fast action potentials during depolarizing current pulses (Fig. 1). Two distinct linear segments of firing can be seen when the repetitive firing is studied by

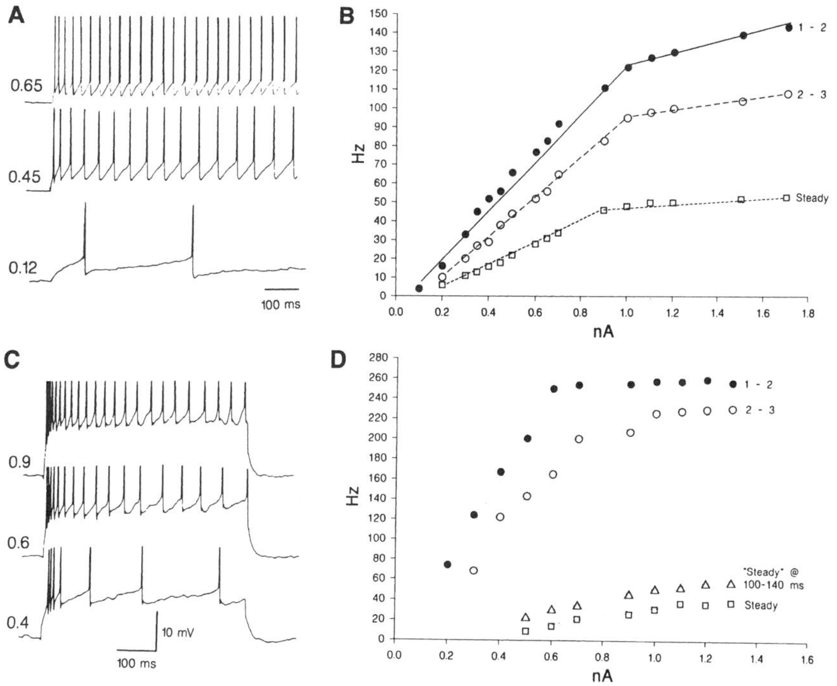

Figure 1. **A**: Repetitive firing observed in human neocortical cell that displayed tonic discharge. Progressive increments of the injected current (0.12–0.65 nA) induce a progressive increase in the rate of firing. **B**: Plot of the firing frequency versus. current intensity (nA) for the first (1–2), the second (2–3) interspike intervals, and the fully-adapted state (steady). Note that for this cell a primary slope and a secondary slope are discernible in the first and second interspike intervals and also occasionally in the steady state. **C**: Repetitive firing in a neuron with initial phasic discharge shows that progressively larger pulses of current generated more action potentials. **D**: Plot of the firing frequency versus current intensity for the first (1–2), the second (2–3) interspike intervals, the adapted state measured 100–140 ms after the pulse onset ("steady" 100–140 ms) and the fully-adapted state (steady). From Avoli et al., 1994a.

injecting depolarizing current pulses of increasing intensities. These two segments are called the "primary" and "secondary" range of firing. Another characteristic that emerges from these experiments is that the firing rate during the depolarizing pulse declines with time i.e., the frequency of the repetitive discharge "adapts". These data indicate, therefore, that pyramidal cells in the human neocortex display firing properties that are similar to those of regularly firing neocortical cells or to hippocampus pyramidal cells of rodents maintained in vitro.[14] These results are also remarkably similar to the firing properties of neocortical long-axon cells studied in situ.[8]

The repetitive firing of action potentials generated by neocortical cells during pulses of depolarizing current is followed by a long-lasting afterhyperpolarization (AHP), which is caused by an increase in K^+ conductance dependent upon Ca^{2+} (Fig. 2). Such a mechanism plays a role in firing adaptation (Fig. 2B), and presumably in the generation of the fast AHP seen after single action potentials (Fig. 2C).

In agreement with findings reported by Knowles et al.,[11] as well as by Schwartzkroin et al.,[18] the comparative analysis of the electrophysiological properties of neurons in slices of "epileptogenic" or "juxta-epileptogenic" brain tissue with slices obtained from "non-epileptogenic" tissue has not revealed any consistent differences in membrane properties

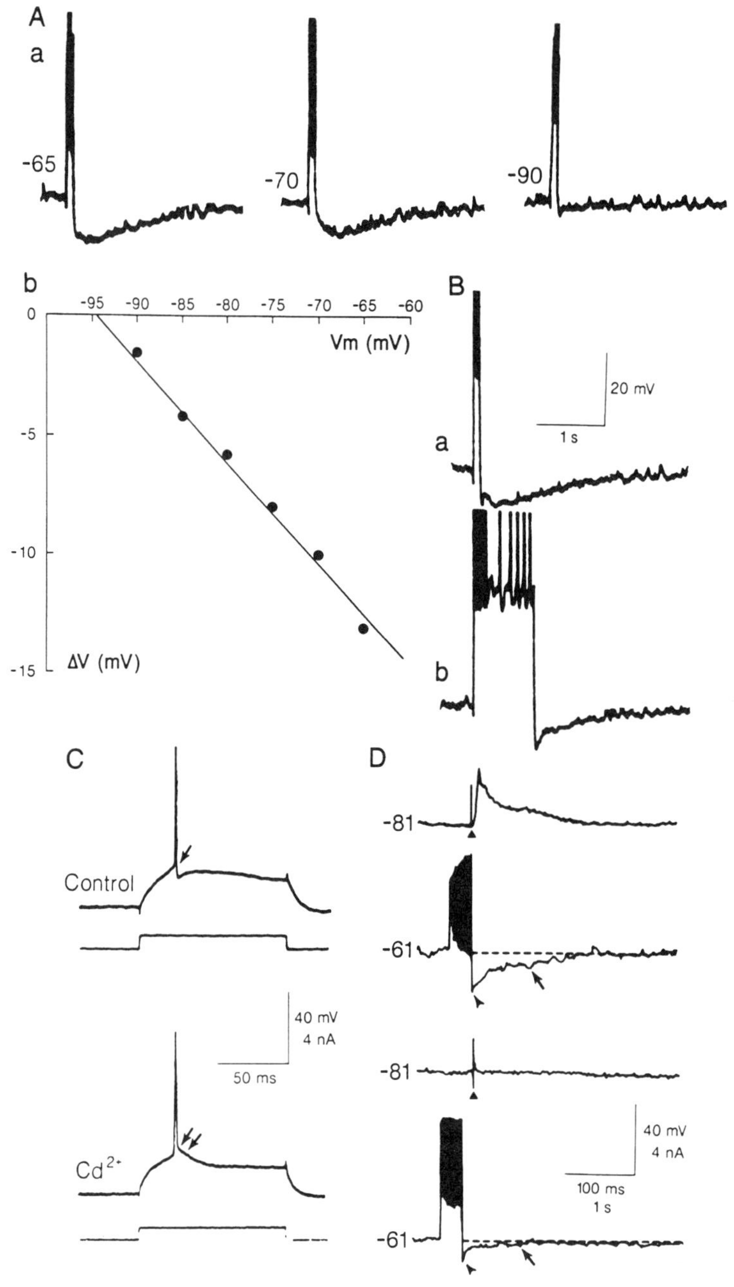

Figure 2. Characteristics of the long-lasting afterhyperpolarization (AHP) that follows repetitive firing. **A**: Sample traces in **a**, showing the effects on the AHP of hyperpolarizing the membrane potential. (Values indicated at left; resting level −65 mV). In **b**, plot of the amplitude of the AHP versus the membrane potential indicates an equilibrium potential of about −94 mV. **B**: The AHP generated in **a** underlies a marked period of adaptation seen 200 ms after the onset of the pulse in **b**. Effects of Cd^{2+} on fast (**C**) and slow (**D**) AHP. In **C**, threshold depolarization (−80 mV) evoked a single action potential followed by a fast AHP in control (single arrow). In Cd^{2+}, the fast AHP is blocked and an underlying depolarizing afterpotential is uncovered (double arrow). In **D**, in control (top two traces), focal electrical stimulation elicits a postsynaptic potential (−81 mV) while a depolarizing current pulse produces a train of action potentials followed by a long-lasting AHP consisting of a medium duration (arrow head) and slow (arrow) components (−61 mV). In Cd^{2+} (bottom two traces), the postsynaptic potential is blocked (−81 mV) and the amplitude of the medium and slow AHP is reduced by more than 50% (−61 mV). From Avoli et al., 1994a.

or intrinsic excitability. Preliminary results indicate that neurons from tissue samples obtained from FCD patients have also normal features.

Synaptic Potentials

Like central neurons of other mammalian species, human neocortical neurons recorded intracellularly respond to focal extracellular stimulation by generating a short-lasting excitatory post-synaptic potential which increases in amplitude when higher-strength focal stimuli are used (Fig. 3A). In most of the cases, high intensity stimuli induce a single action potential, although bursts of action potentials can be recorded in response to *high-strength* extracellular focal stimuli (Fig. 3B). This however represents a graded phenomenon, rather than an all-or-none event as in the case of the paroxysmal

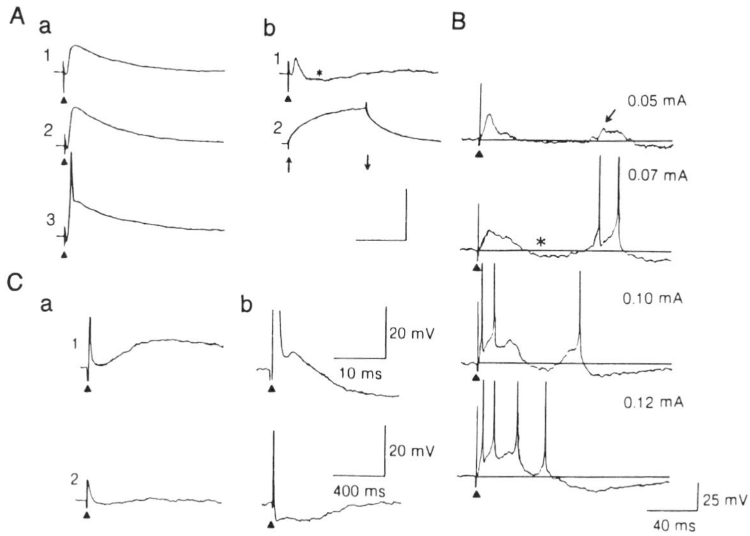

Figure 3. Responses evoked by extracellular focal stimuli in human neocortical neurons obtained from patients with complex partial epilepsy. In **Aa**, single shock stimuli delivered in the white matter at progressively higher strength evoke an EPSP that becomes larger in amplitude and is finally capable of eliciting a single action potential (bottom trace). The stimulus duration in 1–2 was 10, 20, and 90 μs, respectively, whereas the current intensity was kept constant. In **Ab1**, a stimulus of the same intensity as in Aa, was delivered while the resting membrane potential was depolarized by 12 mV; note the presence of an hyperpolarizing IPSP (asterisk) that appears to curtail the initial EPSP. In **Ab2**, response of this neocortical cell to a depolarizing intracellular pulse of intensity sufficient to give a response of the same amplitude as the peak of the EPSP shown in Ab1. **B**: Response obtained from the same slice as the neuron in **A**, to low-strength stimulation (0.05 mA) consists of a short-latency EPSP followed by a low-amplitude (~2 mV) hyperpolarizing IPSP that is interrupted by a late (onset at 180 ms) depolarizing event (arrow). When the stimulus intensity is brought at 0.07 mA both the EPSP and the IPSP (asterisk) become larger in amplitude, whereas the late depolarization is capable of triggering 2 action potentials. By increasing the stimulus strength further, a full-blown burst followed by an hyper polarization appears. **C**: Responses obtained from a different neuron to single shock stimuli delivered within the cortical layers. Steady intracellular, depolarizing current was injected in order to disclose the hyperpolarizing IPSP. Samples shown in **a** and **b** were induced by stimuli of 20 (0.15 mA) and 90 μs in duration (0.2 mA). In each frame, 1 and 2 are successive responses (frequency of stimulation, 0.05 Hz) displayed at different time bases. Note in row 1 that stimulation at maximal strength (**b**) evokes a single action potential at very short latency. In row 2 one can appreciate that the monophasic, hyperpolarizing IPSP (**a**) becomes longer in duration and displays a biphasic shape when stimulus strength is progressively in creased (**b**). From Avoli and Olivier, 1989.

depolarization shift (PDS). Pharmacological analysis of the stimulus-induced depolarizing potential suggests that the EPSP is mediated through non-NMDA receptors, since it is reduced by the AMPA antagonist CNQX.[7] By contrast, the burst is caused by the activation of the NMDA receptor.[5]

An early hyperpolarizing IPSP that is followed by a late IPSP (Figs. 3Ab star, 3Cb) can also be recorded under appropriate conditions. The early IPSP has a reversal potential of −65 to −75 mV (thus, it is close to the resting membrane potential), is sensitive to intracellular injection of Cl^- and is reduced by the $GABA_A$ receptor antagonist bicuculline.[4,5,10] The late part of the hyperpolarizing IPSP appears to be due to the activation of $GABA_B$ receptor subtypes, since it is reduced by $GABA_B$ receptor antagonists.[13] Some evidence also exists for the presence of conductances that are mediated through $GABA_A$ receptors located in the dendrites of human neocortical neurons (*cf.* [1]). Accordingly the shape of the EPSP/IPSP sequence is at times characterized by a transient depolarizing component that occurs between the early and late component of the hyperpolarization. Application of 4-aminopyridine, can disclose long-lasting depolarizations that are due to the activation of dendritic $GABA_A$ receptors[4,6] (see below).

In conclusion, these findings indicate that GABA plays a role as an inhibitory neurotransmitter in the human neocortex. GABAergic mechanisms are present and important for controlling neuron excitability. Accordingly, during application of the $GABA_A$ antagonist bicuculline methiodide, the early IPSP is abolished and the EPSP is transformed into a burst of action potential riding over a depolarizing envelope.[10,13]

The 4-Aminopyridine Model

Recent studies in our laboratory have focused on the effects induced by low doses of 4-aminopyridine (50–100 μM) on cortical slices obtained from patients with temporal lobe epilepsy or FCD. As shown in Fig. 4A, human neocortical slices obtained from a patient with no diagnosis of FCD generated in the presence of 4-aminopyridine (4AP),

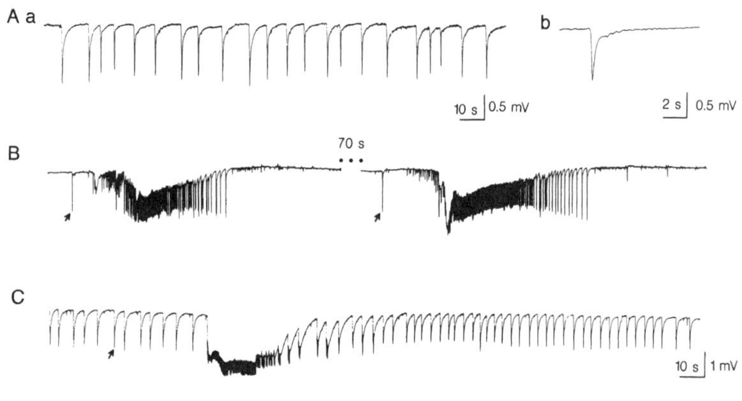

Figure 4. **A**: Synchronous activity recorded in a slice of human temporal neocortex with no dysplastic abnormalities. **a**: Note the long-lasting, negative-going potentials with no epileptiform activity. **b**: One of these events is shown at higher speed. **B**: Epileptiform discharges recorded in a slice of human, dysplastic occipital neocortex. Note the long-lasting seizure-like discharges. **C**: Other example of seizure-like discharge recorded from a temporal neocortical slice, in which the epileptiform discharges were accompanied to short duration, single negative transients. In A and B extracellular recordings were obtained from a level corresponding to the middle layers. Calibration in C applies also to B. From Mattia et al., 1995.

a type of synchronous activity that consisted of long-lasting (>1 s) negative going potentials. These synchronous field potentials are associated with long-lasting depolarizations that appear to be mediated by the postsynaptic activation of GABA receptors.[5]

In contrast, spontaneously occurring epileptiform discharges are seen during application of 4AP (50–100 μM) in neocortical slices obtained from patients with FCD.[12] Some examples of the epileptiform discharges recorded in the middle layers of human neocortical slices are shown in Figs. 4B and C. They consist of prolonged negative-going shifts with superimposed fast, negative transients. These resemble the electrographic pattern associated with the tonic-clonic type of seizure discharges. The intervals between seizure-like discharges show single negative transients (Figs. 1B and C, arrows) that resemble the GABA-mediated field potentials recorded in the temporal-lobe epileptic patients. Epileptiform discharges, similar to those occurring spontaneously, can also be evoked by extracellular, focal single-shock stimuli (Figs. 5A and B).

Simultaneous extracellular and intracellular recordings obtained from regular-spiking neurons in a region corresponding to the superficial to middle strata have demonstrated that the stimulus-induced epileptiform field-potential discharges are correlated with a large depolarization (amplitude of 5–60 mV; duration = 500–600 ms), which is followed by a series of depolarizing and repolarizing sequences with bursts of action potentials. Varying the membrane potential modifies the amplitude of the intracellular epileptiform responses evoked in the presence of 4AP. An increase in amplitude of the evoked depolarization is seen by hyperpolarizing the membrane potential (Fig. 5B, −20 mV), while, at depolarized membrane potentials, the evoked depolarization is smaller and a tonic sequence of action potentials is observed before and after the synaptic stimuli (Fig. 5B, +25 mV). Depolarization of the membrane potential discloses a relatively long-lasting (~5 s) hyperpolarization which terminates the evoked bursting response (Fig. 5B, +25 mV).

FINDINGS RELEVANT TO THE PATHOGENESIS OF SEIZURES IN FCDS

FCDs are often responsible for severe partial epilepsy resistant to pharmacological treatment (for review see[15]). We have found that slices of human dysplastic neocortex do generate *in vitro* long-lasting seizure-like discharges during application of 4AP.[12] It has been shown that non-epileptic human neocortical slices treated with 4AP display only synchronous, GABA-mediated potentials while no sign of epileptiform activity is discernible.[4,6] Intractable temporal lobe epilepsy is a condition that is associated with a morphological profile characterized by a rather selective neuronal loss in the hippocampal formation (so-called "Ammon's horn sclerosis") and only a moderate degree of gliosis of the temporal neocortex, without abnormalities of neocortical organization. In the light of these considerations, it appears that the dysplastic tissue has a peculiar tendency to develop epileptiform activity in vitro. These findings are in keeping with clinical evidence obtained from patients with FCDs, in which epileptogenicity is a common, functional feature of the dysplastic cortical tissue.[9,15,19]

The mechanisms underlying the epileptiform activity generated by dysplastic neocortex in the presence of 4AP remain to be elucidated. It has been suggested that, in the abnormal cortical multilaminar organization that identifies FCDs, neurons would be unable to establish their intended synaptic relations with the neighbors and remain dysfunctional.[16] Intracellular recordings performed in neurons of dysplastic human

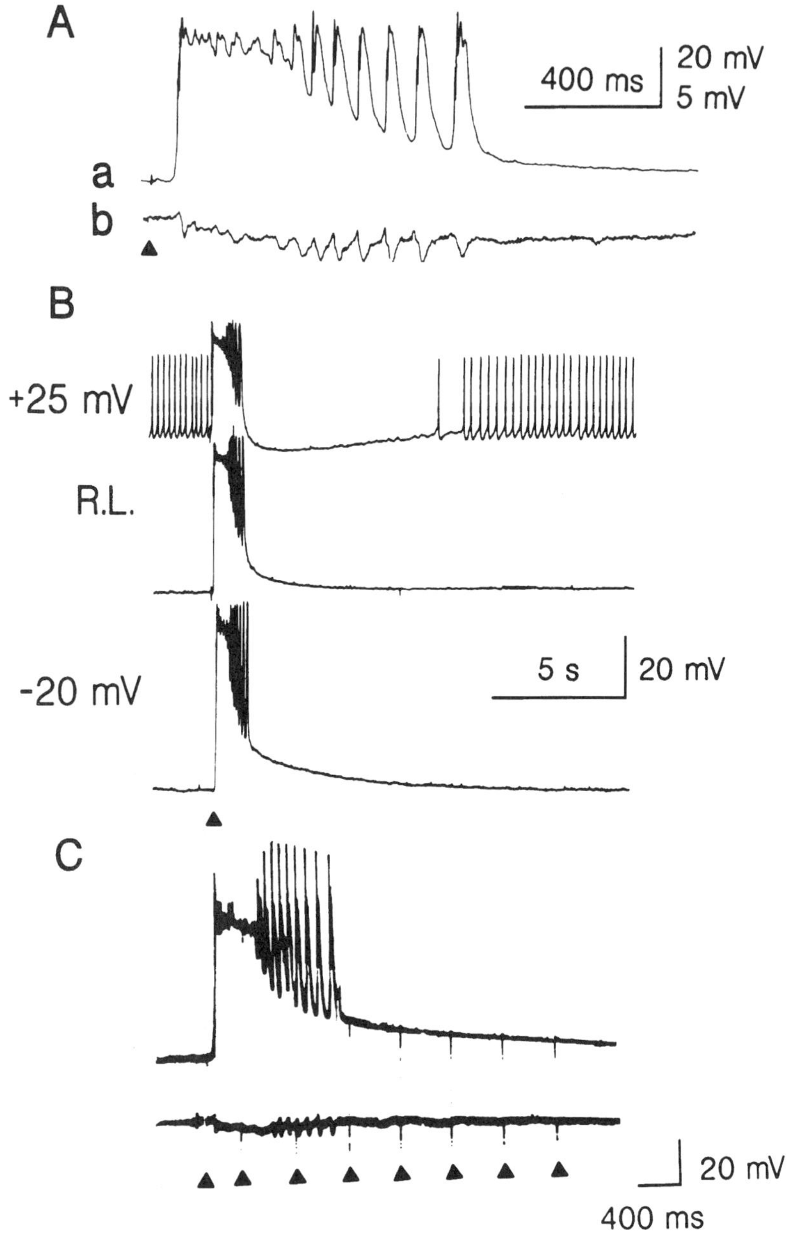

Figure 5. Extra- and intracellular recordings obtained from human dysplastic temporal neocortex. **A** shows an experiment where simultaneous intra- (**a**) and extracellular (**b**) recordings were performed during perfusion with 4AP. A focal single-shock stimulus evoked an initial EPSP followed by a large depolarization associated with a sequence of depolarizing-repolarizing complexes and bursts of action potentials (which are truncated). Note the correlation between the intracellular depolarization and the field potential. **B**: stimulus-induced epileptiform responses at different levels of membrane potential as obtained by intracellular injection of current. The shape of the responses was not modified by this procedure, but the amplitude of the depolarization became larger at hyperpolarized levels (–20 mV). A tonic mode of firing was elicited before and after the stimulus when the cell was depolarized by 25 mV from rest (R.:. = –80 mV). Note the long-lasting post-burst hyperpolarization revealed at +25 mV. **C**: Refractoriness of the field epileptiform activity. Bottom trace shows that the repetitive stimulation at 2.5 Hz lead to a failure of the responses recorded extracellularly. The top trace illustrates the simultaneous intracellular recording. From Mattia et al., 1995.

neocortex have revealed no abnormalities in the membrane properties of single neurons (G. G. C. Hwa, D. Mattia and M. Avoli, unpublished data). These findings appear to be consistent with the hypothesis of a dysfunction at the level of the synaptic circuits that could be responsible for the abnormal synchronization of neuronal populations underlying the genesis of epileptiform activity. The combined morphological-functional changes in synaptic interconnectivity might lead to a facilitation of neuronal recruitment that, in turn, would participate in the abnormal process of neuronal synchronization. Finally, we propose that dysplastic human neocortex maintained in vitro is a unique experimental model for analyzing the cellular mechanisms underlying epileptogenesis related to FCDs.

ACKNOWLEDGMENTS

This study was supported by the Medical Research Council of Canada (MT-8109), the Hospital for Sick Children Foundation and the Savoy Foundation.

REFERENCES

1. Avoli, M. Synaptic activation of $GABA_A$ receptors causes a depolarizing potential under physiological conditions in rat hippocampal pyramidal cells, *Eur. J. Neurosci.*, 4 (1992) 16–26.
2. Avoli, M. Electrophysiology and pharmacology of human neocortex and hippocampus in vitro. In: P.A. Schwartzkroin (Ed.) *Epilepsy: Models, Mechanisms and Concepts*, Cambridge University Press, Cambridge (1993) pp. 244–280.
3. Avoli, M., Hwa, G.G.C., and Lacaille, J.-C. Electrophysiological and repetitive firing properties of neurons in the superficial/middle layers of the human neocortex, *Exp. Brain. Res.*, 98 (1994) 135–144.
4. Avoli, M., Mattia, D., Siniscalchi, A., Perreault, P., and Tomaiuolo, F. Pharmacology and electrophysiology of a synchronous GABA-mediated potential in the human neocortex, *Neuroscience*, 62 (1994) 655–666.
5. Avoli, M. and Olivier, A. Electrophysiological properties and synaptic responses in the deep layers of the human epileptogenic neocortex maintained "in vitro", *J. Neurophysiol.*, 61 (1989) 589–606.
6. Avoli, M., Perreault, P., Olivier, A., and Villemure, J.G. 4-Aminopyridine induces a long-lasting depolarizing GABAergic potential in human neocortical and hippocampal neurons maintained *in vitro*, *Neurosci. Lett.*, 94 (1988) 327–332.
7. Avoli, M. and Williamson, A. Functional and pharmacological properties of human neocortical neurons maintained in vitro, *Prog. Neurobiol.*, 48 (1996) 519–554.
8. Calvin, W.H. and Sypert, G.W. Fast and slow pyramidal tract neurons: an intracellular analysis of their contrasting repetitive firing properties in the cat, *J. Neurophysiol.*, 39 (1976) 420–434.
9. Gambardella, A., Palmini, A., Dubeau, F., Andermann, F., Costa da Costa, J., Andermann, E., and Quesney, L.F. Electroencephalographic patterns in epileptic patients with neuronal migration disorders, *Epilepsia*, 34 (1993) 127.
10. Hwa, G.G.C., Avoli, M., Oliver, A., and Villemure, J.G. Bicuculline-induced epileptogenesis in the human neocortex maintained *in vitro*, *Exp. Brain. Res.*, 83 (1991) 329–339.
11. Knowles, W.D., Lüders, H., Hahn, J.F., and Awad, I.A. In vitro intracellular recordings from epileptic human neocortex and hippocampus, *Epilepsia*, 29 (1988) 711.
12. Mattia, D., Olivier, A., and Avoli, M. Seizure-like discharges recorded in the human dysplastic neocortex maintained *in vitro*, *Neurology*, 45 (1995) 1391–1395.
13. McCormick, D.A. GABA as an inhibitory neurotransmitter in human cerebral cortex, *J. Neurophysiol.*, 62 (1989) 1018–1027.
14. McCormick, D.A., Connors, B.W., Lighthall, J.W., and Prince, D.A. Comparative elctrophysiology of pyramidal and sparsely spiny stellate neurons of the neocortex, *J. Neurophysiol.*, 54 (1985) 782–806.

15. Palmini, A., Andermann, F., Tampieri, D., Andermann, E., Robitaille, Y., and Olivier, A. Epilepsy and cortical cytoarchitectonic abnormalities: an attempt at correlating basic mechanisms with antomoclinical syndromes. In: J., Jr. Engel, C. Wasterlain, E.A. Cavalheiro, U. Heineman, and G. Avanzini (Eds.) *Molecular Neurobiology of Epilepsy*, Elsevier, Amsterdam (1992) pp. 19–30.
16. Sarnat, H.B. Cerebral dysplasias as expressions of altered maturational processes, *Can. J. Neurol. Sci.*, 18 (1991) 196–204.
17. Schwartzkroin, P.A. *Epilepsy: Models, Mechanisms, and Concepts*, Cambridge University Press, Cambridge (1993).
18. Schwartzkroin, P.A., Turner, D.A., Knowles, W.D., and Wyler, A.R. Studies of human and monkey "epileptic" neocortex in the *in vitro* slice preparation, *Ann. Neurol.*, 13 (1983) 249–257.
19. Taylor, D.C., Falconer, M.A., Bruton, C.J., and Corsellis, J.A. Focal dysplasia of the cerebral cortex in epilepsy, *J. Neurol. Neurosurg. Psychiatry*, 34 (1971) 369–387.

15

PSYCHIATRIC ASPECTS OF INTRACTABLE EPILEPSY

Dietrich Blumer

135 North Pauline Street
Memphis, Tennessee 38105
tel: (901) 448-2415
fax: (901) 448-2500

Dr. Blumer argues that intractable temporal lobe epilepsy is often accompanied by a psychiatric syndrome, which he calls "interictal dysphoric disorder". Interictal dysphoric disorder is surprisingly responsive to antidepressant medications, although it is seldom diagnosed or treated.

INTRODUCTION

Epilepsy is a cerebral disorder with both major neurologic and major psychiatric aspects. An earlier era—which was marked by predominant awareness of distinctive psychiatric aspects of intractable epilepsy—has been followed by the modern era which is marked by remarkable progress in the neurologic understanding of all forms of epilepsy, but de-emphasis of the psychiatric aspects of the disorder. Mesial temporal lobe epilepsy, however, the most prominent chronic and often intractable form of epilepsy, requires a unified approach by the disciplines of neurology and psychiatry. Recent neuropsychiatric findings suggest that the psychiatric complications of epilepsy are: 1) of a specific nature; 2) result from the gradual development of inhibitory mechanisms in chronic mesial temporal epilepsy; and 3) tend to be surprisingly well treatable by proconvulsant antidepressive medications. The present chapter will review the evidence—both historical and recent–related to the psychiatric morbidity associated with intractable epilepsy and discuss the symptoms and treatment of this condition.

THE PSYCHIATRIC ERA OF 1850 TO 1950: STUDY OF INTRACTABLE EPILEPSY IN INSTITUTIONS

Beginning with Morel and Jules Falret in France about 1850, psychiatrists maintained a major interest in epilepsy until about 1950.[6,58] During this period, the expert clinical observations of patients admitted for prolonged psychiatric hospitalization led to establishing the modern classification of the major psychiatric disorders. At that time, the psychiatrists' interest in patients with epilepsy matched their interests in those with schizophrenia and manic-depressive illness, and the psychiatric textbooks of that era, from Kraepelin[39] to Bleuler[4] and to Slater and Roth,[53] document that the existence of epilepsy-specific mental disorders—from personality and mood changes to psychotic disorders—was never in doubt. An objective review of the pre-modern literature does not support the bias that these findings were warped by the effects of institutionalization and bromide toxicity.[6] Clearly, however, the pre-modern psychiatric observations of patients with epilepsy were carried out chiefly among those whose disorder had proven intractable and who required hospital confinement. These findings cannot be generalized to patients with more benign forms of epilepsy.

Falret, in 1860,[26] assigned the mental disorders of epileptics to three categories, listed here according to current terminology: 1) peri-ictal (prodromal, ictal, postictal) mental changes; 2) interictal personality and mood changes; and 3) relatively prolonged major or minor psychotic changes (most often postictal, less often prodromal or interictal). He described simple partial and complex partial seizures, and considered loss of consciousness, rather than the relatively rare epileptiform convulsions, as the main element of the epileptic attack. Falret's contemporary Morel[45] listed the intermittent mental changes of excitement and depression, of irritability to the point of fury (for trivial reasons and with amnesia), together with episodes of auditory and visual hallucinations, as the characteristic interictal symptoms that would allow the diagnosis of "larval or masked epilepsy" even in the absence of overt seizures. As Gastaut[29] has documented, the familiarity of French physicians with these concepts accounted for the correct diagnosis of Vincent van Gogh's epilepsy in the latter part of the 19th century.

In the 8th edition of Bleuler's *Textbook of Psychiatry* (1949),[4] *personality changes* characteristic for patients with chronic epilepsy were described as those of prominent intensification of affects, with excessive reactivity and persistence of responses, including not only irritability but affectionate feelings or joy. A sense of justice and religiosity was often particularly enhanced. The patients tended to be orderly, circumstantial, detail-bound and adhesive. If treated gently, they would respond with warm rapport.

Bleuler then emphasized the importance of a variety of *mood changes*, as follows:

> "Of particular importance are the *dysphoric moods* (Verstimmungen), which tend to be ubiquitous among patients with epilepsy: euphoric moods of exceptional well-being, very often with the idea to be now definitely cured of the epilepsy, but without flight of ideas; *depressive* moods with or without anxiety, which may entirely resemble a melancholic mood, are relatively rarely associated with suicidal urge, but fairly often with emphasis of the incurability of their illness; and finally the best known, the *irritable moods*, during which the patients on the one hand may get irritated by whatever happens, on the other hand may comport themselves in a rather demanding and boorish manner, so that conflicts become inevitable. The dysphoric moods may begin suddenly and terminate suddenly or may also begin slowly and last for hours to several days,

> and rarely for a longer period. Common are also simple paraesthesias, in particular headache, pains here and there, and various nervous symptoms . . ."

Bleuler noted that the *psychotic mental changes* of epilepsy, termed "twilight states", were observed mostly after seizures, but at times occurred in the prodromal phase or in place of seizures, lasting hours or days and rarely for longer periods. Hallucinations, particularly visual ones, tended to be the predominant psychotic symptoms. The affective changes of anxiety or fury were often associated, and preoccupation with religious themes was prominent.

Bleuler's textbook concurs with Kraepelin's earlier extensive description of the psychiatric aspects of epilepsy.[6,39] Kraepelin found that the periodic interictal dysphoric moods represented the most common form of all mental changes associated with epilepsy, and specified that their frequency varied from every few days to every few months, with a duration of just a few hours to a couple of days.

Kraepelin also referred to the common neuropathologic finding of usually unilateral Ammon's horn sclerosis among patients who, in the absence of evident cerebral lesions, had been diagnosed as suffering from genuine epilepsy. Bleuler stated that the Ammon's horn sclerosis, reported in about 80% of a rather large series of patients who had suffered from convulsive attacks, was absent when only minor attacks had been present and assumed the anatomic changes of genuine epilepsy were a consequence rather than a cause of the seizure disorder.[4]

GASTAUT: ASSOCIATION OF PSYCHIATRIC MORBIDITY WITH MESIAL TEMPORAL LOBE EPILEPSY

About 1950, at a time when psychiatrists were preoccupied with psychodynamic matters, psychiatry surrendered all interest in epilepsy to the discipline of neurology. All that psychiatrists had learned about intractable patients with epilepsy was judged obsolete once the focus of research centered on the rapidly developing *neurologic* knowledge of epilepsy. Proponents of an improved neurologic treatment of patients with epilepsy assumed that, if they combated the "biased" view of epilepsy as a psychiatric disorder, their patients would become accepted by society and altogether protected from suffering emotional disturbance.

Gastaut was the only one among the neurologically oriented modern epileptologists to explore the significance of the pre-modern psychiatric findings by reviewing the neuropathologic studies among their deceased patients.[28] In the past, the type of epilepsy referred to as "genuine", "idiopathic", "essential", or "chronic" had been listed as opposite to the "symptomatic" type, where seizures resulted from obvious brain disease (infectious or toxic disorders, tumors, abscesses, scars secondary to vascular or traumatic processes). Investigators had concentrated on discovering the lesion specific to genuine epilepsy among the institutionalized patients and indeed had discovered a biologic marker: a lesion consisting of a diffuse hemispheric sclerosis involving predominantly the mesial temporal lobe and the neighboring limbic structures. The first systematic neuropathologic study, performed by Bouchet and Cazauvielh in 1825,[19] had for the first time revealed the prominence of Ammon's horn sclerosis in the postmortem examinations of institutionalized epileptics. This finding was repeatedly confirmed, and Gastaut cited eight subsequent microscopic studies, carried out between 1880 and 1953 on the same populations, that revealed the loss of neurons and reactive gliosis (responsible for the sclerosis) of Ammon's horn in 30–70% of the patients. The significance of these neuropathologic findings for epilepsy had not been realized. They had been looked upon as a puzzle or an oddity at a time when the Ammon's horn was believed to be exclusively an olfactory center.

Gastaut concluded that the idiopathic epileptics from the asylums in times past had been afflicted chiefly with temporal lobe epilepsy. He pointed at studies[41,49,51] that confirmed the very high incidence of patients with temporal lobe epilepsy among the institutionalized. The idiopathic epileptics of modern times, as defined by Lennox, Penfield, and Gastaut tend to be both neurologically and psychiatrically intact. In the absence of any cerebral lesion, they present a genetic-metabolic disorder with a benign course and do not require any hospitalization. They escaped the attention of pre-modern psychiatrist. In the third group of epilepsies corresponding to the old symptomatic group, the patients often show chronic neurologic signs (hemiparesis, aphasia, hemianopsia, etc.), but, in the absence of limbic involvement, they tend to be psychiatrically intact.

Gastaut's own comparison studies of patients with epilepsy, selected from a very large population grouped according to stringent neurologic and electroencephalographic criteria,[30,31,32] documented that hyposexuality and changes in mood and personality were indeed specifically related to mesial temporal epilepsy. Later comparison studies, which purported that patients with a generalized seizure disorder suffered from an equally high percentage of psychiatric changes as those with temporal lobe epilepsy, suffered from the loose definition of what the investigators accepted as generalized seizure disorder.[34,54,55] Pure primary generalized epilepsy tends to be a benign disorder and is under-represented in the tertiary care centers where research is carried out.

Gastaut reported that the mood and personality changes of temporal lobe epilepsy tend to become manifest at least 2 years after onset of the epilepsy.[31] The interictal "schizophrenia-like" psychoses of epilepsy, on the other hand, develop at highly variable intervals, but on the average 15 years after the onset of epilepsy. These are likewise particularly associated with temporal lobe epilepsy.[52] It is then suggested that psychiatric morbidity in epilepsy is related to the chronicity of the seizure disorder and is selectively associated with the partial epilepsy of mesial temporal lobe origin. In view of the major role of the mesial-temporal portion of the limbic system in regulating emotion and behavior, a specific association of mesial temporal epilepsy with personality changes, affective disorders, and psychoses of epilepsy obviously is expected. Gastaut[27] compared his behavioral findings with those of the well-known ablation experiments in animals. The interictal behavior disorders of patients with temporal lobe epilepsy, from viscosity (adherence to detail, hypometamorphosis of attention) to intensification of affects (moods, anger, anxiety) and hyposexuality, represent exact opposites of the flighty attention (hypermetamorphosis of attention), placidity, and hypersexuality observed as characteristic traits of the Klüver-Bucy syndrome after bilateral temporal ablations. Gastaut suggests that this is not a surprising finding if one considers that interictally, owing to the effect of the irritative lesions, patients with temporal lobe epilepsy must present a state of excitation as opposed to the post-ablation state of depression of the temporolimbic system seen in the Klüver-Bucy syndrome. The ictal and interictal discharges in psychomotor epileptics affect the temporolimbic system with its diffuse connections to hypothalamus and mesencephalon and result in specific, profound, behavioral changes.

Studies of psychiatric morbidity in various populations of patients with epilepsy further confirm the link between chronicity of the epilepsy and with temporal lobe epilepsy. Most surveys of epilepsy populations for psychiatric morbidity reveal little about the types of psychiatric disorders associated with epilepsy.[16] At best, one can conclude that depressive and anxiety disorders may be most prevalent,[23] and that psychosis may be a problem in some 7% of the patients at one time or another.[33] Psychiatric morbidity, vaguely defined, ranges from about 20% to 50% of the population,[16] with a clear gradient from low prevalence in a community-based sample of individuals with a history of epilepsy,[60] to a higher prevalence among those who seek the ongoing care of family practitioners, to the highest prevalence in specialty clinics.[16,23,47] The well-documented high psychiatric morbidity among patients surgically treated for temporal lobe epilepsy[25,37,38,57,61] confirms the same trend, as would be expected, because the surgical group is a subgroup of the patients admitted to epilepsy centers. While suicide occurs about five times more frequently among patients with epilepsy, as compared with the general population, this ratio is increased to at least 25-fold among patients with chronic temporal lobe epilepsy.[2,21,43]

RECENT FINDINGS: SPECIFIC PSYCHIATRIC CHANGES ASSOCIATED WITH CHRONIC EPILEPSY, THEIR PATHOGENESIS AND TREATMENT

Specificity of Psychiatric Changes Associated with Chronic Epilepsy

The modern literature on the psychiatric disorders associated with epilepsy tends to be an assembly of fragmentary evidence. In study after study, investigators were content to document, in a cross-sectional approach, the presence of certain mental changes—often of a single type—by using measuring methods that had been certified as useful in the study of various disorders other than epilepsy. Depressive symptoms have been shown to be the most common mental changes associated with epilepsy[1,3,44] and plausibly the most important, given the heightened risk of suicide among epileptic patients. Anxiety, however, has been reported as a problem perhaps as frequent as depression among patients with epilepsy.[3,22,59] The presence of irritability and various other affective and somatoform changes have been investigated less often among patients with epilepsy. This literature has been thoroughly reviewed elsewhere.[14]

Our own studies had shown that patients with chronic epilepsy tend to suffer from a pleomorphic and intermittent pattern of affective and somatoform symptoms that clearly differs from any of the established psychiatric diagnostic categories.[8,15,18] For that reason, the use in epilepsy of psychologic or psychiatric research methods established to assess traditional diagnostic entities appeared inept and useless. We therefore began to develop a more comprehensive method of evaluating patients with epilepsy in the form of an Epilepsy Questionnaire and a Neurobehavioral Inventory. This was to be completed by patient and next of kin and reviewed for completeness and accuracy in semi-structured interviews with the patient and next of kin.[11]

The study, with this method, of 75 patients with epilepsy admitted for long-term monitoring to the Epi-Care Center in Memphis showed that by far the largest group (33 patients; 9 patients had other psychiatric disorders) exhibited the presence of a characteristic polysymptomatic and intermittent affective-somatoform disorder that needed psychiatric treatment.[16] The 33 patients presented with a combination of depressive-somatoform symptoms (depressive mood, 26; anergia, 26; atypical pain, 20; insomnia, 19) and with irritable-explosive affect (24), euphoric mood (21) fear (18) and anxiety (15) (Fig. 1). Most of the symptoms occurred in an episodic fashion for hours or days. Euphoric moods were particularly brief. Depressive symptoms and paroxysmal irritable

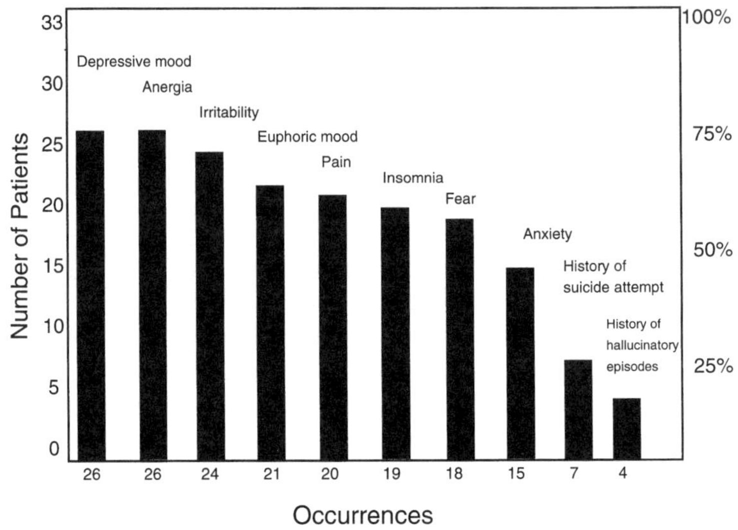

Figure 1. Key symptoms among 33 patients with Interictal dysphoric disorder. The 8 key symptoms of the interictal dysphoric disorder were present in a high percentage of patients (average: five per patient). This constellation accounts for the pleomorphic, yet rather uniform, symptomatology observed in this disorder. Suicide attempts and hallucinations were not concurrent within the period when the evaluation took place.

moods were simultaneously present in most of the patients (24 of 33), and represented the most troublesome complaints requiring treatment. Typically, the patients were genuinely good-natured and conscientious and regretted their temperamental outbursts. This conflict over their own irritability seemed to fuel their depression. The 33 patients averaged 5.1 of the key affective symptoms (range, 3 to 8). A symptom had to be clearly expressed and troublesome to be listed, and the number of symptoms in a given patient reflected the degree of severity of the affective disorder. The need for treatment coincided with the presence of at least three of the 8 key symptoms. Patients from this major group presented with an average of 5 of the 8 key symptoms so commonly that a subgrouping of the population into patients with depressive, irritable-explosive, somatoform, euphoric, anxiety, or phobic mental changes was not feasible.

We have proposed the recognition of this pleomorphic and intermittent affective disorder as the "interictal dysphoric disorder", defined as requiring a minimum of 3 of the 8 key symptoms. Prevalence of suicide attempts (in 7 patients) and hallucinatory episodes (in 4 patients) among this group of patients with interictal dysphoric disorder is also shown in Fig. 1 (a single patient had both histories). The key symptoms were all currently present, while suicide attempts and hallucinatory episodes were recorded as long as they were part of the history.

The interictal dysphoric disorder is identical with the dysphoric disorder recognized by pre-modern psychiatrists. As then, this dysphoric disorder is the most common psychiatric syndrome among patients with chronic epilepsy, and represents the key finding for the understanding of the psychopathology of epilepsy. More or less subtle personality changes (intensified emotionality, viscosity, and global hyposexuality) tend to develop in the same patients who are vulnerable to developing the symptoms of the interictal dysphoric disorder; these patients may not require therapeutic intervention so long as they are not dysphoric.[5,9,12] With increasing chronicity and severity of the interictal dysphoric disorder, patients tend to be at higher risk for developing interictal psychoses.[16] These so-called "schizophrenia-like" psychoses of epilepsy do not occur in the absence of a marked interictal dysphoric disorder and often may be best termed an "interictal dysphoric disorder with psychotic features" (hallucinations or delusions). In stark contrast to what is usually observed in schizophrenic psychoses, the emotionality of patients with interictal psychosis is intense, the pleomorphic affective-somatoform changes of their basic dysphoric disorder are in evidence, and the abilities of the patients to judge reality and to relate to others tend to be intact. According to our experience, pathogenesis and treatment of the interictal dysphoric disorder and the interictal psychosis appear to be identical in nature.[13]

Pathogenesis

Gastaut has emphasized that the seizure threshold, while abnormally low among patients with primary generalized epilepsy, is abnormally high during the interictal phase among patients with partial epilepsy of temporal lobe origin.[31] A significant number of findings suggest that this predominance of inhibitory mechanisms may in fact produce the psychiatric disturbances of the interictal (and peri-ictal) phase of mesial temporal epilepsy.

When Landolt[40] observed that certain patients with epilepsy showed disturbed behavior at the times when their epileptic activity had subsided, he referred to this phenomenon as "forced normalization". Stevens[56] hypothesized early that the mental changes of epilepsy may result not from the seizure activity itself but from the secondary inhibition, and this view is finding increasing support.[24] A number of clinical observations suggest that Landolt's forced normalization may be best ascribed to still vaguely defined, yet clearly evident, inhibitory mechanisms that gradually develop in the course of chronic mesial temporal epilepsy and are forcefully engaged at the time of a seizure to bring it to termination. Personality changes, the intermittent dysphoric symptoms, and episodic and finally chronic psychoses indeed develop gradually in the course of temporal lobe epilepsy, after a delay of at least 2 years from onset of the seizure disorder.[31,52] At the time of the prodromal and postictal phases, when inhibitory mechanisms are markedly engaged, the affective-somatoform changes characteristic for the interictal dysphoric disorder tend to be particularly enhanced.[10,14] When seizure activity is suppressed, many patients experience an exacerbation of their dysphoric disorder, or, in severely dysphoric patients, a psychosis may emerge.

In a series of patients treated at the Epi-Care Center for severe interictal dysphoric disorder, the affective disorder clearly emerged or worsened upon control or significant improvement of the seizure disorder in 9 of the 22 patients (41%).[13] The phenomenon of alternating dysphoric disorders or alternating psychoses is strikingly observed in some patients who are afflicted with a psychiatric disorder whenever the seizures are controlled and remit upon recurrence of the seizures. The availability of more potent antiepileptic drugs has drawn further attention to the fact that certain patients react with emotional and behavioral disorders once seizures are controlled.[50] Similarly, after surgical treatment, extending sometimes to as long as 18 months, there appears to be for some patients a period of enhanced vulnerability toward psychiatric complications,[18,35] coinciding with a predominance of the inhibitory mechanisms that appear to be slow to fade, as they were slow to develop. Overall, the unique intermittent nature of the symptoms of the interictal dysphoric disorder corresponds well with the underlying waxing and waning of seizure activity and seizure-suppressing mechanisms.

The presence of inhibitory mechanisms during the interictal phase can be visualized in positron emission tomography (PET) scans that demonstrate the hypometabolic zones in patients with epilepsy.[24] Successful pharmacologic treatment of the psychiatric complications of epilepsy has to be directed against inhibitory mechanisms, and the proconvulsant antidepressant drugs appear to be most effective. Successful treatment presumably, by diminishing the hypometabolic zones, could be visualized upon repeat PET scan.

Treatment

Ojemann et al.[46] reported in 1983 that of 19 patients treated "for agitated depression or depression with anxiety" with doxepin, 15 demonstrated better seizure control, 2 no change, and 2 increased seizures. Improvement of depression on doxepin (rated in 17 patients) was excellent in 10, good in 2, and mild to moderate in 5 patients. Himmelhoch,[36] commenting on the effectiveness of antidepressants in treating mood disorders related to epilepsy, suggested that the need for their use may be explained partially by the sometimes inverse relationship of interictal behavioral symptoms and seizure control. Blumer and Zielinski[18] documented the effectiveness of treatment with antidepressants in a consecutive series of 19 patients with epilepsy referred for psychiatric treatment. Their patients showed the symptoms of a marked interictal dysphoric disorder, and 12 patients showed psychotic symptoms. Ten achieved complete remission, and only three failed to show significant improvement.

Robertson and Trimble[48] reported a double-blind treatment trial of depression in 42 patients with epilepsy. Their conclusions left an erroneous impression, and their findings need to be scrutinized. Over the initial study period of 6 weeks, placebo was as effective as 75 mg daily of both nomifensine and amitriptyline. Over a subsequent study period of another 6 weeks, 26 patients on active drugs during the initial study period who were judged as nonresponders (i.e., their Hamilton Depression Rating Scale had not fallen by 50% of the initial score) were continued, now on 150 mg daily of nomifensine or amitriptyline. Nomifensine appeared more effective compared with amitriptyline (only 2 of 13 compared with 7 of 13 were nonresponders). No significant worsening of the seizure disorder was reported. For some reason, perhaps because the clinical effect of the (*subtherapeutic*) dose of antidepressant medications prescribed over the initial 6-week period was not distinguishable from the placebo effect, the investigators concluded that "in patients with depression and epilepsy immediate prescription with antidepressants may not be indicated," thereby disregarding the impressive 65% response rate among initial nonresponders at the end of the subsequent 6 weeks on fixed *therapeutic* doses of the antidepressants.

At the Epi-Care Center in Memphis, prescribing a modest amount of tricyclic medication for patients with an interictal dysphoric disorder is done routinely, with imipramine as the drug of first choice, at an average dose of 100 mg at bedtime.[7] If imipramine does not promptly establish sound sleep, it is replaced by a more sedative tricyclic (amitriptyline, doxepin, or trimipramine). In contrast to the treatment of traditional depressive disorders, the therapeutic effects of the tricyclics occur within a couple of days and extend to the full range of symptoms of the interictal dysphoric disorder, including irritability, anxiety, and headaches.

Until recently, if a tricyclic dose up to 150 mg daily was not effective, a small dose of a neuroleptic would be added to the tricyclic (e.g., 4 mg trifluoperazine to 100 mg imipramine). The experience of the last 3 years has shown that enhancing the tricyclic drug (at 100 to 150 mg daily) with a selective serotonin reuptake-inhibitor (10 to 50 mg of paroxetine or fluoxetine daily) represents the most effective treatment for previously intractable interictal dysphoric disorders.

Of about 200 patients treated with tricyclic medication for an interictal dysphoric disorder at the Epi-Care Center over a 20-month period, only 22 proved intractable. These patients required the double antidepressant treatment, with now highly satisfactory response: 15 patients with excellent or good outcome, 4 with partial response, and only 3 with no improvement.[13]

In some particularly difficult interictal dysphoric disorders, the prescription of bupropion (75 mg b.i.d. to t.i.d.) has been effective. In others, associated with marked brain damage, enhancing a tricyclic with a neuroleptic drug or with risperidone has proven most effective.

Experience over the past 12 years has shown us that antidepressant drugs, at a modest dose, prescribed for patients who are taking antiepileptic drugs do not worsen seizure frequency. The proconvulsant effect of the antidepressants appears to make them particularly suitable to mitigate the seizure-suppressing mechanisms that produce the psychiatric disturbances. In view of both the disabling nature of the interictal dysphoric disorder and the associated heightened risk of suicide, withholding antidepressant medication from patients with epilepsy is a critical decision that may have serious, even fatal, consequences. The poor response of interictal psychoses to routinely prescribed neuroleptic drugs compares unfavorably with the good to excellent response of these psychotic disorders to double antidepressant treatment.

Illustrative Cases

Case 1. At age 17, for unknown reasons, this previously healthy, active, and sociable young male began to experience complex partial seizures that, over a few years, reached a frequency of three to five per day. Secondary generalization occurred about twice per month, and he also experienced a few episodes of status or near status. MRI showed no abnormality. At the age of 24, he was referred to the Epi-Care Center, from out of state.

In spite of good intelligence, he dropped out of school after the 10th grade to do menial work, reluctantly married at the age of 17, and had two children before a divorce 5 years later. For a year or more, he was unable to carry out any work. He lived with his mother and his stepfather. (His abusive, alcohol- and drug-addicted father had disappeared from the family when the patient was 5 years old). The family history revealed epilepsy in a maternal aunt, migraine and depression in the mother, depression in the maternal grandfather, and, besides the father's addiction disorder, there was alcoholism in at least two uncles.

Two or 3 years after onset of the epilepsy, he began to be depressed and listless, and at age 22 this condition became severe. He tended to sleep 18 hours a day, yet often felt restless and was troubled by frequent anxieties. In addition, though generally good-natured (he had once aspired to be a preacher), he was often irritable, becoming angry and verbally abusive over little things. On a few occasions he had physically attacked others in a rage. For 2 days a week, perhaps, his depressive state would suddenly bottom out, and it was during such episodes, within the last year before admission to Epi-Care, that he made four serious suicide attempts by overdose, requiring intensive care and psychiatric hospitalization each time. He never received antidepressant medication and continued to be treated in weekly psychotherapy sessions. His mother had become resigned that one day she would find him dead.

Scalp monitoring showed an interictal left frontal-temporal abnormality (sharp activity and slowing) and bilateral activity with right-sided emphasis and secondary generalization during a seizure. Subdural strip recording, carried out 1 month later, showed a right mesial temporal onset of two of his typical seizure events. Within 2 months, a right anterior temporal lobectomy was carried out and, except for two seizures within the first 2 weeks after surgery, he has remained completely seizure free for the entire follow-up period of 1 year after surgery. The histopathology showed a grade II hippocampal sclerosis.

Subsequent to the initial scalp monitoring, he was started on desipramine (100 mg daily) combined with 20 mg of paroxetine. The double antidepressant treatment was chosen because of the severity of his interictal dysphoric disorder, the suicide risk, and the difficulty in monitoring him because he lived out of state. His dysphoric disorder remitted, except for some explosive outbursts, when he did not comply with the antidepressant medication prior to the operation. Following surgery, he was advised to continue the psychotropic medication. Two months after the operation, he returned for a first follow-up visit and reported a remarkable increase of his sexual arousal from about twice per month preoperatively to daily. He had good energy and was sleeping less, but he still experienced depressive and irritable moods, though less frequently and to a lesser degree. He was advised to comply more faithfully with the daily prescription of antidepressants. When seen again 4 months after the operation, he reported stable moods, good energy, normal sleep patterns, and an active sex life. Both family and patient were delighted with the remarkable recovery, described by his brother as "everything is in tune now." Eight months after surgery, we learned that a few weeks earlier he had stopped all his medication, except for the phenytoin, had remained seizure free and in stable moods, and was looking for work.

Comment: The severity of depressive features may have been influenced by the family history of depression. Nevertheless, his remarkable mood swings and episodes of rage were clear evidence of an interictal dysphoric disorder. The failure to prescribe antidepressant medication prior to his referral to the Epi-Care Center could have easily had fatal consequences. As expected, he required continuation of the antidepressants for several months after surgery, in spite of the fact that he was free from any seizures. Restoration of a vigorous sexual arousal results from the surgical elimination of the epileptogenic zone and cannot be accomplished by antidepressant medication. Successful surgery promotes a change from an inverse Klüver-Bucy syndrome towards a mild Klüver-Bucy syndrome—most notably towards placidity and a relative hypersexuality. The initial EEG findings demonstrate how misleading scalp EEG findings can be for localizing mesial temporal epilepsy.

Case 2. At age 53, this housewife from out of state was seen for the first time at the Epi-Care Center. Without significant antecedents, she had started to have complex partial seizures at the age of 21 while she was pregnant with her fourth child. A paternal grandmother had epilepsy, and no psychiatric disorders were reported among family members. The EEG showed an abnormality in the right temporal-frontal-central region, consisting of focal slowing, loss of background, scattered spikes, and sharp and slow-wave discharges. MRI revealed the presence of ectopic gray matter within the right temporal lobe. Because of a psychiatric history, she had been maintained on thioridazine, 75 mg daily, along with phenytoin and phenobarbital. She was started on carbamazepine, maintained on phenytoin and thioridazine, and her phenobarbital was phased out. She seemed mentally stable, and was not referred for psychiatric attention during her initial visit. She was advised to return in 10 weeks.

Subsequent to the change of medication, she became seizure free, but after 3 weeks she began to act strange. Within 10 days she had to be admitted to our psychiatric unit in an acute psychotic state. She was agitated, angry, uncooperative, and hallucinatory, with marked mood swings (crying spells alternated with rapid speech) and insomnia. She accused her husband of being an abusive alcoholic and feared that he was going to kill her. She declared she had found the answers to her prayers: God had spoken to her, and she had to leave spouse and home. She had, in fact, moved their furniture onto the side of the street in order to sell it all. Her history revealed that she had experienced several psychotic episodes in the past, and at one time had been in a mental institution for 4–5 months. In between her psychotic episodes, she had been difficult to live with because of troublesome, irritable and depressive moods with anergia. All four children had moved away from home and maintained little contact with their mother. The husband had remained loyal to her through 35 years, but had gone through a prolonged period of alcohol abuse that he had been able to terminate a couple of years previously.

Her acute psychiatric illness was diagnosed as interictal psychosis precipitated by the arrest of seizures with the medication change. Her thioridazine was replaced by imipramine (100 mg daily) on day 1 of her hospitalization, with a prompt restoration of good sleep and with significant general improvement evident from day 2. On day 3, paroxetine (20 mg daily) was added. By day 4, she was considering a return home, and on day 6, when her husband visited for the first time, she was in full remission and could be discharged home on unchanged doses of phenytoin and carbamazepine plus imipramine 100 mg at bedtime and paroxetine 20 mg a.m.

She has since complied fully with her medication. When the husband returned with her for the first follow-up visit 2 months after discharge from the hospital, he was all smiles, and expressed his delight over the outcome of the new treatment. She still had no seizures, but was now a different person, a better person, in fact she was now again like the woman he had married 35 years ago. The symptoms of her longstanding interictal dysphoric disorder had vanished, and she has maintained a splendid recovery through a follow-up period of 18 months.

Comment: Excellent remissions of longstanding interictal dysphoric disorders, upon antidepressant treatment, can often be observed.[13] Prompt and complete remission of an interictal psychosis with double antidepressant treatment can be achieved in a slight majority of the cases.[13] If a dysphoric disorder stays in remission, upon continuation of the antiepileptic and psychotropic medication, there will be no risk of recurrence of another psychotic episode.

CONCLUSIONS

Pre-modern psychiatrists were very familiar with the psychopathology characteristic of chronic epilepsy, from personality change and dysphoric disorder to psychosis. Premodern and modern findings indicate that this psychopathology is associated specifically with mesial temporal epilepsy, the most common form of chronic epilepsy. Some individuals, with primary or secondary involvement of the mesial temporal zones who are diagnosed by clinical and scalp electroencephalographic evidence as having other forms of epilepsy, as well as individuals with chronic subictal activity of the mesial temporal zone in the absence of overt seizures, experience the same mental changes that are seen in patients with mesial temporal epilepsy.[15,18,31] The precise neurologic identification of all three groups tends to be difficult in the absence of intracranial electroencephalographic recordings and neuroimaging findings. It is important to recognize that there is a specific psychiatric syndrome associated with chronic epileptiform disorders of the mesial temporal zones and that this can be effectively treated—and must be treated—in the absence of complete neurologic evidence.

Two differences between pre-modern and modern observations are noteworthy. First, the high prevalence of postictal psychoses reported by pre-modern observers contrasts with the emphasis on episodic or chronic interictal psychoses in modern observations. This difference can be ascribed to the advent of potent antiepileptic drugs in modern times. Postictal psychoses tend to occur after flurries of seizures, which are nowadays less common; interictal psychoses tend to occur when seizure-suppressing mechanisms, favored by the modern antiepileptic medication, predominate. Secondly, the greater emphasis on the episodes of marked irritability in the past and the greater emphasis on marked depression in the present, may be likewise related to the availability of modern antiepileptic drugs. Suicide was not listed as a significant risk by pre-modern observers. This second difference among pre-modern and modern observers, however, is also related to the tendency of modern observers to depend on cross-sectional assessments and to underrate the intermittent irritability. Both irritability and depressive symptoms can be observed, now as then, as equally important key symptoms of the interictal dysphoric disorder.

An understanding of the pathogenesis of the psychiatric disorders that are specifically associated with epilepsy allows for a rational treatment approach. The effectiveness of proconvulsant antidepressants for the interictal dysphoric disorder and its psychotic features is consistent with the assumption that the seizure-suppressing mechanisms of chronic epilepsy involving the temporal-limbic zone, bring about the psychiatric disorders. Though mesial temporal epilepsy tends to be particularly chronic and intractable, the psychiatric complications of this form of epilepsy in our experience tend to be surprisingly treatable.[13]

Collaboration of neurologist and psychiatrist is urgently needed at tertiary care centers, where about half of the patients with intractable epilepsy need psychiatric intervention.[16] For the majority of interictal dysphoric disorders, the prescription of antidepressants is less complicated than the prescription of antiepileptic drugs, and can be carried out by the neurologist. In their own sphere, psychiatrists need to be aware of the specific psychopathology and treatment needs of their patients with epilepsy. As many as 10% of patients admitted for short-term psychiatric treatment (a figure slightly higher than for bipolar patients) carry the diagnosis of epilepsy,[20] and that percentage is higher in long-term psychiatric treatment units.[42] The persistent neglect by modern biologically oriented psychiatrists of epilepsy and its psychiatric complications contrasts with the major interest in epilepsy on the part of their predecessors, prior to the psychosocial era of psychiatry. With effective psychopharmacologic treatment now available for this group of patients, this neglect cannot be maintained.

REFERENCES

1. Altshuler, L. Depression and epilepsy. In: O. Devinsky and W.H. Theodore (Eds.) Epilepsy and behavior, Wiley-Liss, New York (1991) pp. 47–65.

2. Barraclough, B.M. The suicide rate of epilepsy, *Acta Psychiat. Scand.*, 76 (1987) 339–345.
3. Betts, T.A. Depression, anxiety and epilepsy. In: E.H. Reynolds and M.M. Trimble (Eds.), *Epilepsy and psychiatry*, Churchill Livingstone, Edinburgh (1981) pp. 60–71.
4. Bleuler, E. *Lehrbuch der Psychiatrie*, 8th ed. by Bleuler, M., Springer, Berlin (1949).
5. Blumer, D. Temporal lobe epilepsy and its psychiatric significance. In: D.F. Benson and D. Blumer (Eds.), *Psychiatric aspects of neurologic disease*, Grune & Stratton, New York (1975) pp. 171–197.
6. Blumer, D. The psychiatric dimension of epilepsy: historical perspective and current significance. In: D. Blumer (Ed.), *Psychiatric aspects of epilepsy*, APA Press, Washington, DC, (1984) pp. 26–37.
7. Blumer, D. Diagnosis and treatment of psychiatric problems associated with epilepsy. In: D.B. Smith (Ed.), *Epilepsy: current approaches to diagnosis and treatment*, Raven Press, New York (1990) pp. 193–209.
8. Blumer, D. Epilepsy and disorders of mood. In: D.B. Smith, D.M. Treiman, and M.R. Trimble (Eds.), *Neurobehavioral problems in epilepsy*, Raven Press, New York (1991) pp. 185–195.
9. Blumer, D. Personality in epilepsy. *Seminars in Neurol.*, 11 (1991) 155–166.
10. Blumer, D. Postictal depression: significance for the treatment of the neurobehavioral disorder of epilepsy, *J. Epilepsy*, 5 (1992) 214–219.
11. Blumer, D. Psychiatric evaluation of patients with unilateral temporal lobectomy for epilepsy. In: A. Wyler and B. Hermann (Eds.), *The surgical treatment of epilepsy*, Butterworth-Heineman, London (1994) pp. 90–96.
12. Blumer, D. Personality disorders in epilepsy. In: J. Ratey (Ed.), *Neuropsychiatry of personality disorders*, Blackwell Scientific Publications, Inc. (1995) pp. 230–263.
13. Blumer, D. Antidepressant and double-antidepressant treatment for the affective disorder in epilepsy, *J. Clin. Psychiatry*, 58 (1997) 3–11.
14. Blumer, D. and Altshuler, L. Affective disorders associated with epilepsy. In: J. Engel Jr. and T. Pedley (Eds.) *Epilepsy. A Comprehensive Textbook*, Lippincott-Raven Publishers, Philadelphia (1997) pp. 2083–2099.
15. Blumer, D., Heilbronn, M., and Himmelhoch, J. Indications for carbamazepine in mental illness: Atypical psychiatric disorder or temporal lobe syndrome? *Comprehen. Psy.*, 29(2) (1988) 108–122.
16. Blumer, D., Montouris, G., and Hermann, B. Psychiatric morbidity in seizure patients on a neurodiagnostic monitoring unit, *J. of Neuropsychiatry and Clin. Neurosciences*, 7 (1995) 445–456.
17. Blumer, D., Wakhlu, S., Davies, K., et al. Psychiatric outcome of temporal lobectomy for epilepsy: incidence and treatment of psychiatric complications, *Epilepsia*, 39 (1998) 479–486.
18. Blumer, D. and Zielinski, J.J. Pharmacologic treatment of psychiatric disorders associated with epilepsy, *J. Epilepsy*, 1 (1988) 135–150.
19. Bouchet and Cazauvieilh. De l'épilepsie considéré dans ses rapports avec l'aliénation mentale. Recherches sur la nature et le siège de ces deux maladies, *Arch. Gen. Med.*, 9 (1825) 510–542; 10 (1826) 5–50.
20. Boutros, N.N., Juo-Tzu, Liu, J., Shehata, M., and Millana, R.B. Epileptic psychiatric patients, a special population, *J. Mental Health*, 1 (1995) 79–83.
21. Brent, D.A. Overrepresentation of epileptics in a consecutive series of suicide attempters seen at a children's hospital, 1978–1983, *J. Am. Acad. Child Psychiatry*, 25 (1986) 242–246.
22. Currie, S., Heathfield, K.W.G., Henson, R.A., and Scott, D.F. Clinical course and prognosis of temporal lobe epilepsy: a survey of 666 cases, *Brain*, 94 (1971) 173–190.
23. Edeh, J., Toone, B.K., and Corney, R.H. Epilepsy, psychiatric morbidity, and social dysfunction in general practice. Comparison between hospital clinic patients and clinic nonattenders, *Neuropsychiatry, Neuropsychology, and Behavioral Neurology*, 3 (1990) 180–192.
24. Engel Jr., J. *Seizures and epilepsy.* FA Davis Co., Philadelphia (1989).
25. Falconer, M.A. and Serafetinides, E.A. A follow-up study of surgery in temporal lobe epilepsy, *J. Neurol. Neurosurg. Psychiatry*, 26 (1963) 154–165.
26. Falret, J. De l'état mental des épileptiques, *Arch. Gen. Med.*, 16 (1860) 661–679; 17 (1861) 461–491; 18 (1861) 423–443.
27. Gastaut, H. Interpretation of the symptoms of psychomotor epilepsy in relation to physiological data on rhinencephalic function, *Epilepsia*, 3 (1954) 84–88.
28. Gastaut, H. État actuel des connaissances sur l'anatomie pathologique des épilepsies, *Act. Neurol. Psychiatr. Belg.*, 56 (1956) 5–20.
29. Gastaut, H. La maladie de Vincent Van Gogh envisagée à la lumière des conceptions nouvelles sur l'épilepsie psychomotrice, *Ann. Med. Psychol. (Paris)*, 114 (1956) 196–238.

30. Gastaut, H. and Collomb, H. Étude de comportemente sexual chez les épileptiques psychomoteurs, *Ann. Med. Psychol. (Paris)*, 112 (1954) 657–696.
31. Gastaut, H., Morin, G., and Lesèvre, N. Étude du comportement des épileptiques psychomoteurs dans l'intervalle de leurs crises: les troubles de l'activité globale et de la sociabilité, *Ann Med Psychol (Paris)*, 113 (1955) 1–27.
32. Gastaut, H., Roger, J., and Lesèvre, N. Différenciation psychologique des épileptiques en fonction des formes électrocliniques de leur maladie, *Rev. Psychol. Appl.*, 3 (1953) 237–249.
33. Gudmundsson, G. Epilepsy in Iceland. A clinical and epidemiological investigation, *Acta Neurol. Scand.*, 43 (1966) (Suppl 25) 64–90.
34. Guerrant, J., Anderson, W.W., Fischer, A., Weinstein, M.R., Jaros, R.M., and Deskins, A. *Personality in Epilepsy*, Thomas, Springfield, IL, (1962).
35. Hill, D., Pond, D.A., Mitchell, W., and Falconer, M.A. Personality changes following temporal lobectomy for epilepsy, *J. Mental Science*, 103 (1957) 18–27.
36. Himmelhoch, J.M. Major mood disorders related to epileptic changes. In: D. Blumer (Ed.) *Psychiatric aspects of epilepsy*, American Psychiatric Press, Washington, DC (1984) pp. 271–294.
37. James, J.P. Temporal lobectomy for psychomotor epilepsy, *J. Ment. Sci.*, 106 (1960) 543–558.
38. Jensen, I. and Vaernet, K. Temporal-lobe epilepsy. Follow-up investigation of 74 temporal lobe resected patients, *Acta Neurochir.*, 37 (1977) 173–200.
39. Kraepelin, E. *Psychiatrie.* 8th ed. Lepizig, Barth (1923).
40. Landolt, H. Serial electroencephalographic investigations during psychotic episodes in epileptic patients and during schizophrenic attacks, *Folia. Psychiat. Neurol. Neurochir. Neerlandica.*, S4 (1958) 91–133.
41. Liddell, D.W. Observations on epileptic automatism in a mental hospital population, *J. Ment. Sci.*, 99 (1953) 732–748.
42. Masland, R.L. Commission for the control of epilepsy, *Neurology*, 28 (1978) 861–862.
43. Matthew, W.S. and Barabas, G. Suicide and epilepsy: a review of literature, *Psychosomatics*, 22 (1981) 515–524.
44. Mendez, M.F., Cummings, J., and Benson, F. Depression in epilepsy. Significance and phenomenology, *Arch. Neurol.*, 43 (1986) 766–770.
45. Morel, B.A. D'une forme de délire, suite d'une surexcitation nerveuse se rattachant à une variété non encore décrite d'épilepsie (épilepsie larvée), *Gazette hebdomadaire Med. Chir.*, 7 (1860) 773–775, 819–821, 836–841.
46. Ojemann, L.M., Friel, P.N., Trejo, W.J., and Dudley, D.L. Effect of doxepin on seizure frequency in depressed epileptic patients, *Neurology*, 33 (1983) 646–648.
47. Pond, D.A. and Bidwell, B.H. A survey of epilepsy in fourteen general practices. II. Social and psychological aspects, *Epilepsia*, 1 (1959) 285–299.
48. Robertson, M.M. and Trimble, M.R. The treatment of depression in patients with epilepsy: a double-blind trial, *J. Affective Disorder*, 9 (1985) 127–136.
49. Roger, A. and Dongier, M. Corrélations électrocliniques chez 50 épileptiques internés, *Rev. Neurol.*, 83 (1950) 593–596.
50. Sander, J.W., Hart, Y.M., Trimble, M.R., and Shorvon, S.D. Vigabatrin and psychosis, *J. Neurology, Neurosurgery and Psychiatry*, 54 (1991) 435–439.
51. Sano, K. and Malamud, N. Clinical significance of sclerosis of the cornu Ammonis, *Arch. Neurol. Psychiatry*, 70 (1953) 40–53.
52. Slater, E. and Beard, A.W. The schizophrenia-like psychoses of epilepsy, *Br. J. Psychiatry*, 109 (1963) 95–150.
53. Slater, E. and Roth, M. *Clinical Psychiatry*, 3rd ed., Baillière Tindall, and Cassell, London 1969).
54. Small, J.G., Milstein, V., and Stevens, J.R. Are psychomotor epileptics different? A controlled study, *Arch Neurol*, 7 (1962) 187–194.
55. Stevens, J.R. Psychiatric implications of psychomotor epilepsy, *Arch. Gen. Psychiatry*, 14 (1966) 461–471.
56. Stevens, J.R. Interictal clinical manifestations of complex partial seizures. In: J.K. Penry and D.D. Daly (Eds.), *Advances in neurology, II*, Raven Press, New York (1975) pp. 85–112.
57. Taylor, D.C. and Falconer, M.A. Clinical socio-economic and psychological changes after temporal lobectomy for epilepsy, *Br. J. Psychiatry*, 114 (1968) 1247–1261.
58. Temkin, O. *The falling sickness*, The Johns Hopkins Press, Baltimore (1971).

59. Trimble, M.R. and Perez, M.M. Quantification of psychopathology in adult patients with epilepsy. In: B.M. Kulig, H. Meinardi, and G. Stores (Eds.), *Epilepsy and behavior '79*, Swets and Zeitlinger, Lisse (1980) pp. 118–126.
60. Trostle, J.A., Hauser, W.A., and Sharbrough, F.W. Psychologic and social adjustment to epilepsy in Rochester, Minnesota, *Neurology*, 39 (1989) 633–637.
61. Walker, A.E. and Blumer, D. Behavioral effects of temporal lobectomy for temporal lobe epilepsy. In: D. Blumer (Ed.) *Psychiatric aspects of epilepsy*, American Psychiatric Press, Washington, DC (1984) pp. 295–323.

16

NEURAL GROWTH, NEURAL DAMAGE AND NEUROTROPHINS IN THE KINDLING MODEL OF EPILEPSY

Ronald J. Racine*, Beth Adams*, Philip Osehobo+,
and Margaret Fahnestock+

*Department of Psychology and +Department of Biomedical Sciences
McMaster University
Hamilton, Ontario
CANADA L8S 4K1
tel: (905) 525-9140, ext. 23022
fax: (905) 529-6225

Do seizures change the brain? Studies on the kindling model—a widely used animal model of epilepsy—suggest that they do. Dr. Racine, one of the pioneers in the kindling field, describes the basic phenomena of kindling, and discusses the possible roles of cell growth and cell death in this model.

INTRODUCTION

Kindling is most frequently cited as a model of complex partial seizures with secondary generalization. As pointed out by Racine and Burnham,[119] however, there are a wide variety of clinical events that are suitably modeled by kindling phenomena. We will begin with a brief description of some of these phenomena. Readers familiar with kindling may wish to jump ahead to the section on *mechanisms—recent work*.

Kindling was first described in detail by Graham Goddard.[47] He found that repeated electrical stimulation of the amygdala through implanted electrodes eventually led to the triggering of convulsive activity. Although this development was interesting, Goddard was most intrigued by the apparent permanence of the effect. Electrical stimulation, which initially had little behavioral effect on the animal, eventually came to trigger fully generalized convulsions, and these alterations in reactivity lasted for months in the absence of further stimulation.

KINDLING STAGES AND KINDLING SITES

Kindling effects have now been demonstrated from many different brain sites, and the progression of the electroencephalographic and convulsive activity evoked from these sites is similar, but not identical. When the kindling electrode is located anywhere within the piriform lobe, including the olfactory bulb, the convulsions typically develop through 5 stages.[117] We will use the amygdala as the prototypical piriform lobe structure in the following descriptions.

Amygdala Kindling

If the stimulation intensity is sufficient to trigger an epileptiform discharge, the initial behavioral response is usually a cessation of ongoing behavior (freezing). The first behavioral responses that are clearly epileptic in nature are facial. The mouth, in particular, is involved in this stage (stage 1). Stage 1 often appears early during the kindling process, and the animal may remain in this stage for several stimulations before progressing to the next stage. Once the animal reaches the next stage, however, it generally progresses quite rapidly through the remaining stages: clonic head movements or opisthotonus (stage 2), forelimb clonus (stage 3), clonic rearing (stage 4) and loss of postural control (stage 5). These stages develop in the order described, and, once kindling is complete, the seizures progress through this same sequence in response to each kindling stimulation.

Stage 1 can appear before the evoked afterdischarge has shown any substantial increase in either amplitude or duration. Once stage 2 appears, however, the afterdischarge has generally begun to show considerable growth, often increasing in duration by a factor of 5–10. This growth often accelerates through the development of the final stages of the convulsive response.

It appears that the piriform-lobe structures are closely connected to the brain sites that drive the mouth and facial movements. Additional circuitry must presumably be recruited in order to drive the behavioral components comprising stages 2 to 5. This raises the question of propagation paths, or preferred routes, for the spread of the seizure activity and the recruitment of motor systems. First, however, we will describe the kindling developments seen when other sites are kindled.

Hippocampal Kindling

When the kindling electrodes are placed within the hippocampus, the development of electrographic and convulsive responses is somewhat different than that described above. The hippocampal discharge thresholds are the lowest of any site tested, and the discharges start off 2 to 5 times longer than those in the amygdala. They do not, however, show increases in duration as large as those seen with amygdala kindling. The hippocampus also generally requires 2 to 5 times as many stimulations to kindle to stage 5 behavioral seizures. Although the beginning and end points of hippocampally-induced kindled convulsions are similar to those seen in the amygdala, the intervening stages are often truncated. It is not unusual for hippocampally-stimulated animals to progress abruptly from a stage 1 or 2 to a stage 5 seizure.

Another difference between hippocampus and amygdala is that, even after the appearance of stage 5 seizures, there are often periodic failures of the seizure response in the hippocampally-stimulated animal. The behavioral seizure will often revert to stage 1, and this can occur many times before the response begins to stabilize.

A final striking difference between the hippocampus and the amygdala is that the convulsive response often appears at the apparent *end* of the hippocampal discharge, at which time the piriform lobe discharge may be reaching its maximum amplitude. This observation suggested to us many years ago that the piriform lobe structures might be more critically involved in driving these convulsive responses than the hippocampus. This raises the question of possible generator sites, which we will address, along with propagation patterns, after briefly discussing neocortical kindling.

Neocortical Kindling

The kindling properties of the neocortex have also been thoroughly investigated. Unlike the hippocampus, anterior neocortical sites show convulsive responses during the *first* evoked discharge. These responses, however, are qualitatively different from those evoked from subcortical sites. Although there may be some forelimb clonic movements, these seizures are more clearly characterized by their tonic components. There is generally a tonic turning of the head and twisting of the body trunk during the cortical discharge. The clonic stages described for subcortical sites do not generally appear until 10, 20 or more stimulations have been applied to the cortical site.[121] Coincident with these developments is a clear recruitment of piriform lobe sites into the evoked discharge.

Among the sites that have been tested, it is clear that the neocortex and dorsal hippocampus are among the slowest kindling sites in the brain. The entorhinal cortex and septal area follow with intermediate kindling rates. The most rapid kindling sites are in the anterior piriform lobe in the order (from slowest to fastest) amygdala, piriform cortex, olfactory bulb and perirhinal cortex.[15,48,91,118]

GENERATORS OF SPONTANEOUS ACTIVITY

Kindling is a chronic epilepsy model not only in the sense that there is a permanent change in reactivity to the electrical stimulation, but also because spontaneous epileptic events, both ictal and interictal, develop along with the increased reactivity.

Interictal spikes develop after only a few stimulations. It is not clear how long they persist. They tend to fall off in frequency dramatically over the first few days after the last epileptiform afterdischarge (AD), but we have found that interictal spikes can appear in very relaxed animals up to 6 weeks after the last AD. In addition, all kindled animals will show a resumption of interictal spiking following electrical stimulation that is below threshold for an ictal response.[117]

The development of spontaneous ictal events requires many stimulations, at least in the rat.[114,115] The spontaneous seizures generally fall into one of the traditional 5 stages. Additional seizure stages must be added, however, to describe the *stimulation-induced* seizures that appear after hundreds of kindling stimulations have been applied. The stage 6 seizure consists of multiple stage 5s. Stage 7 is jumping, stage 8 is a running fit, including jumping, stage 9 is running and jumping followed by a stage 4, 5 or 6 limbic seizure, and stage 10 consists of repeated 9's, which include two discrete running and jumping fits and a limbic seizure (N.W. Milgram, personal communication).

One of the authors of this chapter (RJR) attempted to determine the location of the generator(s) for interictal spikes in the kindled preparation. Although several subcortical sites were capable of generating these isolated paroxysmal discharges, the piriform cortex appeared to be the most reactive of the sites monitored.[126] Interictal spikes often appeared first in the piriform cortex, regardless of which structure received the kindling stimulation (although see Zhao and Leung[170]).

McIntyre and co-workers have also found the piriform region to be heavily involved in the kindling process.[93,94] Recently, they have found that the perirhinal cortex may also be involved in the generation or propagation of epileptiform activity.[71] Lesions of the posterior perirhinal cortex appeared to effectively block hippocampal kindled seizures. Other evidence is accumulating that the piriform lobe structures undergo many changes during kindling and related phenomena including c-fos activation,[36,60,81] enhanced neurotrophin levels,[37] activated astrocytes[74,75] and cell loss.[92,98]

PREFERRED ROUTES OF PROPAGATION OF EVOKED SEIZURE DISCHARGE

As described above, there are some differences in the motor responses evoked when stimulation is applied to different sites. Perhaps the most obvious differences are those between neocortical and subcortical "limbic" sites. Between different cortical sites or between different limbic sites, however, the responses are remarkably similar. This similarity would seem to indicate that there are common final pathways for the propagation of the seizure discharge. Although we have not yet been able to determine the final common pathways, we have found that patterned, 3 per second, stimulation of the substantia nigra can drive a stimulation-bound response that looks very much like a limbic convulsion. The fact that stimulation of surrounding tissue did not trigger such responses suggests that there may be some degree of specificity in the brainstem motor circuitry that drives the skeletal seizure response.

There has been even less progress made in the search for preferred routes of discharge propagation in *forebrain* sites. This is not surprising when we consider that the seizure activity appears to be propagating throughout the brain. We, and others, have attempted to determine whether certain structures are critical to the propagation of the discharge into motor nuclei. We made lesions and knife cuts in several different forebrain sites while testing amygdala and septal kindling.[123] Large bilateral lesions of the piriform cortex retarded septal kindling, but the animals still kindled. None of the other knife cuts or lesions had any significant effect on kindling rates. Tsunoda et al.[162] have reported strain differences in dentate gyrus lesion effects. Amygdala kindling was retarded by colchicine lesions of the dentate gyrus in Sprague-Dawley, but not Wistar rats.

Recently, Kelly and McIntyre[71] have used an innovative technique in an attempt to produce more complete piriform lobe lesions. They subjected their animals to the controlled induction of status epilepticus, which led to massive piriform lobe lesions. They found that extensive piriform cortex damage did *not* block kindled responses. They also found, however, that damage to the posterior *perirhinal* cortex reliably blocked the convulsion induced by hippocampal stimulation. Anterograde and retrograde labeling procedures indicated that the frontal cortex was a major target of this perirhinal site. They argued that, at least from hippocampus, the forelimb clonus and clonic rearing are driven by activity that propagates via perirhinal and frontal motor cortices.

While it is true that convulsive responses evoked by direct stimulation of the motor cortex do not look much like those triggered by hippocampal stimulation, this difference may simply be due to the spatio-temporal pattern of activation of the cortical sites. That spatio-temporal patterns may be important in the kindling process is indicated by the work of Spiller and Racine.[152] They tested transfer kindling in the monosynaptic entorhinal cortex-perforant path-dentate gyrus system. Axons alone cannot support epileptiform activity, so it is reasonable to assume that perforant path kindling is mediated via activation of the primary target structure, the dentate gyrus. It is also possible that some or all of the effect is mediated via antidromic activation of the primary source, the entorhinal cortex. If the effect were mediated via dentate gyrus activation, then one might expect that a kindling test of the dentate gyrus following perforant path kindling would yield immediate transfer. It was found, however, that the dentate gyrus still required about 20 additional stimulations to kindle. The same was true for entorhinal cortex transfer.

While this result may initially seem paradoxical, it is perhaps not surprising when considering the probable differences in patterns of activation that arise when the kindling stimulations are applied to different sites. For example, stimulation of the perforant path results in an immediate widespread activation of the hippocampus throughout its longitudinal axis. When the stimulation is applied to a specific site within the dentate gyrus, however, the activity must propagate throughout the hippocampus over quite different, presumably associational, pathways. The original kindling may even have produced changes in the target systems that are inhibitory to propagation over components of the circuitry activated during the transfer tests. This is suggested by the results of McIntyre and Goddard[90] who showed that rekindling in the original site, following transfer kindling of the homologous site in the contralateral hemisphere was often delayed.

In light of these findings, the different types of convulsive response triggered when stimulating the neocortex directly, as opposed to stimulating subcortical sites, may also be attributable to differences in the overall pattern of discharge propagation in the two cases. An epileptiform discharge that originates in a subcortical site, for example, may activate the motor cortex primarily via the perirhinal cortex. The activation would presumably be widespread. When stimulating the motor cortex directly, however, the pattern of propagation throughout the rest of the motor cortex would be primarily via cortico-cortical pathways. A different pattern of convulsive response might result, accounting for a role of motor cortex in both types of seizure response. Alternatively, the limbic forebrain sites might produce their motor effects via their connections to subcortical sites, such as the ventral pallidum and substantia nigra.

WHAT EPILEPTIC PHENOMENA ARE MODELLED BY THE KINDLING EFFECTS?

Racine and Burnham[119] described a number of component phenomena that emerge during the kindling process, and argued that they could be used to model different features of clinical epilepsy. The focal seizure, for example, could be used to model the effects of focal seizures in humans, with different stimulation sites yielding different effects. The progression to a more epileptogenic state with repetition of stimulation models similar phenomena in human epilepsy.[119,136] The fully generalized convulsion is believed to model complex partial seizures with secondary generalization. The spontaneous interictal and ictal responses model similar events in human epilepsy. This theme is more thoroughly developed in the Racine and Burnham paper, but it is sufficient for the purposes of the present chapter to point out that kindling is not, at least at the network level, a unitary phenomenon.

MECHANISMS—THE EARLY WORK

Most of our early work was directed to the investigation of 3 major hypotheses with respect to kindling mechanisms: 1) kindling is based upon an enhanced efficacy in existing excitatory synapses, rendering the affected circuits more excitable and better able to sustain high levels of activation; 2) kindling is based upon a decrement in the function of inhibitory synapses, again rendering the affected circuits more excitable; 3) kindling is based upon a change in the intrinsic response properties of affected neurons, so that they respond to otherwise normal inputs with a paroxysmal, burst mode of responding. Each of these hypotheses will be briefly discussed.

Enhanced Excitation

Briefly, we have obtained partial answers to the questions raised by these hypotheses. We now know that kindling does produce an enhancement at excitatory synapses referred to as kindling-induced potentiation.[31,121,123,124,125,127,134] This enhancement can at least facilitate the kindling process.[127] This facilitation effect is limited, however, so it cannot fully account for kindling. Also, kindling does not always produce a potentiation effect, but the animals always develop convulsions. Finally, kindling-induced potentiation effects, although long-lasting, are shorter-lived than the enhanced epileptogenesis.

Reduced Inhibition

The reduced inhibition hypothesis has run into even more problems. We have run a series of experiments over the years showing that inhibition in the dentate gyrus,[163] and in the piriform cortex,[125] is actually *enhanced*, rather than decreased, by kindling. The only brain site in which inhibition has been shown to be reduced by kindling is area CA1 in the hippocampus.[77] In our hands, this CA1 disinhibition effect is not as robust as the enhanced inhibition in the dentate gyrus. In any case, the hippocampus can be removed without affecting kindling rates in the amygdala or septal area.[123] Consequently, if kindling is based upon a reduction of inhibition, it is likely that this effect occurs in a site as yet untested.

Changed Response Properties

The least thoroughly investigated of the 3 hypotheses mentioned above is that there are changes in the response properties of the affected cells making them more prone to firing in a burst mode. There is at least some evidence in support of this hypothesis. There are some cells in hippocampal slices from kindled animals that respond more strongly than those in slices from non-kindled animals to injected current pulses (which bypass synaptic inputs), indicating that these neurons had become more reactive.[67,120] McIntyre and Wong[94] found larger and more reliable changes in piriform cortex slices taken from kindled rats.

MECHANISMS—RECENT WORK

As might have been expected, the story has become a good deal more complicated in recent years. Although initial studies found no evidence for any degenerative effects induced by kindling, we now know that seizure activity, including that triggered by kindling stimulations, can produce neural damage. We also know that kindling can induce a sprouting effect, including synaptogenesis. These two findings have led to a number of assertions about the role of these events in kindling and epilepsy. For example, it has been proposed that the cell loss compromises inhibitory systems and so contributes to the developing epileptogenesis.[144] It has also been claimed that the sprouting creates recurrent excitatory loops that account for the development of the epileptic condition.[130] As we will discuss below, there are a number of other possible interpretations of the data. We will not be able to provide definitive answers to the questions raised, but we will present our own current interpretations. First, however, we must review this research in more detail.

SEIZURES AND CELL LOSS

It has been known for some time that mesial temporal sclerosis is a common correlate of temporal lobe epilepsy.[3,28,38,46,76,84,103,135] While it has been suggested that this damage may be a causal factor in the epileptic condition (e.g., Falconer[38]), it has also been suggested that the cell loss is a result of the abnormal levels of neuronal activity.[7] Similar forms of damage have now been demonstrated in a number of different epilepsy models. One of the most widely studied is the application of kainic acid, which induces status epilepticus.[7,105] The more severe the status, the more extensive is the neural damage.[111] The structures that are most susceptible include the olfactory bulb, olfactory cortex, amygdala, bed nucleus of stria terminals, septal area, thalamus, medial thalamus, temporal cortex, medial prefrontal cortex, entorhinal cortex and much of the hippocampus.[141]

Kainic acid can produce damage when it contacts cells in high enough concentration. The fact that much of the damage in the status model is distant from the sites of injection, however, suggests that the high level of activation achieved during status can itself produce neural damage. This interpretation was strengthened by the demonstration that administration of an anticonvulsant drug can block the seizure activity and limit the damage to the tissue around the tip of the cannula.[7]

Electrical stimulation produces a more controlled form of epileptogenic activation. If the damage is activation-induced, then seizures driven electrically might also be expected to lead to damage. This has been demonstrated many times, initially with patterns of stimulation that lead, again, to status epilepticus.[92,98,144,150] Recently, however, more subtle forms of cell loss have been demonstrated with standard kindling paradigms. Cavazos and Sutula[19] reported loss of cells in the hilus of the dentate gyrus following kindling of the perforant path. When the animals were kindled to a standard criterion (3 stage 5 seizures), the loss was small but reliable. When kindled to 30 stage 5 seizures, there was a 40% loss of the hilar cells. Spiller and Racine[152] achieved a partial replication of this effect, although the degree of cell loss was considerably less than in the Cavazos and Sutula paper. Cavazos et al.[19] reported cell losses in both the hilus of the dentate gyrus and in area CA1 after only 3 stage 5 seizures. Cell losses were also found in area CA3, entorhinal cortex, and the endopiriform nucleus after 30 stage 5 seizures. In this paper, they described an additional group of animals kindled to 150 stage 5 seizures. Damage was found in all of the sites previously mentioned and also in the granule cell layer of the dentate gyrus and in area CA2. No damage was found in somatosensory cortex. The cell loss was, again, rather extensive, reaching 37%, and 49% in the hilus after 30 and 150 stage 5 seizures, respectively.

These kindling-induced cell loss effects are still somewhat controversial. Bertram and Lothman[11] measured the volume of the hilus of the dentate gyrus and found that it was increased by kindling. The increased volume resulted in a decrease in cell density without an actual loss of cells. J. McNamara (personal communication) has recently found the same effect in mice, again with no cell loss.

In any case, it is now clear that at least intense and persistent seizure activity can induce damage. It is not clear, however, whether that damage further contributes to the developing epileptogenesis. The controversy as to whether damage causes seizures or seizures cause damage is resolved only to the point of realizing that both effects can occur.

One of the main proponents of a sclerosis mechanism for epilepsy is R. Sloviter. He and his co-workers have been looking at the effects of status manipulations on hippocampal anatomy for several years. Although he has also used the kainic acid model, Sloviter is best known for his massed electrical stimulation model. In urethane-anesthetized rats, he applies a continuous pattern of stimulation to the perforant path for a period of 24 hours. This preparation has the advantage of producing a focal activation and a focal region of cell loss. The cell loss tends to be restricted to the region of the hilus, and this damage has been proposed as a model for endfolium sclerosis. Sloviter proposes that endfolium sclerosis is an early stage of hippocampal damage, which can be induced through various insults, including excess activation. This damage then renders the hippocampus more excitable and more likely to generate seizure activity, which can ultimately produce more damage. This view is based partly on the findings of Margerison and Corsellis[84] who found that endfolium sclerosis was always present in the temporal-lobe epileptic human brains that they examined. They also found that endfolium sclerosis was occasionally the *only* pathology evident in these brains.

There have also been reports of reduced numbers of synapses following kindling. Nishizuka et al.[108] found a loss of dendritic synapses in the medial amygdaloid nucleus following kindling in the basolateral amygdala, septal area or corpus callosum. Okada et al.[110] confirmed the loss of dendritic synapses and reported that synapse counts remained low for over 100 days.

Not all reports are consistent with these findings. We will describe conflicting data in a subsequent section in which we address the question of possible causal relationships between these degenerative effects and the developing epileptogenesis. At this point, it is sufficient to conclude that excess activation can induce neuronal damage. The epileptogenic effects of this damage remain to be determined.

SEIZURES, SPROUTING AND SYNAPTOGENESIS

While it is true that seizure activity can lead to cell loss, it is also true that it can lead to neural growth, including axonal sprouting and synaptogenesis. It is well known that electrical activity plays an important role in neural growth and synaptogenesis during development,[39] and it seems likely that this remains true in the adult CNS.[5] There are a number of findings that suggest that growth mechanisms are engaged when seizure activity occurs. Protein synthesis blockers can retard kindling,[17,66,109] and the induction of seizure activity can trigger the activation of astrocytes,[52,53,56,74,75] and the induction of immediate-early genes,[33,34,140,144] neurotrophins[9,10] and GAP-43 mRNA (generally associated with axonal growth cones[95]).

While these findings suggest the possibility of neural growth, there have recently been several demonstrations of *actual* sprouting and synaptogenic effects in both human epileptics and in animals following the administration of epileptogenic agents. It is possible that several systems are affected, but the pathway most thoroughly studied is the mossy fibre projection from the dentate gyrus granule cells to the CA3 pyramidal cells of the hippocampus. The major target for these fibres is the stratum radiatum, which contains the apical dendrites of the CA3 pyramidal cells. This pathway contains very high levels of Zn^{++}, which can be labeled using the Timm stain.[29] This procedure has provided evidence for axonal sprouting in tissue from human epileptics,[59,86,87,88] and in several animal models, including kainic acid induced status[104] and kindling.[130,155,166]

The mossy fiber sprouting shows up in two target sites that usually show quite sparse Timm staining, the basal dendrites, or stratum oriens, in area CA3 and the proximal apical dendrites, or inner molecular layer, of the dentate gyrus. These new inputs appear to form functional synaptic contacts. Represa and Ben-Ari[130] have used electron microscopy to show the development of new giant synapses in stratum oriens. These are characteristic of the mossy fiber inputs. Golairi and Sutula[49] used current source density analyses to show the development of a new sink in the inner molecular layer that is consistent with the development of new mossy fibre contacts at that location. Cronin et al.[27] reported an enhanced response in granule cells to antidromic activation consistent with the development of functional contacts in the sprouted mossy fibre collaterals.

In addition, Chang et al.[22] and Hawrylak et al.[55] have reported an increase in the number of shaft and sessile spine synapses on CA1 pyramidal cells in an *in vitro* hippocampal slice kindling preparation. Hawrylak et al.[55] have found similar changes in hippocampal slices taken from animals kindled *in vivo*. Geinisman et al.[45] report an increase in the number of "perforated" synapses (those with discontinuous postsynaptic densities) in the termination field of the activated fibers in perforant path kindled animals.

Thus, there appears to be little doubt that seizure activity can lead to neural growth and the formation of new synapses. These results raise some interesting issues. First, just as we explored the relationship between seizure-induced cell loss and the developing epileptogenesis, we can query the role of neural growth in promoting epileptic reactivity. Second, it is not clear whether the sprouting effects are triggered directly by the neural activation or by the cell (or synapse) loss. We will address these questions, but first we must introduce a third topic into the discussion.

SEIZURES AND NEUROTROPHINS

If the induction of seizure activity leads to neural growth, then it is reasonable to speculate that neurotrophins, which induce neural growth, might be involved. The neurotrophins comprise a family of neurotrophic factors that have been shown to play a role in the differentiation and survival of CNS and PNS neurons. They share a high degree of structural homology, and include nerve growth factor (NGF), brain-derived neurotrophic factor (BDNF), neurotrophin-3 (NT-3), neurotrophin-4/5, and neurotrophin-6.[50,83] *In situ* hybridization studies have localized the mRNAs for these factors in distinct and overlapping populations of neurons in the brain with the greatest expression in the hippocampus, olfactory cortex, claustrum, neocortex, and amygdala.[37,58,113] The members of the neurotrophin family act on specific populations of neurons to promote cell survival, differentiation, axonal growth, and neuropeptide and neurotransmitter synthesis.

They do so via high-affinity (trk) and low-affinity (p75) receptors. The neurotrophins, particularly BDNF, are highly conserved from fish to mammals.[51]

Neurotrophins and their receptors have been shown to be differentially regulated in several seizure models. Gall and Isackson[44] used electrolytic lesions of the hilus to induce epileptiform activity and found an increase in the levels of mRNA for NGF. Gall et al.,[44] Zafra et al.,[169] Isackson et al.[63] and Dugich-Djordjevic et al.[35] reported increases for NGF and/or BDNF mRNA following kainic acid administration. Hippocampal and amygdala kindling also trigger elevations in NGF protein and in NGF and BDNF mRNA as shown by Bengzon et al.[9,10] The largest effects were found in the hippocampus and in the piriform cortex. They also reported that the levels of NT-3 mRNA were *reduced*, although Ernfors et al.[37] reported no change in NT-3 mRNA levels with amygdala or hippocampal kindling.

The receptors for these neurotrophins are also affected by kindling. Bengzon et al.,[10] found that trkB, the receptor tyrosine kinase that binds BDNF and NT-4, and trkC, which preferentially binds NT-3, were upregulated following the application of kindling stimulations. The NGF receptor, trkA, did not appear to be affected by kindling. Merlio et al.[97] similarly reported a kindling-induced upregulation of trkB with no effect on trkA. They did not, however, find kindling-induced upregulation of trkC.

It is possible that the neurotrophins are involved in the progression or maintenance of kindling epileptogenesis, perhaps via their effects on neural growth. If neurotrophin activity is required for kindling-induced sprouting, then antagonists of the neurotrophins should reduce or block the sprouting and might be expected to suppress kindling. Funabashi[42] tested the effect of intraventricular administration of NGF antibodies on amygdala kindling and found that kindling was retarded. Van der Zee et al.[166] used NGF antibodies that had been assayed for lack of cross-reactivity with the other neurotrophins and found that kindling was delayed and mossy fiber sprouting was reduced. Kindling and sprouting were also inhibited by a bicyclic NGF peptide that blocks the action of NGF.[128]

CAUSAL RELATIONSHIPS

Although there is widespread agreement that cell loss and neural growth can accompany seizure activity, there is considerable controversy about the causal relationships between the developing seizure activity, the cell loss and the synaptogenesis. In the following sections, we will review the data as they relate to these controversies.

Hilar Cell Loss as an Epileptogenic Mechanism (via Reduction of Inhibition)

It has been suggested by a number of epileptologists that the neural degeneration that often accompanies chronic seizure states compromises inhibitory systems to a greater extent than excitatory systems. Although there has been extensive research on this question over the last several decades, the results and their interpretations remain controversial. Various measures of the function of GABA neurons and receptors have been taken during kindling, for example, with highly variable results (see Burnham;[13] Kamphuis and Lopes da Silva[69]). The same is true for measures taken from human epileptic tissue.[132] As outlined earlier, our own physiological measures taken in kindled tissue indicate that the levels of inhibition are either normal or enhanced in most of the sites that we have tested.

The one exception is area CA1 of the hippocampus, where reduced levels of paired-pulse inhibition have been reported following kindling.[77]

One of the current proponents of the hypothesis that epilepsy depends upon a mechanism of reduced inhibition is R. Sloviter. He has monitored the function of inhibitory systems following induction of status epilepticus by administration of kainic acid[145] or by repeated electrical stimulation.[144] Like others (e.g., Shirasaka and Wasterlain[140]), he found that these treatments can lead to a prolonged period of spontaneous seizure generation. He also found that paired-pulse measures often reflect *decreased* levels of inhibition in the dentate gyrus during these periods. Sloviter argues that the loss of inhibition is a primary mechanism underlying the appearance of spontaneous seizure activity.

Initially, it was believed that an actual loss of inhibitory interneurons accounted for the reductions in inhibition. When GABA-containing neurons were directly labeled, however, it was found that their numbers were *not* diminished.[144] Non-GABA neurons, however, were lost following status induction, as described above under the section "Seizures and cell loss." The lost neurons included those staining for somatostatin and neuropeptide Y, but the loss of inhibition was attributed to a loss of mossy cells. Mossy cells (not to be confused with mossy fibers) are excitatory interneurons located in the hilus of the dentate gyrus. They appear to innervate both granule cells and interneurons.[138] To account for an apparent loss of inhibition while inhibitory neurons remained intact, Sloviter[144] proposed the "dormant basket cell" hypothesis. According to this hypothesis, if the mossy cells provide excitatory input to the inhibitory interneurons, then their loss might be expected to reduce the levels of inhibition. More recently, it has been argued that some of the other damaged hilar cells might serve a more direct inhibitory function,[147] but the primary focus has remained on the mossy cells.

The problem with this hypothesis is that we have known for a long time that seizure activity can actually *increase* the levels of inhibition in the dentate gyrus.[163] This has now been shown for normal kindling,[30] kindling to the stage of spontaneous ictal seizures,[99] and it has even been suggested in cases of human epilepsy.[64,165] In all of these cases, there has been an increase in spontaneous epileptogenic activity. Therefore, it is clear that loss of inhibition, at least in the dentate gyrus, cannot be a critical factor for induction of a chronic epileptic state. We also know that this increased inhibition can occur and be maintained, following significant decreases in cell density in the hilus of the dentate gyrus.[152]

A number of studies have shown that there is a *transient* loss of inhibition just prior to the onset of ictal episodes,[8,163] and for some period of time following those episodes. For example, Tuff et al.[163] applied pulse pairs to monitor inhibition. These pairs were applied at a frequency of 5 Hz to trigger epileptiform discharge. This allowed inhibition to be measured before, during and after seizure onset. The inhibition invariably failed just prior to the onset of the ictal discharge. Consequently, when an animal is epileptic (i.e., showing recurrent spontaneous ictal episodes), it is difficult to determine whether chronically reduced levels of inhibition are causing the seizure activity or whether the seizure activity is driving the inhibitory systems into a transient state of failure. In earlier work, Sloviter[149] had reported spontaneous seizures, losses of hilar cells and losses of inhibition following the induction of status epilepticus with kainic acid. Milgram et al.[100] subsequently found that a controlled, one-hour induction of status by systemic kainic acid produced the effects described by Sloviter, but only for one or two days. As soon as the spontaneous seizure activity subsided, inhibition recovered. In fact, the inhibition became stronger on recovery. Sloviter subsequently reported similar results following status. How can the inhibition recover if the mossy cells are dead? Sloviter's argument is that the sprouting that is triggered by the cell loss *re*innervates the inhibitory interneurons, restoring the inhibition. One problem with this explanation is that the inhibition recovered within one to two days in the Milgram et al. experiment, and this is probably too rapid to be accounted for by sprouting. Another problem with Sloviter's speculation is that, in his hands, inhibition appears *not* to recover in his stimulation model, even though sprouting is presumably taking place.

Sloviter (personal communication) claims that a nearly total loss of hilar cells is necessary to achieve a significant loss of inhibition, and the only reliable way to produce this cell loss is with his 24 hour stimulation model.[144] Recently, however, Shirasaka and Wasterlain[140] utilized Sloviter's massed stimulation procedure and found that inhibition recovered in this model as well. Furthermore, there was no correlation between the levels of spontaneous seizure activity and the levels of recorded inhibition. In fact, spontaneous seizures tended to appear at a time when the inhibition had substantially recovered. This late onset of spontaneous epileptogenesis also suggests a poor correlation with hilar cell loss, although Shirasaka and Wasterlain did not report cell counts in their paper. Another problem with the link between hilar cell loss and epileptogenesis is that there have been several reports of kindling *without* any sign of cell damage (e.g., Khurgel et al.[75]).

It is clear that chronic, spontaneous epileptic states can be achieved without a loss of inhibition in the dentate gyrus. Consequently, the usefulness of Sloviter's stimulation model to the study of clinical epilepsy is unclear, although it may provide valuable information about the role of hilar neurons in hippocampal function.

Finally, there are a number of reports indicating that a chronic epileptogenic state can be induced *without* cell loss, or at least without hilar cell loss. Epileptologists at the UCLA School of Medicine, for example, have presented data that appear to be inconsistent with the dormant basket cell hypothesis. First, they have found that dentate gyrus inhibition may be increased, rather than decreased in human epileptics.[73,64,165] Second, they have found that GAD-containing *hilar* interneurons survive and actually sprout in the epileptic hippocampus.[86] Third, they have found that hilar neurons appeared *not* to be damaged in their Mass Lesion and Idiopathic epilepsy groups, even though the amount of seizure activity appeared to be similar to the group showing typical hippocampal sclerosis. The Mass Lesion/Idiopathic groups did, however, show some damage to granule cells and CA3 pyramidal cells showing that the propagated seizure activity was strong enough to damage cells. These authors prefer a mechanism based upon the sprouting and synaptic reorganization that is triggered by cell loss and/or excess activation.

The fact that hilar neuron damage is common to a wide variety of pathologies not necessarily linked to epilepsy (e.g., Alzheimer's type dementia, Chan-Palay;[21] and cerebral ischemia, Johansen[65]), would also seem to weaken the argument for hilar damage as a primary mechanism of epileptogenesis.

In an animal study, Nagao et al.[106] induced status by systemic application of pilocarpine, and then tested hippocampal slices from the treated animals either 3 weeks or 12 weeks later. They found that area CA1 was passively driven in the slices from 3-week animals, but became independently epileptogenic in the slices from 12-week animals. As expected with a status model, cell losses were found in the hippocampus. However, the further increments in epileptogenicity seen in area CA1 at 12 weeks developed without any apparent increases in damage over that seen in the 3-week animals.

Bertram and Lothman[11] kindled up to 1,500 stage 5 seizures in rats (recall that Cavazos et al.[19] only kindled up to 150) and found no evidence for a loss of hilar cells.

Kostopoulos and Antoniadis[79] reported an increased excitability of CA1 pyramidal cells in the epileptic "tottering" mutant mouse, but they found either no change or an increase in paired pulse measures of inhibition in these cells.

Franck and Schwartzkroin[40] used the kainic acid status preparation to investigate the levels of inhibition in area CA1. Two to four weeks following induction of status, slices from the treated animals showed epileptiform activity and reduced levels of inhibition. Two to four months after status, however, the inhibition had recovered to normal levels. While recovery of dentate gyrus inhibition in the kainic acid model was too rapid to be accounted for by sprouting effects of Milgram et al.,[100] recovery of inhibition in the Franck and Schwartzkroin experiment may have been too slow to be readily accounted for by these effects.

Sloviter, and others, have argued that the loss of innervation to the basket cell renders the dentate granule cells hyperexcitable. Mitchell and Barnes,[101] however, have destroyed dentate granule cells with colchicine and found little effect on perforant path kindling. The epileptiform activity, if anything, was actually enhanced.

Qiao and Noebels[116] measured hilar cell density and mossy fiber density in the epileptic mutant mouse, stargazer. They found that the hilar cell loss *followed* the onset of spontaneous seizure activity in this strain of mouse.

Tasker and Dudek[159] cite evidence for *both* increased and decreased levels of inhibition in models of epilepsy and point out that the relationship between excitation and inhibition in these models is likely to be highly complex. Dudek et al.[34] have extended the notion that mossy fiber sprouting, and indeed any sprouting induced by cell loss, is likely to reinnervate both inhibitory and excitatory neurons. They argue that the net effect of this sprouting will depend upon many variables, such as the location and type of cell affected, the amount of damage, the extent of recovery, and so on. They argue that some of the variability in epileptogenic effects may be due to the variability in the effect upon the balance between excitation and inhibition following the damage-induced sprouting.

Hilar Cell Loss vs. Neural Activation as a Trigger for Sprouting

Cavazos and Sutula[20] and others have assumed that the sprouting seen in the mossy fiber system is triggered by the cell loss in the hilus or other areas within the hippocampus. It is possible, however, that the sprouting is purely activation-induced. The most obvious examples of this form of plasticity come from learning and memory and environmental enrichment studies.[5]

Represa and Ben-Ari[130] have reported mossy fiber sprouting and synaptogenesis in the absence of cell damage. We have also recently found mossy fiber sprouting, with no hilar cell loss, in both the amygdala kindled preparation[1,137] and in animals subjected to 8 electroconvulsive shocks (Burnham et al., in preparation). Even more convincing, perhaps, we have found significant mossy fiber sprouting following induction of LTP in the dentate gyrus by *non*-epileptogenic stimulation.[2]

Meberg et al.[95] produced electrolytic lesions in the hilus and monitored both seizure activity and the levels of GAP-43 mRNA expression. GAP-43 is associated with axonal growth cones, and its upregulation may indicate an increase in synaptogenesis. They did not, however, find enhanced GAP-43 expression when the lesions were made with platinum electrodes, which produce damage but not seizures. These results suggest that the sprouting may be activation-induced rather than damage-induced. Gall and Isackson[43] reported similar results for NGF mRNA. Electrolytic lesion-induced seizures triggered increased expression, while hilar lesions with platinum electrodes did not.

Mathern et al.[86,88] reported no correlation between hilar cell loss and mossy fiber sprouting in tissue from human temporal-lobe epileptics.

SPROUTING AS AN EPILEPTOGENIC MECHANISM

The observation of mossy fiber sprouting in several different animal models of epilepsy, as well as in cases of human temporal lobe epilepsy, has led a number of investigators to suggest a causal relationship. It is proposed that new recurrent or feedforward excitatory connections increase the excitability and epileptogenicity of the hippocampus.[130]

The major flaw in this proposal is that spontaneous epilepsy often appears far in advance of mossy fiber sprouting, which becomes measurable at about 5–6 days[155] and reaches near asymptotic levels much later. Spontaneous ictal and interictal episodes, for example, can appear within the first few days after administration of kainic acid.[100]

As mentioned above, Qiao and Noebels[116] found that hilar cell density was reduced in the epileptic mutant mouse, "stargazer", but only *after* spontaneous seizure activity had become well developed. These authors also measured mossy fiber density, which was elevated in the stargazer mouse, and found that the increased innervation of stratum oriens also followed, rather than preceded, the onset of spontaneous seizure activity.

We have administered antibodies to nerve growth factor (NGF) to the ventricles via implanted osmotic pumps, and tested the effects on mossy fiber sprouting and amygdala kindling rates.[166] Sprouting was decreased by 50% overall, and appeared to be completely blocked in some animals. All animals, however, kindled to criterion, although additional stimulations were required. Further work must be done to determine whether we can demonstrate a complete dissociation between mossy fiber sprouting and kindling.

In any case, there appear to be no clear correlates between the induction of mossy fiber sprouting and the appearance of, or further development of, any epileptiform events. It should also be pointed out that we have actually produced bilateral lesions in the hippocampus and found little effect on kindling rates in the intact septal area or amygdala.[123] Tanaka et al.[158] used excitotoxin lesions of the hippocampus and found a similar lack of effect on amygdala kindling. These results show that, whatever the role of cell loss or sprouting within the hippocampus, epileptogenic mechanisms are clearly intact in non-hippocampal structures.

SPROUTING AS A MECHANISM FOR RESTORING INHIBITION

Sloviter[145] has suggested that the recovery of inhibition that he sees in the kainic acid model is due to the mossy fiber sprouting. At least some of the new mossy fiber collaterals, he proposes, are reinnervating the basket cells. This reinnervation then restores some or all of the inhibition that was lost with the destruction of the mossy cells that provided excitatory input to the basket cells.

As described above, however, Milgram et al.[100] found a recovery of inhibition following kainic acid-induced status that was far too rapid to be accounted for by a sprouting mechanism. In their study, spontaneous seizure activity persisted for up to 48 hours during which time inhibition appeared to be reduced. As soon as the spontaneous seizure activity subsided, however, the inhibition rapidly recovered to levels greater than baseline.

Another problem with the idea that recovered or enhanced inhibition is due to reinnervation of inhibitory interneurons is that the new connections formed by the sprouting fibers appears to be very long lasting,[130] while the enhanced inhibition, like LTP, is transient. Although this has not been thoroughly measured in the kainic acid model, it is true for stimulation models, such as kindling, where inhibition has been followed for a period of months. In these models, inhibition is enhanced and sprouting is induced. The enhanced inhibition decays back to baseline, while the sprouted axons appear to retain their connections.

CONCLUSIONS

We now know that seizures can produce many changes in the central nervous system, and that some of these changes are long lasting. There is synaptic enhancement in excitatory systems, both increases and decreases in the function of inhibitory systems, cell loss in the hippocampus and possibly other sites, synapse loss in the amygdala and possibly other sites, and sprouting and synaptogenesis in the hippocampus and possibly other sites.

Unfortunately, we do not yet know the precise relationship between these various phenomena. Is the sprouting activation-induced or damage-induced, or both? Does either the cell loss or the sprouting contribute to epileptogenesis? Do either have any long-term effect on the function of inhibitory systems? What is the role of neurotrophins in these events? Our reading of the literature and our own findings lead us to suspect that all of these phenomena may be relatively independent consequences of excess activation, but a good deal more work needs to be done in order to make this conclusion with any confidence.

REFERENCES

1. Adams, B., Sazgar, M., Osehobo, P., Van derZee, C., Diamond, J., Fahnestock, M., and Racine, R. Nerve growth factor accelerates seizure development, enhances mossy fiber sprouting and attenuates seizure-induced decreases in neuronal density in the kindling model of epilepsy, *J. Neuroscience*, 17 (1997) 5288–5296.
2. Adams, B., Lee, M., Fahnestock, M., and Racine, R.J. Long-term potentiation trains induce mossy fiber sprouting, *Brain Res.*, 775 (1997) 193–197.
3. Babb, T.L., Brown, W.J., Pretorious, J., Davenport, C., Lieb, J.P., and Crandall, P.H. Temporal lobe volumetric cell densities in temporal lobe epilepsy, *Epilepsia*, 25 (1984) 729–740.
4. Babb, T.L., Kupfer, W.R., Pretorius, J.K., Crandall, P.H., and Levesque, M.F. Synaptic reorganization by mossy fibers in human epileptic fascia dentata, *Neuroscience*, 42 (1991) 351–363.
5. Bailey, C.H. and Kandel, E.R. Structural changes accompanying memory storage, *Annu. Rev. Physiol.*, 55 (1993) 397–426.
6. Bekenstein, J.W. and Lothman, E.W. Dormancy of inhibitory interneurons in a model of temporal lobe epilepsy, *Science*, 259 (1993) 97–100.
7. Ben-Ari, Y. Limbic seizure and brain damage produced by kainic acid: mechanisms and relevance to human temporal lobe epilepsy, *Neuroscience*, 14 (1985) 375–403.
8. Ben-Ari, Y., Krnjevic, K., and Reinhardt, W. Hippocampal seizures and failure of inhibition, *Canadian J. Psychol.*, 57 (1979) 1462–1466.

9. Bengzon, J., Soderstrom, S., Kokaia, Z., Kokaia, M., Ernfors, P., Persson, H., Ebendal, T., and Lindvall, O. Widespread increase of nerve growth factor protein in the rat forebrain after kindling-induced seizures, *Brain Research*, 587 (1992) 338–342.
10. Bengzon, J., Kokaia, Z., Ernfors, P., Kokaia, M., Leanza, G., Nilsson, O.G., Persson, H., and Lindvall, O. Regulation of neurotrophin and *trkA*, *trkB* and *trkC* tyrosine kinase receptor messenger RNA expression in kindling, *Neuroscience*, 53 (1993) 433–446.
11. Bertram, E.H. and Lothman, E.W. Morphometric effects of intermittent kindled seizures and limbic status epilepticus in the dentate gyrus of the rat, *Brain Research*, 603 (1993) 25–31.
12. Bertram, E.H., Lothman, E.W., and Lenn, N.J. The hippocampus in experimental chronic epilepsy: a morphometric analysis, *Ann. Neurol.*, 27 (1990) 43–48.
13. Burnham, W.M. The GABA hypothesis of kindling: recent assay studies, *Neuroscience & Biobehavioral Reviews*, 13 (1989) 281–288.
14. Burnham, W.M., Cottrell, G.A., Diosy, D., and Racine, R.J. Long-term changes in entorhinal-dentate evoked potentials induced by electroconvulsive shock seizures in rats, *Brain Research*, 698 (1995) 180–184.
15. Cain, D.P. Seizure development following repeated electrical stimulation of central olfactory structures, *Ann. N.Y. Acad. Sci.*, 290 (1977) 200–216.
16. Cain, D.P. Long-term potentiation and kindling: how similar are the mechanisms? *Trends Neurosci.*, 12 (1989) 6–10.
17. Cain, D.P., Corcoran, M., and Staines, W.A. Effects of protein synthesis inhibition on kindling in the mouse, *Experimental Neurology*, 68 (1980) 409–419.
18. Cavalheiro, E.A., Silva, D.F., Turski, W.A., Calderazzofilho, L.S., Bortolotto, Z.A., and Turski, L. The susceptibility of rats to pilocarpine-induced seizures is age-dependent, *Dev. Brain Res.*, 37 (1987) 43–58.
19. Cavazos, J.E., Das I., and Sutula, T.P. Neuronal loss induced in limbic pathways by kindling: evidence for induction of hippocampal sclerosis by repeated brief seizures, *J. Neuroscience*, 14 (1994) 3106–3121.
20. Cavazos, J.E. and Sutula, T.P. Progressive neuronal loss induced by kindling: a possible mechanism for mossy fiber synaptic reorganization and hippocampal sclerosis, *Brain Research*, 527 (1990) 1–6.
21. Chan-Palay, V., Lang, W., Haesler, U., Kohler, C., and Yasargil, G. Distribution of altered hippocampal neurons and axons immunoreactive with antisera against neuropeptide Y in Alzheimer's-type dementia, *J. Comp. Neurol.*, 248 (1986) 376–394.
22. Chang, F-L., F., Hawrylak, N., and Greenough, W.T. Astrocytic and synaptic response to kindling in hippocampal subfield CA1. I. Synaptogenesis in response to kindling in vitro, *Brain Research*, 603 (1993) 302–308.
23. Clifford, D.B., Olney, J.W., Maniotis, A., Collins, R.C., and Zorumski, C.F. The functional anatomy and pathology of lithium-pilocarpine and high-dose pilocarpine seizures, *Neuroscience*, 23 (1987) 953–968.
24. Clusmann, H., Stabel, J., Stephen, D.N., and Heinemann, U. Alterations in medial perforant path and mossy fiber induced field potentials in amygdala and B-carboline (FG 7142) kindled rats, *Neurosci. Lett.*, 146 (1992) 65–68.
25. Crandall, J.E., Bernstein, J., Boast, C., and Zornetzer, S. Kindling in the rat hippocampus: absence of dendritic alterations, *Behav. Neural Biol.*, 27 (1970) 516–522.
26. Cronin, J. and Dudek, F.E. Chronic seizure and collateral sprouting of dentate mossy fibers after kainic acid treatment in rats, *Brain Research*, 474 (1988) 65–68.
27. Cronin, J., Obenaus, A., Houser, C.R., and Dudek, F.E. Electrophysiology of dentate granule cells after kainate-induced synaptic reorganization of mossy fibers, *Brain Research*, 573 (1992) 305–310.
28. Dam, A.M. Epilepsy and neuron loss in hippocampus, *Epilepsia*, 21 (1980) 617–629.
29. Danscher, G. Histochemical demonstration of heavy metals: a revised version of the sulphide silver method suitable for both light and electron microscopy, *Histochemistry*, 71 (1981) 1–16.
30. de Jonge, M. and Racine, R.J. The development and decay of kindling-induced increases in paired-pulse depression in the dentate gyrus, *Brain Research*, 412 (1987) 318–328.
31. Douglas, R.M. and Goddard, G.V. Long-term potentiation of the perforant path-granule cell synapse in the rat hippocampus, *Brain Research*, 86 (1975) 205–215.
32. Dragunow, M. and Robertson, H.A. Kindling stimulation induces c-fos protein(s) in granule cells of the rat dentate gyrus, *Nature*, 329 (1987) 441–442.
33. Dragunow, M., Currie, R.W., Faull, R.L.M., Robertson, H.A., and Jansen, K. Immediate-early genes, kindling and long-term potentiation, *Neuroscience & Biobehavioral Reviews*, 13 (1989) 301–313.

34. Dudek, F.E., Obehaus, A., Jeffrey, S.S., and Wuarin, J-P. Functional significance of hippocampal plasticity in epileptic brain: Electrophysiological changes of the dentate granule cells associated with mossy fiber sprouting, *Hippocampus*, 4 (1994) 259–265.
35. Dugich-Djordjevic, M.M., Tocco, G., Lapchak, P.A., Pasinetti, G.M., Najm, I., Baudry, M., and Hefti, F. Regionally specific and rapid increases in brain-derived neurotrophic factor messenger RNA in the adult rat brain following seizures induced by systemic administration of kainic acid, *Neuroscience*, 47 (1992) 303–315.
36. Ebert, U. and Loscher, W. Strong induction of c-fos in the piriform cortex during focal seizures evoked from different limbic brain sites, *Brain Research*, 671 (1995) 338–344.
37. Ernfors, P., Bengzon, J., Kokaia, Z., Persson, H., and Lindvall, O. Increased levels of messenger RNAs for neurotrophic factors in the brain during kindling epileptogenesis, *Neuron*, 7 (1991) 165–176.
38. Falconer, M.A. Mesial temporal (Ammon's horn) sclerosis as a common cause of epilepsy-etiology, treatment and prevention, *Lancet* (1974) September 28 (1974) 767–770.
39. Fields, R.D. and Nelson, P.G. Activity-dependent development of the vertebrate nervous system, *Int. Rev. Neurobiol.*, 34 (1992) 133–214.
40. Franck, J.E. and Schwartzkroin, P.A. Do kainate-lesioned hippocampi become epileptogenic? *Brain Research*, 329 (1985) 309–313.
41. Freund, T.F., Ylinen, A., Miettinen, R., Pitkanen, A., Lahtinen, H., Baimbridge, K.G., and Riekkinen, P.J. Pattern of neuronal death in the rat hippocampus after status epilepticus. Relationship to calcium binding protein content and ischemic vulnerability, *Brain Research*, 28 (1991) 27–38.
42. Funabashi, T., Sasaki, H., and Kimura, F. Intraventricular injection of antiserum to nerve growth factor delays the development of amygdaloid kindling, *Brain Research*, 458 (1988) 132–136.
43. Gall, C.M. and Isackson, P.J. Limbic seizures increase neuronal production of messenger RNA for nerve growth factor, *Science*, 245 (1989) 758–761.
44. Gall, C.M., Murray, K., and Isackson, P.J. Kainic acid-induced seizures stimulate increased expression of nerve growth factor mRNA in rat hippocampus, *Mol. Brain Res.*, 9 (1991) 113–123.
45. Geinisman, Y., Morrell, F., and deToledo-Morrell, L. Increase in the relative proportion of perforated axospinous synapses following hippocampal kindling is specific for the synaptic field of stimulated axons, *Brain Research*, 507 (1990) 325–331.
46. Gloor, P. Mesial temporal sclerosis: historical background and overview from a modern perspective. In: H. Luders, (Ed.) *Epilepsy surgery*, New York, Raven Press (1991) pp. 689–703.
47. Goddard, G.V. Development of epileptic seizures through brain stimulation at low intensity, *Nature*, 214 (1967) 1020–1021.
48. Goddard, G.V., McIntrye, D.C., and Leech, C.K. A permanent change in brain function resulting form daily electrical stimulation, *Exp. Neurol.*, 25 (1969) 295–330.
49. Golarai, G. and Sutula, T.P. Functional alterations in the dentate gyrus after induction of long-term potentiation, kindling, and mossy fiber sprouting, J. *Neurophysiology*, 75 (1996) 343–354.
50. Gotz, R., Koster, R., Winkler, C., Raulf, F., Lottspeich, F., Schartl, M., and Thoenen, H. Neurotrophin-6 is a new member of the nerve growth factor family, *Nature*, 372 (1994) 266–268.
51. Hallbook, F., Ibanez, C.F., and Persson, H. Evolutionary studies on the nerve growth family reveal a novel member abundantly expressed in *Xenopus* ovary, *Neuron*, 68 (1991) 845–858.
52. Hansen, A., Jorgensen, O.S., Bolwig, T.G., and Barry, D.I. Hippocampal kindling in the rat is associated with time-dependent increases in the concentration of glial fibrillary acidic protein, *J. Neurochemistry*, 57 (1991) 1716–1720.
53. Hansen, A., Jorgensen, O.S., Bolwig, T.G., and Barry, D.I. Hippocampal kindling alters the concentration of glial fibrillary acidic protein and other marker proteins in rat brain, *Brain Research*, 531 (1990) 307–311.
54. Hauser, W.A. Status epilepticus: frequency, etiology, and neurological sequelae, *Adv. Neurol.*, 34 (1983) 3–14.
55. Hawrylak, N., Chang, F-L.F., and Greenough, W.T. Astrocytic and synaptic response to kindling in hippocampal subfield CA1. II. Synaptogenesis and astrocytic process increases to in vivo kindling, *Brain Research*, 603 (1993) 309–316.
56. Heinemann, U., Clussmann, H., Dreier, J., and Stabel, J. Changes in synaptic transmission in the kindled hippocampus, *Adv. Exp. Med. Biol.*, 268 (1990) 445–450.
57. Hikiji, M., Tomia, H., Ono, M., Fujiwara, Y., and Akiyama, K. Increase of kainate receptor mRNA in the hippocampal CA3 of amygdala-kindled rats detected by *in situ* hybridization, *Life Sciences*, 53 (1993) 857–864.

58. Hofer, M., Pagliusi, S.R., Hohn, A., Leibrock, J., and Barde, Y.A. Regional distribution of brain-derived neurotrophic factor mRNA in the adult mouse brain, *EMBO J.*, 9 (1990) 2459–2464.
59. Houser, C.R., Miyashiro, J.E., Swartz, B.E., Walsh, G.O., Rich, J.R., and Delgado-Escueta, A.V. Altered patterns of dynorphin immunoreactivity suggest mossy fiber reorganization in human hippocampal epilepsy, *J. Neuroscience*, 10 (1990) 267–282.
60. Hughes, P. and Dragunow, M. Muscarinic receptor-mediated induction of Fos protein in rat brain, *Neuroscience Letters*, 150 (1993) 122–126.
61. Humpel, C., Wetmore, C., and Olson, L. Regulation of brain-derived neurotrophic factor messenger RNA and protein at the cellular level in phenylenetetrazol-induced epileptic seizures, *Neuroscience*, 53 (1993) 909–918.
62. Ip, N.Y., Li, Y., Yancopoulos, G.D., and Lindsay, R.M. *J. Neuroscience*, 13 (1993) 3394–3405. (Note: Report that BDNF elicits expression of c-fos in hippocampus—check).
63. Isackson, P.J., Huntsman, M.M., Murray, K.D., and Gall, C.M. BDNF mRNA expression is increased in adult rat forebrain after limbic seizures: temporal patterns of induction distinct form NGF, *Neuron*, 6 (1991) 937–948.
64. Isokawa-Akesson, M., Wilson, C.L, and Babb, T.L. Inhibition in synchronously firing human hippocampal neurons, *Epilepsy Res.*, 3 (1989) 236–247.
65. Johansen, F.F. Interneurons in rat hippocampus after cerebral ischemia, *Acta Neurologica Scandinavica*, 88, Suppl 150 (1993) 1–32.
66. Jonec, V. and Wasterlain, C.G. Effect of inhibitors of protein synthesis on the development of kindled seizures in rats, *Exp. Neurol.*, 66 (1979) 524–532.
67. Kairiss, E.W. Hippocampal slice studies of kindling-induced epilepsy, Ph.D. Thesis, McMaster University, Hamilton, Ontario, 1985.
68. Kamphuis, W., Wadman, W., Buijs, R.M., and Lopes da Silva, F.H. Decrease in number of hippocampal gamma-aminobutyric acid (GABA) immunoreactive cells in the rat kindling model of epilepsy, *Exp. Brain Res.*, 64 (1986) 491–495.
69. Kamphuis, W. and Lopes da Silva, F.H. The kindling model of epilepsy: the role of GABAergic inhibition, *Neuroscience Research Communications*, 6 (1990) 1–9.
70. Kamphuis, W., Monyer, H., De Rijk, T.C., and Lopes da Silva, F.H. Hippocampal kindling increases the expression of glutamate receptor -A Flip and -B Flip mRNA in dentate granule cells, *Neuroscience Letters*, 148 (1992) 51–54.
71. Kelly, M.E. and McIntyre, D.C. The effects of piriform and perirhinal cortex damage on convulsive seizures in kindled rats. *Epilepsia*, 36, (suppl. 4) (1995) p. 87.
72. Khan, S.U., Wilson, C.L., Isokawa-Akesson, M., Babb, T.L., and Levesque, M.E. Increased paired-pulse inhibition in the epileptogenic human temporal lobe, *Soc. Neurosci. Abstra.*, 15 (1989) 236.
73. Khan, S.U., Wilson, C., Engel Jr. J., and Levesque, M.F. Contrasting changes in EPSP slope of associational and perforant pathways in epileptic patients, *Epilepsia*, 31 (1990) 651.
74. Khurgel, M., Racine, R.J., and Ivy, G.O. Kindling causes changes in the composition of the astrocytic cytoskeleton, *Brain Research*, 592 (1992) 338–342.
75. Khurgel, M., Ivy, G.O., and Racine, R.J. Seizure-induced activation of astrocytes in the absence of neuronal degeneration. *Neurobiology of Disease*, 2 (1995) 23–35.
76. Kim, J.H., Guimaraes, P.O., Shen, M.Y., Masukawa, L.M., and Spencer, D.D. Hippocampal neuronal density in temporal lobe epilepsy with and without gliomas, *Acta Neuropathol.*, 80 (1990) 41–45.
77. King, G.L., Dingledine, R., Giacchino, J.L., and McNamara, J.O. Abnormal neuronal excitability in hippocampal slices from kindled rats, *J. Neurophysiol.*, 54 (1985) 1295–1304.
78. Kohr, G., de Koninck, Y., and Mody, I. Properties of NMDA receptor channels in neurons acutely isolated from epileptic (kindled) rats, *J. Neuroscience*, 13 (1993) 3612–3627.
79. Kostopoulos, G. and Antoniadis, G. A comparison of recurrent inhibition and of paired-pulse facilitation in hippocampal slices from normal and genetically epileptic mice, *Epilepsy Res.*, 9 (1991) 184–194.
80. Kraus, J.E., Yeh, G-C., Bonhaus, D.W., Nadler, J.V., and McNamara, J.O. Kindling induces the long-lasting expression of a novel population of NMDA receptors in hippocampal region CA3, *J. Neuroscience*, 14 (1994) 4196–4205.
81. Labiner, D.M., Butler, L.S., Cao, Z., Hosford, D.A., Shin, C., and McNamara, J.O. Induction of c-fos mRNA by kindled seizures: complex relationship with neuronal burst firing, *J. Neuroscience*, 13 (1993) 744–751.
82. Leung, L.S. Spontaneous hippocampal interictal spikes following local kindling: Time-course of change and relations to behavioural seizures, *Brain Research*, 513 (1990) 308–314.

83. Lindsay, R.M., Wiegand, S.J., Alter, C.A., and DiStefano, P.S. Neurotrophic factors: from molecule to man, *Trends Neurosci.*, 17 (1994) 182–190.
84. Margerison, J.H. and Corsellis, J. Epilepsy and the temporal lobes, *Brain*, 89 (1966) 499–530.
85. Masukawa, L.M., Uruno, K., Sperling, M., O'Connor, M.J., and Burdette, L.J. The functional relationship between antidromically evoked field response of the dentate gyrus and mossy fiber reorganization in temporal lobe epileptic patients, *Brain Research*, 579 (1992) 119–127.
86. Mathern, G.W., Pretorius, J.K., and Babb, T.L. Quantified patterns of mossy fiber sprouting and neuron densities in hippocampal and lesional seizures, *J. Neurosurgery*, 82 (1995) 211–219.
87. Mathern, G.W., Babb, T.L., Pretorius, J.K., and Leite, J.P. Reactive synaptogenesis and neuron densities for neuropeptide Y, somatostatin, and glutamate decarboxylase-immunoreactivity in the epileptogenic human fascia dentata, *J. Neuroscience*, 15 (1995) 3990–4004.
88. Mathern, G.W., Pretorius, J.K., Babb, T.L., and Quinn, B. Unilateral hippocampal mossy fiber sprouting and bilateral asymmetric neuron loss with episodic postictal psychosis, *J. Neurosurgery*, 82 (1995) 228–233.
89. Mathern, G.W., Pretorius, J.K., and Babb, T.L. Influence of the type of initial precipitating injury and at what age it occurs on course and outcome in patients with temporal lobe seizures, *J. Neurosurgery*, 82 (1995) 220–227.
90. McIntyre, D.C. and Goddard, G.V. Transfer, interference and spontaneous recovery of convulsions kindled from the rat amygdala, *Electroencephalogr. Clin. Neurophysiol.*, 35 (1973) 533–543.
91. McIntyre, D.C., Kelly, M.E., and Armstrong, J.N. Kindling in the perirhinal cortex, *Brain Research*, 615 (1993) 1–6.
92. McIntyre, D.C., Nathanson, D., and Edson, N. A new model of partial status epilepticus based on kindling, *Brain Research*, 250 (1982) 53–63.
93. McIntyre, D.C. and Plant, J.R. Pyriform cortex involvement in kindling, *Neuroscience & Behavioral Reviews*, 13 (1989) 277–280.
94. McIntyre, D.C. and Wong, R.K.S. Cellular and synaptic properties of amygdala-kindled pyriform cortex in vitro, *J. Neurophysiol.*, 55 (1986) 1295–1307.
95. Meberg, P.J., Gall, C.M., and Routtenberg, A. Induction of F1/Gap-43 gene expression in hippocampal granule cells after seizures, *Molecular Brain Research*, 17 (1993) 295–297.
96. Meldrum, B., Vigouroux, R., and Brierley, J. Systemic factors and epileptic brain damage: prolonged seizures in paralyzed artificially ventilated baboons, *Arch. Neurol. Psychol.*, 29 (1973) 82–87.
97. Merlio, J.P., Ernfors, P., Kokaia, Z., Middlemas, D.S., Bengzon, J., Kobaia, M., Smith, M.-L., Siesjo, B.K., Hunter, T., Lindvall, O., and Persson, H. Increased production of the trkB protein tyrosine kinase receptor after brain insults, *Neuron*, 10 (1993) 151–164.
98. Milgram, N.W., Green, I., Siberman, M., Riexinger, K., and Petit, T.L. Establishment of status epilepticus by limbic system stimulation in previously unstimulated rats, *Exp. Neurol.*, 88 (1985) 253–264.
99. Milgram, N.W., Michael, M., Cammisuli, S., Head, E., Ferbinteanu, J., Reid, C., Murphy, M.P., and Racine, R. Development of spontaneous seizures over extended electrical kindling. II. Persistence of dentate inhibitory suppression, *Brain Research*, 670 (1995) 112–120.
100. Milgram, N.W., Yearwood, T., Khurgel, M., Ivy, G.O., and Racine, R. Changes in inhibitory processes in the hippocampus following recurrent seizures induced by systemic administration of kainic acid, *Brain Research*, 551 (1991) 236–246.
101. Mitchell, C.L. and Barnes, M.I. Effect of destruction of dentate granule cells on kindling induced by stimulation of the perforant path, *Physiology & Behavior*, 53 (1993) 45–49.
102. Mody, I., Otis, T.S., Staley, K.J., and Kohr, G. The balance between excitation and inhibition in dentate granule cells and its role in epilepsy, In J. Engel, C. Wasterlain, E.A. Cavalheiro, U. Heinemann, and G. Avanzini (Eds.), *Molecular Neurobiology of Epilepsy* (*Epilepsy Res. Suppl.*, 9), Elsevier, 1992, pp. 331–339.
103. Mouritzen-Dam, A. Epilepsy and neuron loss in the hippocampus, *Epilepsia*, 21 (1980) 617–629.
104. Nadler, J.V., Perry, B.C., and Cotman, C.W. Selective reinnervation of hippocampal area CA1 and the fascia dentata after destruction of CA3–CA4 afferents with kainic acid, *Brain Research*, 192 (1980) 387–403.
105. Nadler, J. and Cuthbertson, G. Kainic acid neurotoxicity toward the hippocampal formation: dependence on specific excitatory pathways. *Brain Research*, 195 (1980) 47–56.
106. Nagao, R., Avoli, M., and Gloor, P. Interictal discharges in the hippocampus of rats with long-term pilocarpine seizures, *Neurosci. Lett.*, 1974 (1994) 160–164.
107. Nevander, G., Ingvar, M., Auer, R., and Siesjo, B.K. Status epilepticus in well-oxygenated rats causes neuronal necrosis, *Ann. Neurol.*, 18 (1985) 281–290.

108. Nishizuka, M., Okada, R., Seki, K., Arai, Y., and Iizuka, R. Loss of dendritic synapses in the medial amygdala associated with kindling, *Brain Research*, 552 (1991) 351–355.
109. Ogata, N. Effects of cycloheximide on experimental epilepsy induced by daily amygdaloid stimulation in rabbits, *Epilepsia*, 18 (1977) 101–108.
110. Okada, R., Nishizuka, M., Iizuka, R., and Arai, Y. Persistence of reorganized synaptic connectivity in the amygdala of kindled rats, *Brain Res. Bull.*, 31 (1993) 631–635.
111. O'Shaugnessy, D. and Gerber, G.J. Damage induced by systemic kainic acid in rats is dependent upon seizure activity—a behavioral and morphological study, *Neurotoxicology*, 7 (1986) 187–202.
112. Patel, M.N. and McNamara, J.O. *Soc. Neuroscience Abstra.*, 19 (1993) 257. (Note: Report that BDNF triggers dendritic branching in cultures of dentate gyrus granule cells—check)
113. Phillips, H.S., Hains, J.M., Laramee, G.R., Rosenthal, A., and Winslow, J.W. Widespread expression of BDNF but not NT-3 by target areas of basal forebrain cholinergic neurons, *Science*, 250 (1990) 290–294.
114. Pinel, J.P.J., Mucha, R.F., and Phillips, A.G. Spontaneous seizures generated in rats by kindling: a preliminary report, *Physiol. Psychol.*, 3 (1975) 127–129.
115. Pinel, J.P.J. and Rovner, L.I. Electrode placement and kindling-induced experimental epilepsy, *Exp. Neurol.*, 58 (1978) 335–346.
116. Qiao, X. and Noebels, J.L. Developmental analysis of hippocampal mossy fiber outgrowth in a mutant mouse with inherited spike-wave seizures, *J. Neuroscience*, 13 (1993) 4622–4635.
117. Racine, R.J. Modification of seizure activity by electrical stimulation: II. Motor seizure, *Electroenceph. Clin. Neurophysiol.*, 32 (1972) 281–294.
118. Racine, R.J. Kindling: The first decade, *Neurosurgery*, 3 (1978) 234–252.
119. Racine, R.J. and Burnham, W.M. The Kindling Model, In: H. Wheal, and P. Schwartzkroin (Eds.) *Electrophysiology of Epilepsy*, Academic Press, London (1984) pp. 153–171.
120. Racine, R.J., Burnham, W.M., Gilbert, M., and Kairiss, E.W. Kindling mechanisms: I. Electrophysiological studies, *Kindling 3*, In J.A. Wada (Ed.) Raven Press, New York (1986) pp. 263–282.
121. Racine, R.J., Chapman, C.A., Teskey, G.C., and Milgram, N.W. Post-activation potentiation in the neocortex. III. Kindling-induced potentiation in the chronic preparation, *Brain Research*, 702 (1995) 77–86.
122. Racine, R.J., Gartner, J.G., and Burnham, W.M. Epileptiform activity and neural plasticity in limbic structures, *Brain Research*, 47 (1972) 262–268.
123. Racine, R.J., Paxinos, G., Mosher, J.M., and Kairiss, E.W. The effects of various lesions and knife-cuts on septal and amygdala kindling in the rat, *Brain Research*, 454 (1988a) 264–274.
124. Racine, R.J., Milgram, W.N., and Hafner, S. Long-term potentiation phenomena in the rat limbic forebrain, *Brain Research*, 260 (1983) 217–231.
125. Racine, R.J., Moore, K.-A., and Evans, C. Kindling-induced potentiation in the piriform cortex, *Brain Research*, 556 (1991) 218–225.
126. Racine, R.J., Mosher, M., and Kairiss, E.W. The role of the piriform cortex in the generation of interictal spikes in the kindled preparation, *Brain Research*, 454 (1988b) 251–263.
127. Racine, R.J., Newberry, F., and Burnham, W.M. Post-activation potentiation and the kindling phenomenon, *Electroencephalgr. Clin. Neurophysiol.*, 39 (1975) 261–271.
128. Rashid, K., Van der Zee, C.E.E.M., Ross, G.M., Chapman, C.A., Stanisz, J., Riopelle, R.J., Racine, R.J., and Fahnestock, M. A nerve growth factor peptide retards seizure development and inhibits neuronal sprouting in a rat model of epilepsy, *Proc. Natl. Acad. Sci. U.S.A.*, 92 (1995) 9495–9499.
129. Represa, A., Jorquera, I., Le Gal La Salle, G., and Ben-Ari, Y. Epilepsy induced collateral sprouting of hippocampal mossy fibers: Does it induce the development of ectopic synapses with granule cell dendrites? *Hippocampus*, 3 (1993) 257–268.
130. Represa A. and Ben-Ari Y. Kindling is associated with the formation of novel mossy fibre synapses in the CA3 region, *Exp. Brain Res.*, 92 (1992) 69–78.
131. Represa, A., Robain, O., Tremblay, E., and Ben-Ari, Y. Hippocampal plasticity in childhood epilepsy, *Neurosci. Lett.*, 99 (1989) 351–355.
132. Ribak, C.E. Epilepsy and the cortex, In: A. Peters (Ed.) *Cerebral Cortex, Vol. 9*, Plenum Press, New York (1991) pp. 427–483.
133. Rocamora, N., Palacios, J.M., and Mengod, G. Limbic seizures induce a differential regulation of the expression of nerve growth factor, brain-derived neurotrophic factor and neurotrophin-3, in the rat hippocampus, *Mol. Brain Res.*, 13 (1992) 27–33.
134. Russell, R.D. and Stripling, J.S. Effect of olfactory bulb kindling on evoked potentials in the piriform cortex, *Brain Research*, 361 (1985) 61–69.
135. Sagar, H.J. and Osbury, J.M. Hippocampal neuron loss in temporal lobe epilepsy: correlation with early childhood convulsions, *Ann. Neurol.*, 22 (1987) 334–340.

136. Sato, M., Racine, R.J., and McIntyre, D.C. Kindling: basic mechanisms and clinical validity, *Electroencephalgr. Clin. Neurophysiol.*, 76 (1990) 459–472.
137. Sazgar, M., Chick, B.A., Rashid, K., Van der Zee, C.E.E.M., Diamond, J., Fahnestock, M., and Racine, R.J. Role of nerve growth factor in kindling and kindling-induced mossy fiber sprouting, *Soc. Neurosci. Abstr.*, 21 (1995) 1973.
138. Scharfman, H.E. EPSPs of dentate gyrus granule cells during epileptiform bursts of dentate hilar "mossy" cells and area CA3 pyramidal cells in disinhibited rat hippocampal slices, *J. Neurosci.*, 14 (1994) 6041–6057.
139. Shin, C., McNamara, J.O., Morgan, J.L., Curran, T., and Cohen, D.R. Induction of c-fos mRNA expression by afterdischarge in the hippocampus of naive and kindled rats, *J. Neuroschemistry*, 55 (1990) 1050–1055.
140. Shirasaka, Y. and Wasterlain, C.G. Chronic epileptogenicity following focal status epilepticus, *Brain Research*, 655 (1994) 33–44.
141. Schwob, J.E., Fuller, T., Price, J.L., and Olney, J.W. Widespread patterns of neuronal damage following systemic or intracerebral injections of kainic acid: a histological study, *Neuroscience*, 5 (1980) 991–1014.
142. Scharfman, H.E. Electrophysiological evidence that dentate hilar mossy cells are excitatory and innervate both granule cells and interneurons, *J. Neurophysiol.*, 74 (1995) 179–194.
143. Simonato, M., Hosford, D.A., Labiner, D.M., Shin, C., Mansbach, H.H., and McNamara, J.O. Differential expression of immediate early genes in the hippocampus in the kindling model of epilepsy. *Molecular Brain Research*, 11 (1991) 115–124.
144. Sloviter, R.S. Permanently altered hippocampal structure, excitability, and inhibition after experimental status epilepticus in the rat: the "dormant basket cell" hypothesis and its possible relevance to temporal lobe epilepsy, *Hippocampus*, 1 (1991) 41–66.
145. Sloviter, R.S. Possible functional consequences of synaptic reorganization in the dentate gyrus of kainate-treated rats, *Neurosci. Lett.*, 137 (1992) 91–96.
146. Sloviter, R.S. On the relationship between neuropathology and pathophysiology in the epileptic hippocampus of humans and experimental animals, *Hippocampus*, 4 (1994) 250–253.
147. Sloviter, R.S. The functional organization of the hippocampal dentate gyrus and its relevance to the pathogenesis of temporal lobe epilepsy. *Ann. Neurol.*, 35 (1994) 640–654.
148. Sloviter, R.S. and Damiano, B.P. Sustained electrical stimulation of the perforant path duplicates kainate-induced electrophysiological effects and hippocampal damage in rats, *Neurosci. Lett.*, 24 (1981) 279–284.
149. Sloviter, R. Decreased hippocampal inhibition and a selective loss of interneurons in experimental epilepsy, *Science*, 235 (1987) 73–75.
150. Sloviter, R. Epileptic brain damage in rats induced by sustained electrical stimulation of the perforant path. I. Acute electrophysiological and light microscopic and Golgi study, *J. Anat.*, 10 (1983) 675–697.
151. Sperber, E.F., Haas, K.Z., Stanton, P.K., and Moshe, S.L. Resistance of the immature hippocampus to seizure-induced synaptic reorganization, *Dev. Brain Res.*, 60 (1991) 321–330.
152. Spiller, A. E. and Racine, R.J. The effect of kindling beyond the "stage 5" criterion on paired-pulse depression and hilar cell counts in the dentate gyrus, *Brain Research*, 635 (1994) 139–147.
153. Sundstrom, L.E., Mitchell, J., and Wheal, H.V. Bilateral reorganization of mossy fibers in the rat hippocampus after a unilateral intracerebroventricular kainic acid injection, *Brain Research*, 609 (1993) 321–326.
154. Sutula, T., Cascino, G., Cavazos, J., Parada, I., and Ramirez, L. Mossy fiber synaptic reorganization in the epileptic human temporal lobe, *Ann. Neurol.*, 26 (1989) 321–330.
155. Sutula, T., He, XX., Cavazos, J., and Scott, G. Synaptic reorganization in the hippocampus induced by abnormal functional activity, *Science*, 239 (1988) 1147–1150.
156. Swanson, L.W., Wyss, J.M., and Cowan, W.M. An autoradiographic study of the organization of intrahippocampal association pathways in the rat, *J. Comp. Neurol.*, 181 (1978) 681–710.
157. Swanson, L.W., Wawchenko, P.E., and Cowan, W.M. Evidence for collateral projections by neurons in Ammon's horn, the dentate gyrus, and the subiculum: a multiple retrograde labelling study in the rat, *J. Neurosci.*, 1 (1981) 548–559.
158. Tanaka, T., Kondo, S., Hori, T., Tanaka, S., and Yonemasu, Y. Various hippocampal lesions induced by multi-fractional ibotenic acid injections and amygdala kindling in rats, *Brain Research*, 559 (1991) 154–158.
159. Tasker, J.G. and Dudek, E. Electrophysiology of GABA-mediated synaptic transmission and possible roles in epilepsy, *Neurochemical Res*, 16 (1991) 251–262.

160. Tauck, D.L. and Nadler, J.V. Evidence of functional mossy fiber sprouting in hippocampal formation of kainic acid treated rats, *J. Neurosci.*, 5 (1985) 1016–1022.
161. Titulaer, M.N.G., Kamphuis, W., Pool, C.W., van Heerikhuize, J.J., and Lopes da Silva, F.H. Kindling induces time-dependent and regional specific changes in the [^{3}H]muscimol binding in the rat hippocampus: a quantitative autoradiographic study, *Neuroscience*, 59 (1994) 817–826.
162. Tsunoda, K., Mori, N., Osonoe, M., Ariga, K., Saitoh, H., Kittaka, H., and Ogata, S-I. Different effect of hippocampal granule cell destruction on amydgaloid kindling in Sprague-Dawley and Wistar rats, *Brain Research*, 691 (1995) 18–24.
163. Tuff, L.P., Racine, R.J., and Adamec, R. The effects of kindling on GABA-mediated inhibition in the dentate gyrus of the rat: I. Paired pulse depression, *Brain Research*, 277 (1983) 79–90.
164. Turski, L., Cavalheiro, E.A., Sieklucka-Dziuba, M., Ikonomidou-Turski, C., Czuczwar, S.J., and Turski, W.A. Seizures produced by pilocarpine: neuropathological sequelae and activity of glutamate decarboxylase in the rat forebrain, *Brain Research*, 398 (1986) 37–48.
165. Uruno, K., O'Connor, M.J., and Masukawa, L.M. Effects of bicuculline and baclofen on paired-pulse depression in the dentate gyrus of epileptic patients, *Brain Research*, 695 (1995) 163–172.
166. Van der Zee, C.E.E.M., Rashid, K., Le, K., Moore, K-A., Stanisz, J., Diamond, J., Racine, R.J., and Fahnestock, M. Intraventricular administration of antibodies to nerve growth factor retards kindling and blocks mossy fiber sprouting in adult rats, *J. Neuroscience*, 15 (1995) 5316–5323.
167. Vicedomini, J.P. and Nadler, J.V. A model of status epilepticus based on electrical stimulation of hippocampal afferent pathways, *Exp. Neurol.*, 96 (1987) 681–691.
168. Wenzel, H.J., Buckmaster, P.S., Kunkel, D.D., Anderson, N.L., and Schwartzkroin, P.A. Electron microscopic analysis of rat hippocampal mossy cell axon: synaptic relationship with postsynaptic neurons, *Soc. Neurosci. Abstra.*, 20 (1994) 350.
169. Zafra, F., Hengerer, B., Leibrock, J., Thoenen, H., and Lindholm, D. Activity dependent regulation of BDNF and NGF mRNAs in the rat hippocampus is mediated by non-NMDA glutamate receptors, *EMBO J.*, 9 (1990) 3545–3550.
170. Zhao, D. and Leung, L.S. Effects of hippocampal kindling on paired-pulse response in CA1 in vitro, *Brain Research*, 564 (1991) 220–229.

17

EFFECTS OF UNCONTROLLED SEIZURES

Neural Changes in Animal Models

Carl E. Stafstrom[1] and Gregory L. Holmes[2]

[1] Department of Neurology
H4/614 CSC
600 Highland Avenue
Madison, WI 53792
tel: (608) 262-2154
fax: (608) 262-5543
[2] Division of Clinical Neurophysiology
Children's Hospital
Harvard Medical School
Boston, MA02115

Doctors Stafstrom and Holmes have spent several years studying the brain damage and behavioral deficits caused by repeated seizures or status epilepticus. They discuss possible mechanisms and the effects of age.

INTRODUCTION

Children with epilepsy are at significant risk for a variety of problems involving cognition and behavior. The distribution of IQ scores is skewed toward lower values[43,77] and the number of children requiring special education services varies from 10 to 33 per cent.[77,78,79] Several authors have observed that seizures beginning in early childhood are associated with a higher risk of intellectual impairment than seizures beginning in late childhood or in the teenage years.[36,54,55] Behavioral and psychiatric disorders in children with epilepsy are also higher than in the general population, with surveys demonstrating that the prevalence of psychiatric disease is two to four times greater among children with epilepsy than in controls.[77,78,79] While the cognitive and behavioral abnormalities may often be explained by the etiological factors responsible for the epilepsy, there is evidence that some children with poorly controlled epilepsy have progressive declines of IQ on serial intelligence tests[60,65,170] and behavioral and psychiatric deterioration over time. Whether this decline is secondary to antiepileptic medications, progression of the underlying encephalopathy responsible for the seizures, or the seizures *per se*, is not certain.

It is very difficult to determine from clinical studies whether seizures adversely affect cognitive functioning, since there are always a large number of variables that must be considered. Etiology of the seizures, age of onset, antiepileptic drugs, and seizure type, frequency and duration, all may play a critical role in outcome. It is for this reason that investigators have turned to animal studies, where many of these variables can be controlled. In this chapter we will review those animal studies which bear on the question of how prolonged or recurrent seizures alter brain function.

EFFECTS OF PROLONGED SEIZURES ON THE BRAIN

Although partial seizures can arise from anywhere in the cerebral cortex, the temporal lobe, and particularly the hippocampus, is the most common site. Temporal-lobe epilepsy is common in both children and adults.[77] Among patients with temporal lobe epilepsy, mesial temporal sclerosis (MTS) is the sole pathology found in two-thirds or more.[7,216] MTS is characterized by selective neuronal loss and reactive gliosis in the hippocampus and other mesial temporal structures.[7,112] Loss of dentate hilar neurons (endfolium sclerosis) is a common feature, and in some patients is the only apparent hippocampal lesion. In the most severe form of hippocampal sclerosis, there is severe cell loss of hilar neurons and CA1 and CA3 pyramidal cells.

Prolonged febrile seizures[4,58,160,169,213] or status epilepticus[2,31,45,214] have been implicated as major predisposing factors in the development of temporal lobe epilepsy and MTS. Postmortem studies of patients who died as a consequence of prolonged seizures have shown the typical pattern of neuronal loss in the hippocampus that characterizes MTS.[160,217] While it is recognized that status epilepticus may be associated with severe metabolic aberrations, including hypoxia, hypotension, and acidosis, there is compelling evidence from both human and animal studies to indicate that the seizures, *per se*, are responsible for the majority of the cell loss. The current consensus is that prolonged seizures cause neuronal damage secondary to excessive release of the excitatory amino acids (EAAs), glutamate and aspartate, which cause cell death by over-activation of neurons. Since the observation by Lucas and Newhouse[116] that glutamate had a toxic effect on newborn mouse retinal neurons, evidence has accumulated demonstrating that exposure of neurons to glutamate for even a brief period causes immediate or delayed cell death.[39,155,166]

Glutamate Receptors and Epileptogenesis

Because of their importance in a variety of neurological disorders, including epilepsy, glutamate receptors have been the subject of intense investigation.[33,109,177] Glutamate receptor subtypes are distinguishable by biochemical, electrophysiological, and pharmacological criteria.[124,125] Molecular cloning studies have shown that there are multiple receptor gene families that mediate the versatile and widespread function of glutamate signaling.[179] There are two distinct groups of glutamate receptors, metabotropic and ionotropic.[139,194,202] Both the ionotropic[131,139] and metabotropic[173] receptors have been implicated in the initiation and propagation of seizures, as well as in seizure-related cell death. Furthermore, the different distribution and sensitivity of the various glutamate receptor subtypes in the developing and mature brain could provide information on age-specific mechanisms of epileptogenesis.[119,124]

The ionotropic glutamate receptor (GluR) family is composed of closely related subunits that combine to form receptors that are selectively activated by agonists. Ionotropic receptors can be further divided into three major groups, based on the agonist which preferentially activates the receptor: N-methyl-D-aspartate (NMDA), a-amino-3 hydroxy-5 methyl-4 isoxazol propionic acid (AMPA), and kainic acid (KA) (Fig. 1a–c). Glutamate binds to all of these receptors. The NMDA receptor binds NMDA or glutamate, the AMPA receptor binds AMPA, KA, or glutamate, and the KA receptor binds KA or glutamate. Binding of an agonist to any of these glutamate receptor subtypes leads to a conformational change in the ionic channel linked to the receptor, with subsequent flow of cations into the neuron. Channels of the NMDA receptor allow influx of sodium and calcium ions and AMPA and kainate receptors admit Na^+ (and to a lesser extent, Ca^{++}) ions. Another important property of NMDA receptors is their voltage-sensitive block by Mg^{++}. Activation and subsequent increase in intracellular Ca^{++} require ligand stimulation coupled with the removal of the Mg^{++} by the depolarization of the membrane. The NMDA receptor contains several modulatory sites, both within the cation channel and at extra-channel sites near the ligand binding domain.[126] For example, there are glycine and polyamine binding sites near the ligand binding site. Glycine is a "co-agonist" since its binding is required for function of the NMDA receptor complex. Function of the channel itself can be modulated by Mg^{++} (which blocks the channel in a voltage-dependent manner), phencyclidine, Zn^{++} and redox agents.[32,107,108] The redox site consists of one or more thiol (sulfhydryl) groups. Oxidation appears to down-regulate channel activity. The redox modulatory site can also be affected by nitric oxide, which can facilitate disulfide bond formation and lead to less NMDA-induced Ca^{++} entry and thus confer neuroprotection.[110] It is possible that agents could be developed to counteract NMDA receptor function at any of these sites, potentially leading to protection against seizure-induced neuronal damage.

The other major class of glutamate receptors is termed "metabotropic", because they are coupled via guanosine triphosphate (GTP) binding proteins to intracellular enzymes.[23,172,180,193,202] A receptor family comprised of at least eight different subtypes of metabotropic glutamate receptor (mGluR) has been identified.[149] Two major subtypes of mGluR exist with different functions in different types of neurons. One type of mGluR, the L-AP4 (L-2-amino-4-phosphonoproprionic acid) type, are presynaptic inhibitory receptors that may control glutamate release.[53] Both L-AP4 and glutamate hyperpolarize neurons by activating a G protein, which acts via phophodiesterase to convert cyclic AMP or cyclic GMP to AMP or GMP. This hydrolysis allows the nucleotides to alter channel function.

Another mGluR subtype, perhaps of greater interest in epileptogenesis,[23] selectively binds the agonist ACPD (aminocyclopentyl dicarboxylic acid). ACPD then activates the intracellular enzyme phospholipase C (PLC; Fig. 1d), which catalyzes the conversion of phosphoinositides (especially phophatidyl inositol-4,5-biphosphate, PIP_2) to inositol 1,4,5-triphosphate (IP_3) and diacylglycerol (DAG). IP_3 causes prompt release of calcium ions from intracellular stores, particularly from smooth endoplasmic reticulum. This increased $[Ca^{++}]_i$, in conjunction with Ca^{++} entry via other pathways (e.g., NMDA receptors, voltage-sensitive calcium channels), can activate protein kinases (such as protein kinase C [PKC]) to phosphorylate membrane proteins and thus modulate excitability. In different cell types, this modulation can have either a net excitatory or inhibitory effect.[13,48,67] Metabotropic receptors can modulate either glutamatergic or

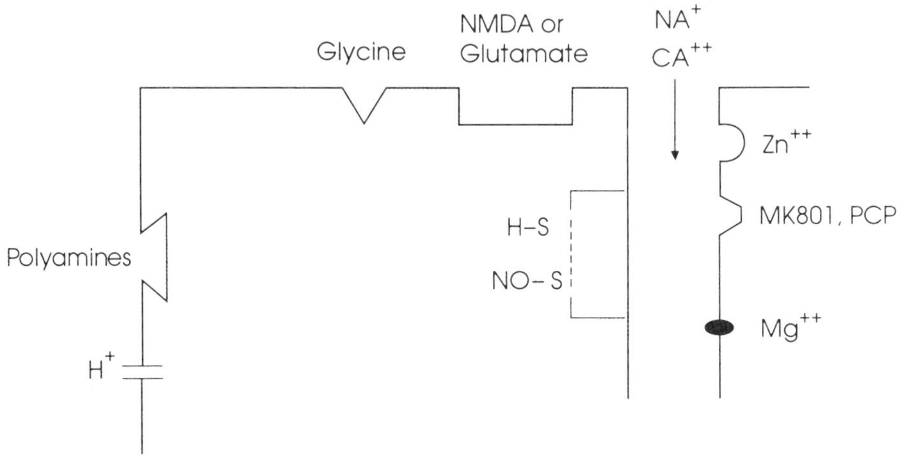

a NMDA Receptor Complex

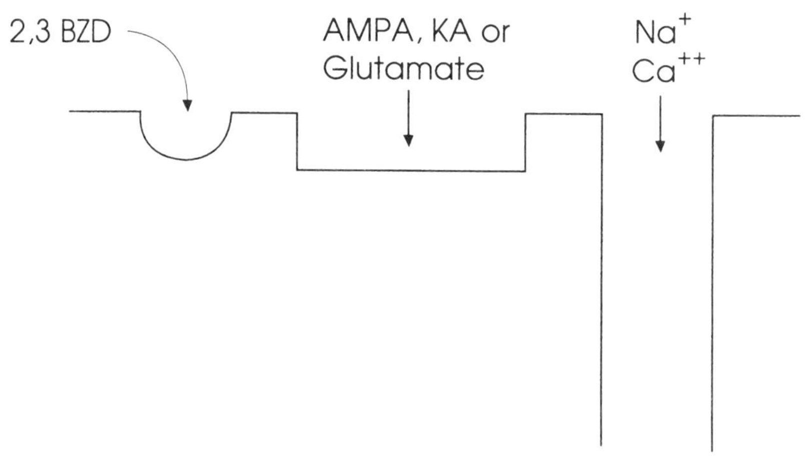

b AMPA Receptor Complex

Figure 1. Schematics of glutamate receptors. Three types of inotropic (a–c) and one type of metabotropic glutamate receptor (d) are illustrated. See text for details. a. NMDA receptor complex, showing agonist binding site, ionic channel and modulators of each. b. AMPA receptor complex. c. Kainate receptor complex. d. one type of metabotropic glutamate receptor that may be important in hippocampus with glutamate or ACPD as agonists. Abbreviations: NMDA—N-methyl, D-aspartate; MK801—(+)-5-methyl-10,11-dihydro-5H-dibenzo[a,d]cycloheptan-5,10-imine maleate; PCP—phencyclidine; NO—nitric oxide; BZD—benzodiazepine; AMPA—amino-3-hydroxy-5-methyl-4-isoxazoleproprionic acid; KA—kainic acid; ACPD—amino-cyclopentane-1,3-dicarboxylate; G—G-protein, PLC—phospholipase C; IP3—inositol-1.4.5-triphosphate,; DAG—diacyl glycerol; PIP_2—phosphatidyl inositol-4,5-diphosphate; PKC—phosphokinase C.

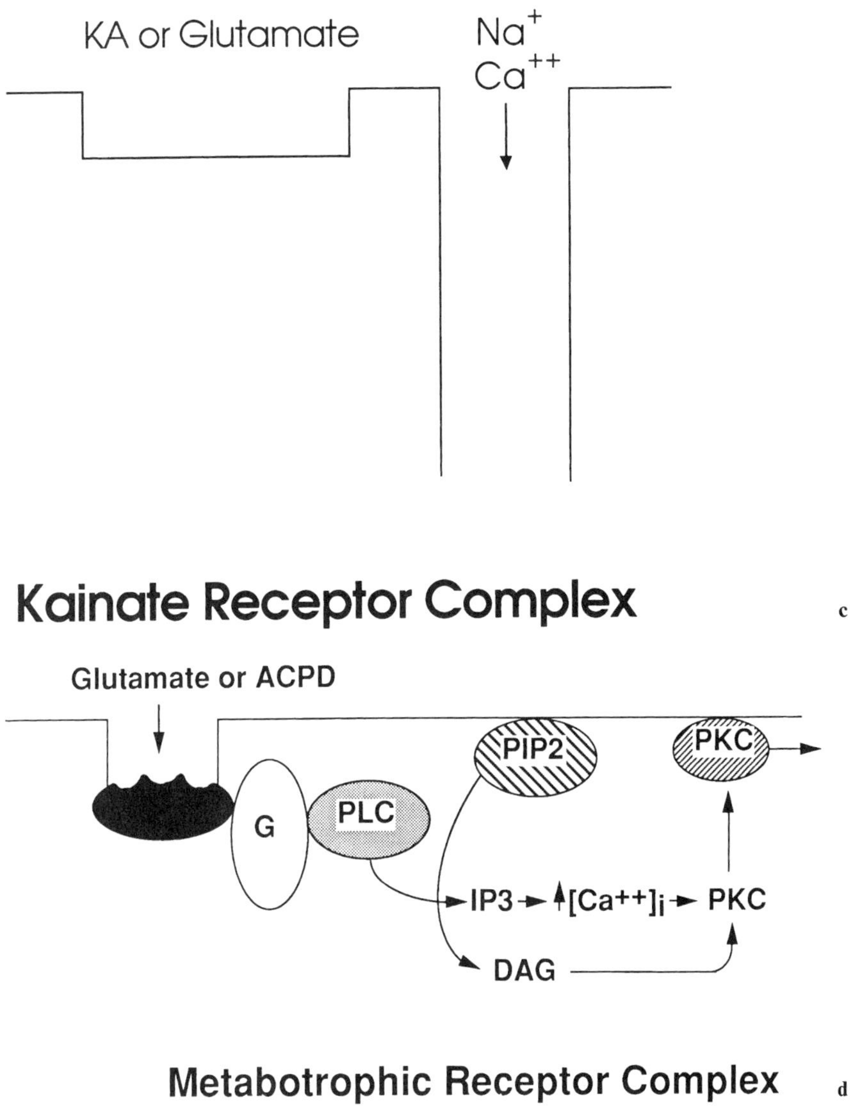

Figure 1. *Continued.*

GABAergic synaptic transmission. In hippocampal pyramidal neurons stimulation of metabotropic receptors increases intracellular Ca^{++} which causes a slow, persistent depolarization, activating high-threshold voltage and ligand-gated channels.[48] Metabotropic receptors can be crucial for the establishment of long-term potentiation (LTP) in the hippocampus,[12,24] a phenomenon with potential importance to learning and memory as well as epileptogenesis.

Prolonged Seizures Can Lead to Excitotoxicity and MTS

Several lines of experimental evidence suggest that MTS can be caused by prolonged seizures. Systemic and intracerebral injections of the excitotoxin kainic acid (KA) or cholinergic agonists (i.e., pilocarpine) in rats and mice result in intense seizure activity in limbic pathways, which resembles the behavioral and EEG features of human complex partial seizures. The morphological sequelae of these seizures are reminiscent of MTS in human temporal lobe epilepsy: selective cell loss and gliosis in sectors CA1 and CA3 of the hippocampus with variable sparing of sector CA2 and granule cells of dentate gyrus.[7,20,146,147,156,157,208,209]

Experimental studies have demonstrated that seizures can cause MTS in the absence of hypotension, hypoxemia, and hypoglycemia.[64,132] When these systemic metabolic disturbances are prevented by mechanical ventilation and neuromuscular blockade in a model of drug-induced status epilepticus in primates, neuronal damage is still observed in the hippocampus, thalamus, and the cortex.[132,133,134] Intense epileptic activity in the hippocampus induced by repeated activation of afferent pathways with high-frequency stimulation in anesthetized, paralyzed rats also induces neuronal loss in the hippocampus and hilus of the dentate gyrus.[186] These results suggest that seizure-induced neuronal damage can be dissociated from the systemic metabolic disturbances related to behavioral seizures. The role of energy and metabolic defects in the pathogenesis of MTS, however, should not be discounted. It is probable that excitotoxic and defects in metabolism are synergistic in producing MTS.

Sloviter and Dempster[186] have demonstrated that neuronal injury induced in the hippocampus by stimulation of the perforant path shares many of the characteristics of excitotoxic injury. Immediately after 24 hours of intermittent, unilateral perforant-path stimulation, acute cell injury could be detected in the ipsilateral dentate hilus and CA3 pyramidal cells. High concentrations of glutamate or aspartate injected directly into the brain replicated all of the qualitative features of acute seizure-induced damage,[186] supporting the hypothesis that the sustained release of an excitatory amino acid is the initiating factor in seizure-associated cell injury. Holmes et al.[88] demonstrated that intrahippocampal infusion of glutamate and quisqualate produces similar neuronal injury as that following seizures induced by KA and pilocarpine. Liu et al.[113] showed that direct equimolar injections of glutamate into immature hippocampus caused cell death in an age-dependent fashion, with little damage below P20 and progressively more neuronal death at older ages. These results support the notion that excessive amounts of glutamate and aspartate released during seizures result in neuronal damage and that this damage is age-dependent.

Based on this excitotoxic hypothesis, it would be expected that hippocampal subfields CA1 and CA3 would be vulnerable to damage because both are rich in EAA receptors. Indeed, quantitative autoradiographic studies have demonstrated that CA1 contains the highest density of L-(^{3}H) glutamate binding sites in the brain,[140,141] and CA3 and hilus regions contain abundant KA receptors.[47,206] Since these regions are most vulnerable to damage, it can be surmised that seizure-induced neuronal damage is associated with powerful activation of NMDA receptors in CA1 and KA receptors in CA3. As described above, these receptors, when powerfully activated by the appropriate ligands during seizures, allow a considerable influx of Ca^{2+} ions into postsynaptic pyramidal neurons.[89,98] Non-NMDA receptor activation is also important in glutamate-induced neuronal damage.[62,97] The increased intracellular Ca^{2+} will trigger a cascade of reactions resulting in activation of proteases, phospholipases, and endonucleases, ultimately leading to the generation of active and potentially toxic metabolites.[38,40,159,182]

Several lines of evidence have shown that seizures are associated with increased extracellular levels of excitatory amino acids. In experimental studies, Dodd et al.[56] reported that enhancement of glutamate release from cobalt epileptic lesions was approximately 3-fold greater than that from non-epileptic lesions. In addition, epileptogenic lesions showed a marked reduction in glutamine release compared to controls. Lehmann[103] found that the seizures induced by kainic acid or folic acid were accompanied by pronounced elevations of extracellular alanine and phosphoethanolamine in the hippocampus. Extracellular glutamate was slightly elevated (50–70% over baseline values). Lallement et al.[100,101] noted a sustained increase in the extracellular glutamate level in the amygdala and the hippocampus following seizures precipitated by soman, an inhibitor of acetylcholinesterase. Minamoto et al.[137] reported that the extracellular concentrations of glutamate and GABA were increased during early stages of kindling in comparison with the values in control animals. After full kindling, with secondarily generalized seizures, these values were increased even further.

One of the difficulties in measuring changes in EAAs during seizures is the rapid uptake of glutamate.[104] Lehmann et al.[104] found no change in extracellular glutamate in rabbits treated with bicuculline or KA. Millan et al.[135] demonstrated that there were large and sustained increases in extracellular hippocampal aspartate and glutamate levels following treatment with the glutamate uptake inhibitor, 1-trans-pyrrolidine-2,4-dicarboxylic acid (PDC).[135] After pilocarpine administration, however, only a transient increase in aspartate and glutamate occurred before the onset of seizures. Aspartate and glutamate levels were not significantly increased (relative to PDC controls) during seizures.

Additional supporting evidence that EAAs are responsible for the damage induced by prolonged seizures comes from the observation that administration of glutamate antagonists significantly reduces cell loss.[34,59,130] In models of status epilepticus, protection from neuronal necrosis was obtained with both NMDA and non-NMDA antagonists.[93,102,142]

MTS Can Lead to Subsequent Seizures: The Synaptic Reorganization Hypothesis

Experimental studies also support the view that MTS is an important factor for the development of subsequent seizures. Spontaneous recurrent seizures (SRS) are frequently observed following seizure-induced structure damage in several animal models.[29,85,86,111,191] These seizures are typically partial in onset with secondary generalization, and, based on EEG and behavioral features, they begin in the hippocampus or amygdala.[3,18,144,164,191]

How could MTS lead to epilepsy? Several possible mechanisms have been proposed. Among them, the sprouting hypothesis has attracted considerable recent attention.[49,57,121,154,161,200,201,204] To understand sprouting, it is necessary to review the primary afferent pathways into the hippocampus. Associational regions of the neocortex project to the lateral areas of the entorhinal cortex, which in turn project to the dentate gyrus via the perforant pathway. The entorhinal cortex fibers terminate on dendritic trees located in the molecular layer of the dentate granule cells. The dentate granule cells give rise to axons, termed "mossy fibers", which project to the CA3 field of the hippocampus. CA3 neurons have widespread associational projections to various levels of CA1. Other collaterals, namely the Schaeffer collaterals, provide the major input from CA3 to CA1. CA1 projects to the subiculum, which in turn projects to the pre- and parasubiculum and to the entorhinal cortex thus completing the circuit. Cells located deep in the entorhinal cortex project back to the cortical field from which they receive input.

Following prolonged seizures there is a selective loss of vulnerable neurons, especially in CA3 and the hilus. Since the mossy fibers have lost their target, there is an apparent aberrant growth of granule cell axons back into the inner molecular layer of the dentate. The spouting of mossy fiber axons is readily observed with the Timm method, a histochemical technique that selectively stains mossy fiber axon terminals because of their high zinc content.[50] Since the neurotransmitter of the mossy fibers is glutamate, these aberrant synaptic conditions may contribute to the state of hyperexcitability that either provokes or facilitates abnormal discharges (Fig. 2).[204] Reorganization of mossy fiber synapses is also observed in models of generalized seizures,[68] in the tottering mouse mutant[192] and during kindling of limbic structures, including perforant path, amygdala, and olfactory bulb.[197] Evidence of synaptic reorganization in epileptic human temporal lobe has also been found with the use of several histologic markers of the mossy fiber pathway, including Timm histochemistry.[9,120,196,199] The net effect of such synaptic reorganization has been shown to be excitatory in the kindling model,[199,201] although sprouting of inhibitory GABAergic neurons could make the functional effect of such synaptic reorganization more complicated.[8]

While sprouting suggests that recurrent seizures might be secondary to excessive excitation, there is also evidence that the recurrent seizures are related to a lack of inhibition. Some investigators have reported that GABA-containing basket cells remain intact in both animal models[183,184] and human specimens[8,10,181,210] from patients with temporal lobe epilepsy. Sloviter,[185] using the perforant pathway stimulation model, found that the most vulnerable neurons in the hilus were the so-called "mossy cells", which are believed to synapse with GABA-containing basket cells. If there is a loss of mossy cells the GABA-containing basket cells will be essentially non-functional, i.e., "dormant", and this lack of functional inhibition may be responsible for the expression of seizures. This dormant basket cell hypothesis has been supported by recent electrophysiological data.[17] Other evidence favors a contribution of diminished inhibitory circuitry. Obenaus and colleagues[152] found that a period of severe seizure activity resulted in loss of glutamic acid decarboxylase (GAD) mRNA-containing neurons in the hilus. The loss of GAD, the synthetic enzyme for GABA, may contribute to hyperexcitability and epileptogenesis in the hippocampus. Many other factors likely participate in seizure-induced synaptic reorganization, including dentate granule cell neurogenesis,[162] neurotrophins,[66] immediate early genes[27,96,148] and alterations in glutamate receptor expression.[163]

Prolonged Seizures in the Developing Brain

Considerable clinical data suggest that prolonged febrile seizures are a major predisposing factor for the development of temporal lobe epilepsy and MTS.[4,58,160,169,213] Postmortem studies in children who died as a consequence of febrile seizures have shown the typical pattern of neuronal loss in the hippocampus that characterizes MTS.[160,217] In a series of 67 patients with mesial temporal lobe epilepsy followed by French and colleagues,[63] 54 (81%) had a history of convulsions during early childhood or infancy, 52 of which were associated with fever. Grattan-Smith et al.,[70] in a review of MRI scans from 53 children with medically refractory temporal lobe epilepsy, found that hippocampal sclerosis was present in 30 children (57%). MTS was frequently associated with a history of prior neurological insults, primarily febrile seizures, while other MRI-detected lesions in patients with temporal lobe epilepsy were not highly related to prior neurological insults.

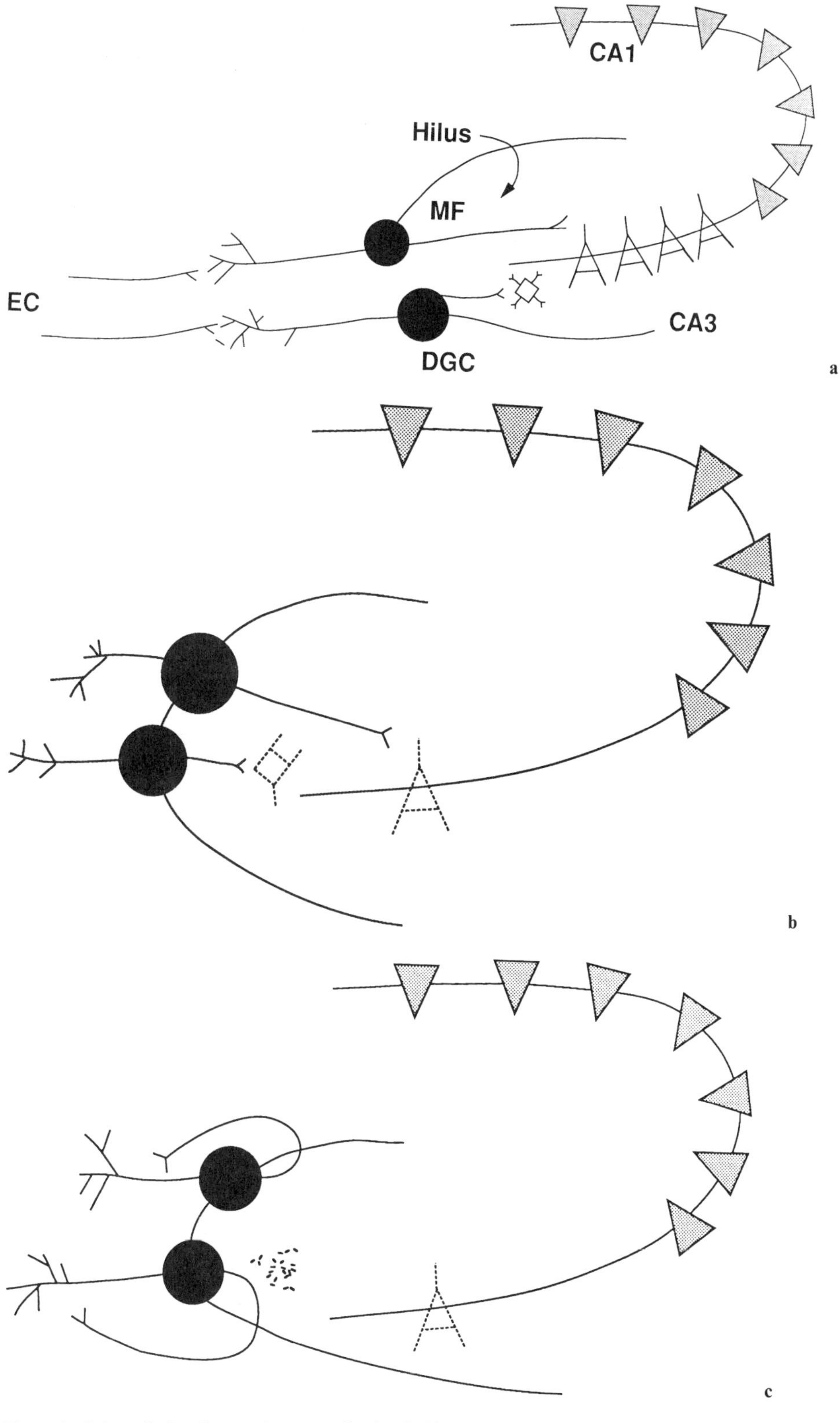

Figure 2. Seizure-induced synaptic reorganization in hippocampus (a–c). a. Normal situation. Dentate granule cells innervate CA3 pyramidal neurons and neurons of the dentate hilus via their "mossy fiber" axons. b. After seizures. Death of CA3 pyramidal and hilar neurons leave mossy fibers (which contain zinc) without sites to innervate. c. Synaptic reorganization. Aberrant dendritic innervation in the inner molecular layer of the dentate, sets up a potentially hyperexcitable circuit that may lead to enhanced epileptogenicity.

Despite the considerable evidence implicating prolonged febrile seizures in infancy to MTS, laboratory studies have suggested that the immature animal may be less susceptible to seizure-induced brain damage than the mature animal. In a series of studies using the excitotoxin kainic acid (KA), we found that long-term cognitive and behavioral sequelae are highly dependent on the age at which an animal receives KA.[3,86,189,191] Rats receiving KA at postnatal (P) ages 30 and 60 days, and tested at P80 had significant hippocampal lesions and deficits in learning, memory, activity level, and behavior when compared to controls.[189] When KA is administered to P5 and P10 rats, severe seizures occur, but no major histological lesions are noted nor do the animals have any long-term deficits in cognition or behavior. In addition, while subsequent seizure threshold is not altered by KA in P5 and P10 rats, animals receiving KA at P30 or P60 have a significantly lower threshold to flurothyl seizures than age-matched controls.[191]

The long-term age-dependent effects of status epilepticus are not limited to the KA model of epilepsy. In another study, animals receiving KA at P15 and P21 developed kindled seizures at the same rate as controls, while rats receiving KA at P27 kindled more quickly than controls.[86] Thurber et al.[205] found similar results using continuous hippocampal stimulation to produce prolonged limbic seizures in rats on P20, P30 and P60, and assessed spatial behavior using a water maze beginning at P80. Limbic status epilepticus on P60 produced impairment in water maze performance. However, rats which seized on P20 and P30 showed only slightly impaired spatial responses. Thus, the age at which status epilepticus occurs appears to be an important determinant of long-term impairment in visual spatial discrimination and memory.

The lack of behavioral effects in young animals may depend upon the particular task. Holmes and colleagues recently used an auditory discrimination method to evaluate the long-term effects of status epilepticus on the developing brain.[150] Rats were injected with convulsant dose of pilocarpine on either P20 or P45, then were trained as adults on two auditory discrimination tasks: sound source location and sound-silence discrimination. Control rats acquired both discriminations immediately. Status epilepticus on P45 produced spontaneous recurrent seizures (SRS), CA3 cell loss, periods of unresponsiveness, and failure to acquire either discrimination task. Status epilepticus on P20 did not cause SRS or a hippocampal lesion; however, acquisition of the sound source location discrimination was slow and erratic. We concluded that pilocarpine-induced status epilepticus in the rat causes an age-dependent long-term impairment in auditory discrimination. It should be cautioned that impairment in acquisition and performance of auditory location discrimination may also be due to damage in structures other than the hippocampus. Harrigan et al.[73] administered lithium plus pilocarpine to induce status epilepticus in mature rats. The status epilepticus caused damage in many diencephalic and subcortical telencephalic structures, and caused significant deficits in acquisition and recall of a radial arm maze. Damage within the medial dorsal thalamic group of nuclei was significantly correlated with the number of errors.

In addition to fewer behavioral deficits following prolonged seizures in the immature brain, there are also fewer histological lesions. Status epilepticus produced by continuous electrical hippocampal stimulation in the immature animal is associated with fewer histological lesions and behavioral abnormalities in immature animals than mature animals.[205] We have not detected hippocampal lesions, using light microscopy, in animals that receive KA before P20.[191] Likewise, pilocarpine administration, which produces severe seizures in both immature and mature animals, results in fewer behavioral abnormalities and less histological damage in the immature animal than the mature animal.[30,76] In addition, Sperber et al.[188] found that while KA caused more severe seizures in the P15 rats than in adults, these seizures did not produce mossy fiber sprouting or changes in paired-pulse facilitation in hippocampal slices from rats tested 2–4 weeks after status epilepticus. However, the immature brain is less resistant to seizure-induced histologic damage if there has been a prior brain injury.[187] This finding has clinical implications, suggesting that damage emerges according to an ontogenetic sequence, which can be modified by preexisting neurologic compromise.

More recent work suggests that, although the immature brain is more resistant to seizure induced damage than the mature brain, histological deficits can be identified.[80] Wasterlain and Sankar[215] have reported that two hours of stimulation of the perforant pathway resulted in the loss of paired-pulse inhibition and damage to the inner granule cells of the dentate gyrus and hilar interneurons in P14–P16 rats. There was remarkable sparing of CA1 and CA3 pyramidal cells. The authors concluded that "the pattern of neuronal necrosis seen after perforant pathway stimulation in immature rats would be unlikely to lead to Ammon's horn sclerosis, and calls into question the hypothesis that hippocampus sclerosis and Ammon's horn sclerosis can result from complicated febrile seizures or from status epilepticus in infancy or childhood". The same group has also shown that in the lithium-pilocarpine model,[175] status epilepticus damages CA1, CA3 and dentate granule cells in an age-specific pattern, with features of both apoptotic and necrotic cell death. Supragranular sprouting has been reported following statues epilepticus caused by corticotrophin-releasing hormone in neonatal rats.[167]

While the immature rat brain appears to be less sensitive to the effects of prolonged seizures, this may not be the case in all species. The animal model may be an important factor in determining the degree of brain damage following seizures. Franck and Schwartzkroin[61] found that systemic administration of KA in P7 rabbits led to cell loss in the CA1 region, but not the CA3 region. While intraventricular injections of KA produced cell loss in CA3 of mature rabbits, it failed to produce lesions in the hippocampus of immature animals. The reasons for the different pathological patterns with the two routes of administration were not determined. Mathern and colleagues,[120] in a study of surgically resected hippocampi from children with extrahippocampal seizures, found significant cell loss (by neuronal counting) and aberrant supragranular mossy fiber sprouting. While it is likely that prolonged seizures or recurrent seizures can lead to some degree of brain damage in the immature brain, experimental data indicate that when compared to the mature brain, the developing brain is less susceptible to the adverse effects of prolonged seizures.

One reason that the immature brain may be less vulnerable to prolonged seizures is that the immature brain has fewer EAA receptors or has functionally immature EAAs. However, KA and NMDA receptors are present at birth and increase dramatically during the first few weeks of life.[22,207] Miller and colleagues[136] studied the ontogeny of KA binding sites in rat forebrain using *in vitro* receptor autoradiography. Specific binding was detectable in the hippocampus by postnatal day one. In the CA3 region, binding increased progressively with age, peaking at P21. Insel et al.[90] also found that specific binding to NMDA and AMPA receptors could be detected at P1 in the hippocampus and striatum with the adult pattern of binding to NMDA receptors emerging by P14, with high densities of binding in CA1 and the dentate gyrus. Physiological development of EAAs parallels the biochemical maturation. The population spike elicited by NMDA application to CA1 pyramidal neurons peaks at 10 days and long-term potentiation peaks around 15 days of age in the rat.[19,72,74] Besides their role as neurotransmitters, EAAs in the developing brain are involved in plasticity and excitotoxicity.[42,44,52,75,211] Administration of NMDA to immature rats (17- and 35-day-old), for example, results in an increase in synaptic density within hours of injection[26] and 8 day-old rats given 10 days of subcutaneous NMDA have a 15% increase in total number of synapses.[25] Small amounts of glutamate in the extracellular space promote the growth and differentiation of cultured neurons.[6]

In some circumstances, EAAs produce *greater* neurotoxicity in the immature brain than in the mature brain.[104,107,109,110] McDonald and colleagues[126] found that NMDA neurotoxicity is greater following injection in the corpus striatum of 7 day-old rats than in adults. NMDA toxicity *in vivo* transiently peaks near postnatal day 7 in rats, with the vulnerability of the developing brain to direct intrastriatal infusion of equimolar NMDA being approximately 60 times greater at this age, as compared to adults. The severity of brain injury produced by infusion of NMDA at age 1, 14, 21, and 28 days was comparable to that produced by NMDA in adults, whereas intermediate levels of injury were present at 4 and 10 days of age and peak levels were present at age 7 days. Likewise, when given by a single injection, AMPA and quisqualic acid have greater toxicity in immature animals than mature animals.[128,129] In addition, we have found that KA, when infused by osmotic pump into the immature brain, causes lesions similar in size to those of the mature brain.[203]

While the immature brain is as sensitive, if not more sensitive, to the direct effects of EAAs, brain damage following prolonged seizures is significantly less in immature animals than in mature animals. We hypothesize that one reason for this decreased vulnerability in the immature brain might be that less glutamate is released from presynaptic terminals during prolonged seizures. This hypothesis is supported by the finding that glutamate concentrations in the immature rat brain are low,[28] as are concentrations of glutamate-related enzymes.[11,99,171] Hippocampal slices from immature rats have low presynaptic release of EAAs[138] and low Na^+-dependent high-affinity uptake of glutamate and aspartate into presynaptic terminals and into glial cells.[178] Ribak and Navetta,[168] in a study of mossy fiber development, have found that an adult pattern of Timm-labeled mossy fiber innervation is not observed until age 25 days. These authors also found that at age 7 days, the deep hilus showed only a few asymmetric excitatory synapses formed by the developing mossy fibers. At 12 and 15 days, the number of asymmetric synapses was less than half of that found in the adult. The subgranular zone of the hilus reached adult levels by 21 days, while the deep hilus attained adult values at 30 days.

LTP and Excitotoxicity

Since brain damage in mature animals following seizures appears to be related to excitotoxicity secondary to release of EAAs, there has been considerable interest in developing agents that block the NMDA or non-NMDA receptor. NMDA antagonists can block long-term potentiation (LTP), the most thoroughly characterized example of synaptic plasticity. Following LTP, which is induced by a short burst of high-frequency electrical stimuli, there is enhanced synaptic transmission that can last hours, weeks or longer.[1,12,41,51,143] For this reason, LTP has been considered to be the cellular basis for learning and memory. Despite decades of study, the mechanism of LTP, its molecular mediators and even its pre-versus postsynaptic localization are still hotly debated.[174] While initially described and most extensively studied in the hippocampus,[21] LTP has also been described in several other brain regions including the visual system,[5,153] neocortex,[14,91,195] piriform cortex,[94] and amygdala.[35] Studying LTP in the CA1 region of the hippocampus, an area rich in NMDA receptors,[46] has increased our understanding of its mechanisms. LTP requires a temporal coupling of presynaptic transmitter release and postsynaptic depolarization.[71,95,118,176] This results in an activation of postsynaptic receptors that allows Ca^{++} influx and increases intracellular Ca^{++} in postsynaptic spines. LTP can be observed *in vitro* as early as P5; by P15 the duration of LTP approaches adult values.[74,92,145] *In vivo* studies have demonstrated the presence of LTP during the first week of life.[15,16,51,143]

The relationship between learning and LTP has been observed in *in vivo* behavioral studies. For example, Davis et al.[51,143] administered the NMDA receptor antagonist D-2-amino-5-phosphonopentanoate (D-AP5) via osmotic pumps during water maze testing. D-AP5 impaired spatial learning in a linear dose-dependent manner, correlating with its corresponding impairment of hippocampal LTP *in vivo*.

As noted above, postsynaptic EAAs, particularly of the NMDA and AMPA receptors, are transiently enhanced early in life. This transient enhancement is presumably beneficial to the immature brain because of the role that physiological activation of the EAAs, and presumably LTP, plays in early learning and morphogenesis. Because of the involvement of EAAs in learning and brain plasticity, blocking these receptors may impair learning and memory. For example, administration of 2-amino-5-phosphonovalerate (APV), a competitive NMDA antagonist, and MK-801 ((+)-5-methyl-10,11-dihydro-5H-dibenzo[a,d]cycloheptan-5,10-imine maleate), a non-competitive NMDA antagonist, intracortically in kittens prevents or reduces ocular dominance shift.[165] Even in mature animals, NMDA receptor antagonists impair performance in a variety of behavioral tests.[37,51,143,211,212]

While there have been few studies on the long-term effects of EAA antagonists on the developing brain, there are some indications that these agents cause permanent changes in the brain. Gorter et al.[69] chronically treated rat pups with MK-801 from postnatal day 8 through 19. Hippocampal kindling at age 70 days demonstrated that the early stages of kindling were reached more quickly in the MK-801 treated rats as compared to controls, showing that neonatal interference with NMDA receptor function leads to long-lasting increases in hippocampal excitability, perhaps through up-regulation of NMDA receptors. Support for this notion is provided by McDonald and colleagues,[127] who showed that neonatal rats become more sensitive to NMDA-mediated brain injury after pretreatment with MK-801. This NMDA antagonist can also exacerbate KA-induced seizures in the neonatal period.[190]

The anticonvulsant action of NMDA receptor antagonists initially generated considerable excitement about the potential clinical role of these compounds. Most of these agents, however, cause excessive sedation at clinically effective doses, and no such medications have yet been released for treatment of epilepsy.

EFFECTS OF RECURRENT SEIZURES ON THE BRAIN

There is increasing evidence that brief recurrent seizures, like prolonged seizures, can lead to brain damage, although the changes are not as dramatic as those following status epilepticus. Newer studies are revealing subtle structural and physiological consequences of brief, recurrent seizures in the immature brain that were previously undetected. These results have profound clinical implication. Clinicians may not be as inclined to treat brief seizures, even if recurrent, yet their consequences may be equally detrimental as with those following prolonged seizures.

Holmes et al.,[87] subjected genetically epilepsy prone rats (GEPRs) to 66 audiogenic seizures between 45–75 days of age, administered at a rate of 2–3 stimulations a day.

Compared to controls, GEPRs reached criteria less frequently in the T-maze (a test in which water-deprived rats have to learn to go to alternate arms of a T-shaped maze in order to receive a water reward), required longer times to find the platform in the Morris water maze, were less active in the open field, less aggressive in the home cage intruder test, and more irritable and aggressive in the handling test. These results demonstrated that frequent, brief seizures could lead to long-term deficits in learning, memory, activity level, and behavior. Since GEPRs do not demonstrate the audiogenic seizure trait consistently until after 20 days of age, however, prepubescent animals could not be studied.

Kindling is another model that has been used to assess the long-term effects of recurrent seizures. Leung and colleagues[105,106] kindled rats in the hippocampus and then measured relearning in the radial 8-arm maze, another measure of hippocampal function.[158] Kindled rats were found to perform worse than control, non-stimulated rats[105] or rats subjected to low-frequency stimulations.[106] The abnormal performance in the maze persisted for 3–4 weeks after the kindling. Lopes da Silva et al.[115] also performed hippocampal kindling following training in the 8-arm maze. The authors found that both working (short-term) and reference (long-term) memory were impaired. In another study, adult rats' deficits in the radial arm maze were directly correlated with the number of previous kindled seizures,[198] supporting the role of repeated seizures in cognitive impairment.

While kindling appears to adversely alter performance on a task previously learned, Holmes et al.[81] found that kindling had little effect on subsequent learning of a new task. The authors kindled 20, 40, and 60 day-old rats to stage 5 seizures and then elicited an additional 15 seizures using the same kindling stimulation. At age 80 days, all animals that reached stage 5 kindling and their respective age-matched controls underwent behavioral testing using the Morris water maze, open field test (a test of activity), and handling test (a test of emotionality). Prior to euthanasia the animals had their seizure thresholds tested using flurothyl inhalation. No differences were noted in time to platform in the water maze or activity level in the open field test between the kindled rats and controls in any of the three age groups. Rats kindled at age 20 and 40 days, however, were more emotionally labile than the controls in the handling test. In the flurothyl inhalation test, rats kindled at 40 and 60 days of age had a shorter latency to all seizure stages than the controls. These results demonstrate that kindling alters emotionality and activity level in immature animals and reduces seizure threshold in pubescent and mature rats.

In addition to kindling studies, investigators have used other seizure models to evaluate the effects of recurrent seizures on cognition. In P15 rats subjected to three flurothyl seizures on each of five consecutive days, deficits were found in spatial learning in the water maze and in auditory location learning when the rats were later tested as adults.[151] Therefore, serial seizures in the developing brain caused long-term deficits in learning and memory, even though no hippocampal cell loss was seen.

Morphological changes also occur following recurrent seizures. As introduced above, synaptic reorganization in the from of mossy fiber sprouting has been observed in many experimental epilepsy models. Such sprouting occurs following both prolonged and brief, recurrent seizures. Neonatal rats were subjected to 25 flurothyl-induced seizures over a five day period, starting on the day of birth.[82] When later studied as adults, compared to controls, the rats with multiple flurothyl neonatal seizures had several neurologic abnormalities, including impaired learning and memory in the water maze, lower thresholds to pentylenetetrazole seizures, sprouting of mossy fibers in the dentate supragranular region and infrapyramidal layer of CA3, and newly formed dentate granule cells.

In another study, young rats were subjected to 50 brief flurothyl seizures over a twelve day period, beginning on P11. Again, the early seizures were associated with increased supragranular and CA3 sprouting, plus progressively increased brain excitability assessed by c-fos staining.[114]

Seizure-induced alterations in cell number and axonal growth during development are not limited to the flurothyl model. Beginning at P1, P10 or P60, rats underwent one or fifteen daily pentylenetetrazole seizures.[83,84] Compared to controls, rats with fifteen seizures had increased CA3 sprouting, whereas those with a single seizure did not. These findings were felt to represent new axonal growth rather than failure of normal synaptic regression. None of these immature rats had cell loss under these conditions, whereas CA1 cell loss was seen in P60 rats. In P10–25 rats undergoing two daily seizures, an increase in dentate granule cells was noted. These findings suggest that during development, brief, recurrent seizures can result in permanent changes in neuron connectivity, which might account in part for the increased seizure susceptibility seen after seizures early in life.

Importantly, in these recent studies, synaptic reorganization occurred in the absence of significant neuronal loss, pointing to the role of factors other than cell death in the pathogenesis of the epileptic state. The hypothesis that granule cell neurogenesis plays a critical role in the epileptogenic process is currently being investigated.

CONCLUSIONS

Our understanding of the mechanisms of seizure-induced brain damage has increased dramatically during the last decade. Prolonged or recurrent seizures cause excessive activation of glutamate receptors, which leads to a cascade of events, eventually resulting in cell death. Since glutamate and other EAAs are important in learning, brain development and plasticity, a future challenge for investigators will be to develop methods that reduce the excitotoxicity that occurs with seizures without impairing normal brain function. In evaluating the effects of prolonged or recurrent seizures on the brain, investigators must consider the animal model, the age of the animal, the task required, and the timing of the testing in relationship to the seizures.

SUMMARY

The majority of children with epilepsy have a good prognosis. Seizures are often adequately controlled with antiepileptic drugs and normal mental development occurs. However, approximately a third of children with epilepsy will develop medically intractable epilepsy, and tragically, in some of these patients the recurrent seizures are associated with cognitive decline. Whether this decline occurs because of the recurrent seizures, drug therapy, or the underlying etiology of the seizures is unclear. However, it is recognized that prolonged or recurrent seizures can lead to cell loss in a number of brain regions with resulting long-term deficits in memory and learning. There is substantial evidence that this brain damage is secondary to excitotoxicity caused by excessive release of excitatory amino acids (EAAs) during seizures. This release of EAAs stimulates ionotropic and metabotropic glutamate receptors and leads to a cascade of events: excessive Ca^{++} entry into neurons activates several Ca^{++} dependent processes, such as the proteases, phospholipases, and endonucleases, ultimately leading to the generation of toxic metabolites. The brain damage associated with seizures appears to be age-related; immature animals have fewer histological lesions and cognitive disturbances following prolonged seizures than do mature animals. While the mechanism for these age-related differences is not known, some studies suggest that the immature brain may be relatively less vulnerable to the adverse effects of prolonged seizures than the mature brain.

ACKNOWLEDGMENTS

Supported by an NINDS grant (NS27984) to GLH.

REFERENCES

1. Abraham, W.C. and Mason, S.E. Effects of the NMDA receptor/channel antagonists CPP and MK801 on hippocampal field potentials and long-term potentiation in anesthetized rats, *Brain Res*, 462 (1988) 40–46.
2. Aicardi, J. and Chevrie, J.-J. Consequences of status epilepticus in infants and children. In: A.V. Delgado-Escueta, C.G. Wasterlain, D.M. Treiman, and R.J. Porter (Eds.): *Advances in Neurology. Vol. 34. Status Epilepticus: Mechanisms of Brain Damage and Treatment*, Raven Press, New York (1983) pp. 115–125.
3. Albala, B.J., Moshé, S.L., and Okada, R. Kainic-acid-induced seizures: A developmental study, *Dev. Brain Res.*, 13 (1984) 139–148.
4. Annegers, J.F., Hauser, W.A., Elveback, L.R., and Kurland, L.T. The risk of epilepsy following febrile convulsions, *Neurology*, 29 (1979) 297–303.
5. Artola, A. and Singer, W. Long-term potentiation and NMDA receptors in rat visual cortex, *Nature*, 330 (1987) 649–652.
6. Aruffo, C., Ferszt, R., Hildebrandt, A.G., and Cervos-Navarro, J. In vitro neuronal plasticity triggered by subtoxic doses of monosodium glutamate, *Prog. Clin. Biol. Res.*, 253 (1987) 229–237.
7. Babb, T. and Brown, W. Pathological findings in epilepsy. In: J. Engel (Ed.) *Surgical Treatment in Epilepsy*, Raven Press, New York (1987) pp. 511–540.
8. Babb, T.L. and Brown, W.J. Neuronal, dendritic and vascular profiles of human temporal lobe epilepsy correlated with cellular physiology "in vivo", *Adv. Neurol.*, 44 (1986) 949–966.
9. Babb, T.L., Kupfer, W.R., Pretorius, J.K., Crandall, P.H., and Levesque, P.F. Synaptic reorganization by mossy fibers in human epileptic fascia dentata, *Neurosci.*, 42 (1991) 351–363.
10. Babb, T.L., Pretorius, J.K., Kupfer, W.R., and Crandall, P.H. Glutamate decarboxylase-immunoreactive neurons are preserved in human epileptic hippocampus, *J. Neurosci.*, 9 (1989) 2562–2574.
11. Barca, M.A. and Toledano, A. Histochemical electron microscopic study of the enzyme glutamate dehydrogenase (GD) in post-natal developing cerebellum, *Cell Mol. Biol.*, 28 (1982) 187–195.
12. Bashir, Z.I., Bortolotto, Z.A., Davies, C.H., Berretta, N., Irving, A.J., Seal, A.J., Henley, J.M., Jane, D.E., Watkins, J.C., and Collingridge, G.L. Induction of LTP in the hippocampus needs synaptic activation of glutamate metabotropic receptors, *Nature*, 363 (1993) 347–350.
13. Batchelor, A.M., Madge, D.J., and Garthwaite, J. Synaptic activation of metabotropic glutamate receptors in the parallel fibre-purkinje cell pathway in rat cerebellar slices, *Neuroscience*, 63 (1994) 911–915.
14. Bear, M.F. and Kirkwood, A. Neocortical long-term potentiation, *Curr. Opin. Neurobiol.*, 3 (1993) 197–202.
15. Bekenstein, J.W. and Lothman, E.W. An in-vivo study of the ontogeny of long-term potentiation (LTP) in the CA1 region and in the dentate gyrus of the rat hippocampal formation, *Dev. Brain Res.*, 63 (1991) 245–251.
16. Bekenstein, J.W. and Lothman, E.W. A comparison of the ontogeny of excitatory and inhibitory neurotransmission in the CA1 region and dentate gyrus of the rat hippocampal formation, *Dev. Brain Res.*, 63 (1991) 237–243.

17. Bekenstein, J.W. and Lothman, E.W. Dormancy of inhibitory interneurons in a model of temporal lobe epilepsy, *Science*, 259 (1993) 97–100.
18. Ben-Ari, Y. Limbic seizure and brain damage produced by kainic acid: mechanisms and relevance to human temporal lobe epilepsy, *Neuroscience*, 14 (1985) 375–403.
19. Ben-Ari, Y., Cherubini, E., and Krnjevic K. Changes in voltage dependence of NMDA currents during development, *Neurosci. Lett.*, 94 (1988) 88–92.
20. Ben-Ari, Y., Tremblay, E., Ottersen, O.P., and Meldrum, B.S. The role of epileptic activity in hippocampal and "remote" cerebral lesions induced by kainic acid, *Brain Res.*, 191 (1980) 79–97.
21. Bliss, T.V. and Collingridge, G.L. A synaptic model of memory: Long-term potentiation in the hippocampus, *Nature*, 362 (1993) 31–39.
22. Bode-Greuel, K.M. and Singer, W. The development of N-methyl-D-aspartate receptors in cat visual cortex, *Dev. Brain Res.*, 46 (1989) 197–204.
23. Bordi, F. and Ugolini, A. Group I metabotropic glutamate receptors: implications for brain diseases, *Progr. Neurobiol.*, 59 (1999) 55–79.
24. Bortolotto, Z.A., Bashir, Z.I., Davies, C.M., and Collingridge, G.L. A molecular switch activated by metabotropic glutamate receptors regulates induction of long-term potentiation, *Nature*, 368 (1994) 740–743.
25. Brooks, W.J., Petit, T.L., and LeBoutillier, J.C. Neural development following NMDA administration in the rat: an electron microscopic examination of the occipital neocortex layer I, *Brain Res. Bull.*, 33 (1994) 621–624.
26. Brooks, W.J., Petit, T.L., LeBoutillier, J.C., and Lo, R. Rapid alteration of synaptic number and postsynaptic thickening length by NMDA: an electron microscopic study in the occipital cortex of postnatal rats, *Synapse*, 8 (1991) 41–48.
27. Burazin, T.C. and Gundlach, A.L. Rapid and transient increases in cellular immediate early gene and neuropeptide mRNAs in cortical and limbic areas after amygdaloid kindling seizures in the rat, *Epilepsy Res.*, 26 (1996) 281–293.
28. Campochiaro, P. and Coyle, J.T. Ontogenic development of kainate neurotoxicity: Correlates with glutamatergic innervation, *Proc. Natl. Acad. Sci. USA*, 75 (1978) 2025–2029.
29. Cavalheiro, E.A., Czuzwar, S.J., Kleinrok, Z., Turski, L., and Turski, W.A. Intracerebral cholinomimetics produce seizure-related brain damage in rats, *Brit. J. Pharmacol.*, 79 (1983) 284P.
30. Cavalheiro, E.A., Silva, D.F., Turski, W.A., Calderazzo-Filho, L.S., Bortolotto, Z.A., and Turski, L. The susceptibility of rats to pilocarpine-induced seizures is age-dependent, *Dev. Brain Res.*, 37 (1987) 43–58.
31. Cavanagh, J.B. and Meyer, A. Aetiological aspects of Ammon's horn sclerosis associated with temporal lobe epilepsy, *Brit. Med. J.*, 2 (1956) 1403–1407.
32. Chahal, H., D'Souza, S.W., Barson, A.J., and Slater, P. Modulation by magnesium of N-methyl-D-aspartate receptors in developing human brain, *Arch. Dis. Child Fetal Neonatal Ed.*, 78 (1998) F116–F120.
33. Chapman, A.G. Glutamate receptors in epilepsy, *Progr. Brain Res.*, 116 (1998) 371–383.
34. Chapman, A.G., Smith, S.E., and Meldrum, B.S. The anticonvulsant effect of the non-NMDA antagonists, NBQX and GYKI 52466, in mice, *Epilepsy Res.*, 9 (1991) 92–96.
35. Chapman, P.F., Kairiss, W.W., Keenan, C.L., and Brown, T.H. Long-term synaptic potentiation in the amygdala, *Synapse*, 6 (1990) 271–278.
36. Chevrie. J.J. and Aicardi, J. Childhood epileptic encephalopathy with slow spike-wave. A statistical study of 80 cases, *Epilepsia*, 13 (1972) 259–271.
37. Chiamulera, C., Costa, S., and Reggiani, A. Effect of NMDA- and strychnine-insensitive glycine site antagonists on NMDA-mediated convulsions and learning, *Psychopharmacology*, 102 (1990) 551–552.
38. Choi, D.W. Ionic dependence of glutamate neurotoxicity, *J. Neurosci.*, 7 (1987) 369–379.
39. Choi, D.W. Calcium-mediated neurotoxicity: Relationship to specific channel types and role in ischemic damage, *Trends Neurosci.*, 11 (1988) 465–469.
40. Choi, D.W. Glutamate toxicity and diseases of the nervous system, *Neuron.*, 1 (1988) 623–624.
41. Colley, P.A. and Routtenberg, A. Long-term potentiation as synaptic dialogue, *Brain Res. Rev.*, 18 (1993) 115–122.
42. Collingridge, G.L. and Lester, R.J. Excitatory amino acid receptors in the vertebrate central nervous system, *Pharmacol. Rev.*, 41 (1989) 143–210.
43. Collins, A.L. and Lennox, W.G. The intelligence of 300 private epileptic patients, *Proc. Assoc. Res. Nerv. Ment. Dis.*, 26 (1947) 586–603.
44. Constantine-Paton, M., Cline, H.T., and Debski, E. Patterned activity, synaptic convergence, and the NMDA receptor in developing visual pathways, *Ann. Rev. Neurosci.*, 13 (1990) 129–154.

45. Corsellis, J.A.N. and Bruton, C.J. Neuropathology of status epilepticus in humans. In: A.V. Delgado-Escueta, C.G. Wasterlain, D.M. Treiman, and R.J. Porter (Eds.): *Status Epilepticus*, Raven Press, New York (1983) pp. 129–139.
46. Cotman, C.W., Bridges, R.J., Taube, J.S., Clarke, A.S., Geddes, J.W., and Monanhan, D.T. The role of the NMDA receptor in central nervous system plasticity and pathology, *J. NIH Res.*, 1989:65–74.
47. Cotman, C.W. and Monaghan, D.T. Anatomical organization of excitatory amino acid receptors and their properties. In: R. Schwarcz and Y. Ben-Ari (Eds.): *Excitatory Amino Acids and Epilepsy*, Plenum Press, New York (1986) pp. 237–252.
48. Crépel, V., Aniksztejn, L., Ben-Ari, Y., and Hammond, C. Glutamate metabotropic receptors increase a Ca^{2+}-activated nonspecific cationic current in CA1 hippocampal neurons, *J. Neurophysiol.*, 72 (1994) 1561–1569.
49. Cronin, J. and Dudek, F.E. Chronic seizures and collateral sprouting of dentate mossy fibers after kainic acid treatment in rats, *Brain Res.*, 474 (1988) 181–184.
50. Danscher, G. Histochemical demonstration of heavy metals. A revised version of the sulphide silver method suitable for both light and electron microscopy, *Histochem.*, 71 (1981) 1–16.
51. Davis, S., Butcher, S.P., and Morris, R.G.M. The NMDA receptor antagonist D-2-amino-5-phophonopentanoate (D-AP5) impairs spatial learning and LTP in vivo at intracerebral concentrations comparable to those that block LTP in vitro, *J. Neurosci.*, 12 (1992) 21–34.
52. Daw, N.W., Stein, P.S.G., and Fox, K. The role of NMDA receptors in information processing, *Ann. Rev. Neurosci.*, 16 (1993) 207–222.
53. Dietrich, D., Kral, T., Clusmann, H., Friedl, M., and Schramm, J. Reduced function of L-AP4-sensitive metabotropic glutamate receptors in human epileptic sclerotic hippocampus, *Eur. J. Neurosci.*, 11 (1999) 1109–1113.
54. Dikmen, S., Matthews, C.G., and Harley, J.P. Effect of early versus late onset of major motor epilepsy upon cognitive-intellectual performance, *Epilepsia*, 16 (1975) 73–81.
55. Dikmen, S., Matthews, C.G., and Harley, J.P. Effect of early versus late onset of major motor epilepsy upon cognitive-intellectual performance: Further considerations, *Epilepsia*, 18 (1977) 31–36.
56. Dodd, P.R. and Bradford, H.F. Release of amino acids from the maturing cobalt-induced epileptic focus, *Brain Res.*, 111 (1976) 377–388.
57. Dudek, F.E. and Spitz, M. Hypothetical mechanisms for the cellular and neurophysiologic basis of secondary epileptogenesis: proposed role of synaptic reorganization, *J. Clin. Neurophysiol.*, 14 (1997) 90–101.
58. Falconer, M.A., Serafetinides, E.A., and Corsellis, J.A.N. Etiology and pathogenesis of temporal lobe epilepsy, *Arch. Neurol.*, 10 (1964) 233–248.
59. Fariello, G.G., Golden, G.T., Smith, G.G., and Reyes, P.F. Potentiation of kainic acid epileptogenicity and sparing from neuronal damage by an NMDA receptor antagonist, *Epilepsy Res.*, 3 (1989) 206–213.
60. Farwell, J.R., Dodrill, C.B., and Batzel, L.W. Neuropsychological abilities of children with epilepsy, *Epilepsia*, 26(5) (1985) 395–400.
61. Franck, J.E. and Schwartzkroin, P.A. Immature rabbit hippocampus is damaged by systemic but not intraventricular kainic acid, *Dev. Brain Res.*, 13 (1984) 219–227.
62. Frandsen, A., Drejer, J., and Schousboe, A. Direct evidence that excitotoxicity in cultured neurons is mediated via N-methyl-D-aspartate (NMDA) as well as non-NMDA receptors, *J. Neurochem.*, 53 (1989) 297–299.
63. French, J.A., Williamson, P.D., Thadani, V.M., Darcey, T.M., Mattson, R.H., Spencer, S.S., and Spencer, D.D. Characteristics of medial temporal lobe epilepsy: I. Results of history and physical examination, *Ann. Neurol.*, 34 (1993) 774–780.
64. Fujikawa, D.G., Dwyer, B.E., and Wasterlain, C.G. Preferential blood flow to brainstem during generalized seizures in the newborn marmoset monkey, *Brain Res.*, 397 (1986) 61–72.
65. Funakoshi, A., Morikawa, T., Muramatsu, R., Yagi, K., and Seino, M. A propsective WISC-R study in children with epilepsy, *Jpn. J. Psychiatry Neurol.*, 42(3) (1988) 562–564.
66. Gall, C.M. Seizure-induced changes in neurotrophin expression: implications for epilepsy, *Exp. Neurol.*, 124, (1993) 150–166.
67. Glaum, S.R. and Miller, R.J. Metabotropic glutamate receptors mediate excitatory transmission in the nucleus of the solitary tract, *J. Neurosci.*, 12 (1992) 2251–2258.
68. Golarai, G., Cavazos, J., and Sutula, T. Activation of dentate gyrus by pentylenetetrazol evoked seizures induces mossy fiber synaptic reorganization, *Brain Res.*, 593 (1992) 257–264.

69. Gorter, J.A., Titulaer, M., Bos, N.P.A., and Huisman, E. Chronic neonatal MK-801 administration leads to a long-lasting increase in seizure sensitivity during the early stages of hippocampal kindling, *Neurosci. Lett.*, 134 (1991) 29–32.
70. Grattan-Smith, J.D., Harvey, A.S., Desmond, P.M., and Chow, C.W. Hippocampal sclerosis in children with intractable temporal lobe epilepsy: detection with MR imaging, *AJR*, 161 (1993) 1045–1048.
71. Gustafsson, B., Wigstrom, H., Abraham, W.C., and Huang, Y.Y. Long-term potentiation in the hippocampus using depolarizing current pulses as the conditioning stimulus to single volley synaptic potentials, *J. Neurosci.*, 7 (1987) 774–780.
72. Hammon, B. and Heinemann, U. Developmental changes in neuronal sensitivity to excitatory amino acids in area CA1 of the rat hippocampus, *Dev. Brain Res.*, 38 (1988) 286–290.
73. Harrigan, T., Peredery, O., and Persinger, M.A. Failure to acquire an inhibitory task following seizure-induced brain damage, *Percep. Motor Skills*, 70 (1990) 268–270.
74. Harris, K.M. and Teyler, T.J. Developmental onset of long-term potentiation in area CA1 of the rat hippocampus, *J. Physiol.*, 346 (1984) 27–48.
75. Hattori, H. and Wasterlain, C.G. Excitatory amino acids in the developing brain: ontogeny, plasticity, and excitotoxicity, *Pediatr. Neurol.*, 60 (1990) 219–228.
76. Hirsch, E., Baram, T.Z., and Snead, O.C., III. Ontogenic study of lithium-pilocarpine-induced status epilepticus in rats, *Brain Res.*, 583 (1992) 120–126.
77. Holmes, G.L. *Diagnosis and Management of Seizures in Children.* W.B. Saunders Company, Philadelphia, (1987) pp. 1–293.
78. Holmes, G.L. The long-term effects of seizures on the developing brain: clinical and laboratory issues, *Brain Dev.*, 13 (1991) 393–409.
79. Holmes, G.L. Do seizures cause brain damage? *Epilepsia*, 32 (Suppl. 5) (1991) S14–S28.
80. Holmes, G.L. and Ben-Ari, Y. Seizures in the developing brain: perhaps not so benign after all, *Neuron*, 21 (1998) 1231–1234.
81. Holmes, G.L., Chronopoulos, A., Stafstrom, C.E., Mikati, M.A., Thurber, S.J., Hyde, P.A., and Thompson, J.L. Effects of kindling on subsequent learning, memory, behavior, and seizure susceptibility, *Develop. Brain Res.*, 73 (1993) 71–77.
82. Holmes, G.L., Gaiarsa, J.L., Chevassus-Au-Louis, N., and Ben-Ari, Y. Consequences of neonatal seizures in the rat: morphological and behavioral effects, *Ann. Neurol.*, 44 (1998) 845–857.
83. Holmes, G.L., Sarkisian, M., Ben-Ari, Y., and Chevassus-Au-Louis, N. Effects of recurrent seizures in the developing brain. In: A. Nehlig, J. Motte, S.L. Moshe, and P. Plouin (Eds.): *Childhood Epilepsies and Brain Development*, John Libbey, London (1999) pp. 263–276.
84. Holmes, G.L., Sarkisian, M., Ben-Ari, Y., and Chevassus-Au-Louis, N. Mossy fiber sprouting after recurrent seizures during early development in rats, *J. Comp. Neurol.*, 404 (1999) 437–553.
85. Holmes, G.L. and Thompson, J.L. Effects of kainic acid on seizure susceptibility in the developing brain, *Brain Res.*, 467 (1988) 51–59.
86. Holmes, G.L., Thompson, J.L., Marchi, T., and Feldman, D.S. Behavioral effects of kainic acid administration on the immature brain, *Epilepsia*, 29 (1988) 721–730.
87. Holmes, G.L., Thompson, J.L., Marchi, T.A., Gabriel, P.S., Hogan, M.A., Carl, F.G., and Feldman, D.S. Effects of seizures on learning, memory, and behavior in the genetically epilepsy-prone rat, *Ann. Neurol.*, 29 (1990) 24–32.
88. Holmes, G.L., Thurber, S.T., Liu, Z., Stafstrom, C.E., Gatt, A.M., and Mikati, M.A. Effects of quisqualic acid and glutamate on subsequent learning, emotionality, and seizure susceptibility in the immature and mature animal, *Brain Res.*, 623 (1993) 325–328.
89. Hori, N., Ffrench-Mullen, J.M.H., and Carpenter, D.O. Kainic acid response and toxicity show pronounced Ca^{2+} dependence, *Brain Res.*, 358 (1985) 380–384.
90. Insel, T.R., Miller, L.P., and Gelhard, R.E. The ontogeny of excitatory amino acid receptors in rat forebrain—I. N-Methyl-D-Aspartate and quisqualate receptors, *Neuroscience*, 35 (1990) 31–43.
91. Iriki, A., Pavlides, C., Keller, A., and Asanuma, H. Long-term potentiation in the motor cortex, *Science*, 245 (1989) 1385–1387.
92. Jackson, P.S., Suppes, T., and Harris, K.M. Stereotypical changes in the pattern and duration of long-term potentiation expressed at postnatal days 11 and 15 in the rat hippocampus, *J. Physiol.*, 70 (1993) 1412–1419.
93. Jensen, F.E., Alvarado, S., Firkusny, I.R., and Geary, C. NBQX blocks the acute and late epileptogenic effects of perinatal hypoxia, *Epilepsia*, 36 (1995) 996–972.
94. Kanter, E.D. and Haberly, L.B. NMDA-dependent induction of long-term potentiation in afferent and association fiber systems of piriform cortex in vitro, *Brain Res.*, 525 (1990) 175–179.

95. Kelso, S.R. and Brown, T.H. Differential conditioning of associative synaptic enhancement in hippocampal brain slices, *Science*, 232 (1986) 85–87.
96. Kiessling, M. and Gass, P. Immediate early gene expression in experimental epilepsy, *Brain Pathol.*, 3 (1993) 381–393.
97. Koh, J.-Y., Goldberg, M.P., Hartley, D.M., and Choi, D.W. Non-NMDA receptor-mediated neurotoxicity in cortical culture, *J. Neurosci.*, 7 (1990) 693–705.
98. Kudo, Y. and Ogura, A. Glutamate-induced increase in intracellular Ca++ concentration in isolated hippocampal neurons, *Brit. J. Pharmacol.*, 89 (1986) 191–198.
99. Kvamme, E., Svenneby, G., Torgner, I.A.A., Drejer, J., and Schousboe, A. Postnatal development of glutamate metabolizing enzymes in hippocampus from mice, *Int. J. Development Neurosci.*, 3 (1985) 359–364.
100. Lallement, G., Carpentier, P., Collet, A., Pernot-Marino, I., Baubichon, D., and Blanchet, G. Effects of soman-induced seizures on different extracellular amino acid levels and on glutamate uptake in rat hippocampus, *Brain Res.*, 563 (1991) 234–240.
101. Lallement, G., Carpentier, P., Collet, A., Pernot-Marino, I., Baubichon, D., Sentenac-Roumanou, H., and Blanchet, G. Involvement of glutamatergic system of amygdala in generalized seizures induced by soman: comparison with the hippocampus, *Comptes Rendus de l'Academie des Sciences*, 313 (1991) 421–426.
102. Lees, G.J. and Leong, W. The non-NMDA glutamate antagonist NBQX blocks the local hippocampal toxicity of kainic acid, but not the diffuse extrahippocampal damage, *Neurosci. Lett.*, 143 (1992) 39–42.
103. Lehmann, A. Alterations in hippocampal extracellular amino acids and purine catabolites during limbic seizures induced by folate injections into the rabbit amygdala, *Neuroscience*, 22 (1987) 573–578.
104. Lehmann, A., Hagberg, H., Jacobson, I., and Hamberger, A. Effects of status epilepticus on extracellular amino acids in the hippocmpus, *Brain Res.*, 359 (1985) 147–151.
105. Leung, L.S., Boon, K.A., Kaibara, T., and Innis, N.K. Radial maze performance following hippocampal kindling, *Behav. Brain Res.*, 40 (1990) 119–129.
106. Leung, L.S. and Shen, B. Hippocampal CA1 evoked response and radial 8-arm maze performance after hippocampal kindling, *Brain Res.*, 555 (1991) 353–357.
107. Lewin, A.H. Sun, G., Fudala, L., Navarro, H., Zhou L.M., Popik, P., Faynsteyn, A., and Skolnick, P. Molecular features associated with polyamine modulation of NMDA receptors, *J. Med. Chem.*, 41 (1998) 988–995.
108. Lipton, S.A. Prospects for clinically tolerated NMDA antagonists: open-channel blockers and alternative redox states of nitric oxide, *Trends Neurosci.*, 16 (1993) 527–532.
109. Lipton, S.A. and Rosenberg, P.A. Excitatory amino acids as a final common pathway for neurologic disorders, *N. Engl. J. Med.*, 330 (1994) 613–622.
110. Lipton, S.A., Singel, D.J., and Stamler, J.S. Nitric oxide in the central nervous system, *Progr. Brain Res.*, 103 (1994) 359–364.
111. Liu, Z., Gatt, A., Mikati, M., and Holmes, G.L. Long-term behavioral deficits following pilocarpine seizures in immature rats, *Epilepsy Res.*, 19 (1995) 191–204.
112. Liu, Z., Mikati, M., and Holmes, G.L. Mesial temporal sclerosis: pathogenesis and significance, *Pediatr. Neurol.*, 12 (1995) 5–16.
113. Liu, Z., Stafstrom, C.E., Sarkisian, M., Tandon, P., Yang, Y., Hori, A., and Holmes, G.L. Age-dependent effects of glutamate toxicity in the hippocampus, *Dev. Brain Res.*, 97 (1996) 178–184.
114. Liu, Z., Yang, Y., Silveira, D.C., Sarkisian, M.R., Tandon, P., Huang, L.T., Stafstrom, C.E., and Holmes, G.L. Consequences of recurrent seizures during early brain development, *Neuroscience*, 92 (1999) 1443–1454.
115. Lopes da Silva, F.H., Gorter, J.A., and Wadman, W.J. Kindling of the hippocampus induces spatial memory deficits in the rat, *Neurosci. Lett.*, 63 (1986) 115–120.
116. Lucas, D.R. and Newhouse, J.P. The toxic effects of sodium L-glutamate on the inner layers of the retina, *Arch. Opthalmol.*, 58 (1957) 193–201.
117. MacDonald, J.F., Xiong, X.G., Lu, W.Y., Raouf, R., and Orser, B.A. Modulation of NMDA receptors, *Progr. Brain Res.*, 116 (1998) 191–208.
118. Malinow, R. and Miller, J.P. Postsynaptic hyperpolarization during conditioning reversibly blocks induction of long-term potentiation, *Nature*, 321 (1986) 529–530.
119. Mares, P. Excitatory amino acids and epileptogenesis during ontogenesis. In: A. Nehlig, J. Motte, S.L. Moshe, and P. Plouin (Eds.): *Childhood Epilepsies and Brain Development*, John Libbey, London, (1999) pp. 157–160.

120. Mathern, G.W., Leite, J.P., Pretorius, J.K., Quinn, B., Peacock, W.J., and Babb, T.L. Children with severe epilepsy: evidence of hippocampal neuron losses and aberrant mossy fiber sprouting during postnatal granule cell migration and differentiation, *Dev. Brain Res.*, 78 (1994) 70–80.
121. Mathern, G.W., Leite, J.P., Pretorius, J.K., Quinn, B., Peacock, W.J., and Babb, T.L. Severe seizures in young children are associated with hippocampal neuronal losses and aberrant mossy fiber sprouting during fascia dentata postnatal development, *Epilepsy Res. Supplement*, 12 (1996) 33–43.
122. McDonald, J.W., Fix, A.S., Tizzano, J.P., and Schoepp, D.D. Seizures and brain injury in neonatal rats induced by 1S,3R-ACPD, a metabotropic glutamate receptor agonist, *J. Neurosci.*, 13 (1993) 4445–4455.
123. McDonald, J.W. and Johnston, M.V. Pharmacology of N-methyl-D-aspartate-induced brain injury in an in vivo perinatal rat model, *Synapse*, 6 (1990) 179–188.
124. McDonald, J.W. and Johnston, M.V. Physiological and pathophysiological roles of excitatory amino acids during central nervous system development, *Brain Res. Rev.*, 15 (1990) 41–70.
125. McDonald, J.W., Silverstein, F.S., Cardona, D., Hudson, C., Chen, R., and Johnston, M.V. Systemic administration of MK-801 protects against N-methy-D-aspartate- and quisqualate-mediated neurotoxicity in perinatal rats, *Neuroscience*, 36 (1990) 589–599.
126. McDonald, J.W., Silverstein, F.S., and Johnston, M.V. Neurotoxicity of N-methyl-D-aspartate is markedly enhanced in developing rat central nervous system, *Brain Res.*, 459 (1988) 200–203.
127. McDonald, J.W., Silverstein, F.S., and Johnston, M.V. MK-801 pretreatment enhances N-methyl-D-aspartate-mediated brain injury and increases brain N-methyl-D-aspartate recognition site binding in rats, *Neuroscience*, 38 (1990) 103–113.
128. McDonald, J.W., Trescher, W.H., and Johnston, M.V. The selective ionotropic-type quisqualate receptor agonist AMPA is a potent neurotoxin in immature rat brain, *Brain Res.*, 526 (1990) 165–168.
129. McDonald, J.W., Trescher, W.H., and Johnston, M.V. Susceptibility of brain to AMPA induced excitotoxicity transiently peaks during early postnatal development, *Brain Res.*, 583 (1992) 54–70.
130. Meldrum, B. Excitatory amino acid transmitters in epilepsy, *Epilepsia*, 32 (1991) S1–S3.
131. Meldrum, B. and Garthwaite, J. Excitatory amino acid neurotoxicity and neurodegenerative disease, *Trends Pharmacol. Sci.*, 11 (1990) 379–387.
132. Meldrum, B.S. and Brierley, J.B. Prolonged epileptic seizures in primates: Ischaemic cell change and its relation to ictal physiological events, *Arch. Neurol.*, 28 (1973) 10–17.
133. Meldrum, B.S. and Horton, R.W. Physiology of status epilepticus in primates, *Arch. Neurol.*, 28 (1973) 1–9.
134. Meldrum, B.S., Vigouroux, R.A., and Brierley, J.B. Systemic factors and epileptic brain damage. Prolonged seizures in paralysed artificially ventilated baboons, *Arch. Neurol.*, 29 (1973) 82–87.
135. Millan, M.H., Chapman, A.G., and Meldrum, B.S. Extracellular amino acid levels in hippocampus during pilocarpine-induced seizures, *Epilepsy Res.*, 14 (1993) 139–148.
136. Miller, L.P., Johnson, A.E., Gelhard, R.E., and Insel, T.R. The ontogeny of excitatory amino acid receptors in the rat forebrain—II. Kainic acid receptors, *Neuroscience*, 35 (1990) 45–51.
137. Minamoto, Y., Itano, T., Tokuda, M., Matsui, H., Janjua, N.A., Hosokawa, K., Okada, Y., Murakami, T.H., Negi, T., and Hatase, O. In vivo microdialysis of amino acid neurotransmitters in the hippocampus in amygdaloid kindled rat, *Brain Res.*, 573 (1992) 345–348.
138. Minc-Golomb, D., Levy, Y., Kleinberger, N., and Schramm, M. D-[^{3}H]Aspartate release from hippocampus slices studied in a multiwell system: controlling factors and postnatal development of release, *Brain Res.*, 402 (1987) 255–263.
139. Monaghan, D.T., Bridges, R.J., and Cotman, C.W. The excitatory amino acid receptors: their classes, pharmacology, and distinct properties in the function of the central nervous system, *Ann. Rev. Pharmacol. Toxicol.*, 29 (1989) 365–402.
140. Monaghan, D.T. and Cotman, C.W. Distribution on N-methyl-D-aspartate sensitive L-[3H]glutamate binding sites in rat brain, *J. Neurosci.*, 5 (1985) 2909–2919.
141. Monaghan, D.T., Yao, D., and Cotman, C.W. L-[^{3}H] glutamate binds to kainate-, NMDA- and AMPA-sensitive binding sites: an autoradiographic analysis, *Brain Res.*, 340 (1985) 378–383.
142. Moncada, C., Arvin, B., Le Peillet, E., and Meldrum, B.S. Non-NMDA antagonists protect against kainate more than AMPA toxicity in the rat hippocampus, *Neurosci. Lett.*, 133 (1991) 287–290.
143. Morris, R.G.M., Anderson, E., Lynch, G.S., and Baudry, M. Selective impairment of learning and blockade of long-term potentiation by an N-methyl-D-aspartate receptor antagonist, AP5, *Nature*, 319 (1986) 774–776.
144. Moshé, S.L. Epileptogenesis and the immature brain, *Epilepsia*, 28 (Suppl. 1) (1987) S3–S15.
145. Muller, D., Oliver, M., and Lynch, G. Developmental changes in synaptic properties in hippocampus of neonatal rats, *Dev. Brain Res.*, 49 (1989) 105–114.

146. Nadler, J.V. Kainic acid as a tool for the study of temporal lobe epilepsy, *Life Sci.*, 29 (1981) 2031–2042.
147. Nadler, J.V., Perry, B.W., and Cotman, C.W. Intraventricular kainic acid preferentially destroys hippocampal pyramidal cells, *Nature*, 271 (1978) 676–677.
148. Nahm, W.K. and Noebels, J.L. Nonobligate role of early or sustained expression of immediate early gene proteins c-fos, c-jun, and Zif/286 in hippocampal mossy fiber sprouting, *J. Neurosci.*, 18 (1998) 9245–9255.
149. Nakanishi, S. Metabotropic glutamate receptors: synaptic transmission, modulation, and plasticity, *Neuron.*, 13 (1994) 1033–1037.
150. Neill, J.C., Liu, Z., Gatt, A., Mikati, M., and Holmes, G.L. Pilocarpine-induced seizures impair acquisition rate of auditory discrimination using location cues, *Soc. Neurosci. Abstr.*, 19 (1993) 394.
151. Neill, J.C., Liu, Z., Sarkisian, M., Tandon, P., Yang, Y., Stafstrom, C.E., and Holmes, G.L. Recurrent seizures in immature rats: effect on auditory and visual discrimination, *Dev. Brain Res.*, 95 (1996) 283–292.
152. Obenaus, A., Esclapez, M., and Houser, C.R. Loss of glutamate decarboxylase mRNA-containing neurons in the rat dentate gyrus following pilocarpine-induced seizures, *J. Neurosci.*, 13 (1993) 4470–4485.
153. Okada, Y. and Miyamoto, T. Formation of long-term potentiation in superior colliculus slices from the guinea pig, *Neurosci. Lett.*, 96 (1989) 108–113.
154. Okazaki, M.M., Evenson, D.A., and Nadler, J.V. Hippocampal mossy fiber sprouting and synapse formation after status epilepticus in rats: visualization after retrograde transport of biocytin, *J. Comp. Neurol.*, 352 (1995) 515–523.
155. Olney, J.W., Collins, R.C., and Sloviter, R.S. Excitotoxic mechanisms of epileptic brain damage, *Adv. Neurol.*, 44 (1986) 857–877.
156. Olney, J.W., De Gubareff, T., and Labruyere, J. Seizure-related brain damage induced by cholinergic agents, *Nature*, 301 (1983) 520–522.
157. Olney, J.W., De Gubareff, T., and Sloviter, R.S. "Epileptic" brain damage in rats induced by sustained electrical stimulation of the perforant path. II. Ultrastructural analysis of acute hippocampal pathology, *Brain Res. Bull.*, 10 (1983) 699–712.
158. Olton, D.S., Becker, J.T., and Handelman, G.E. Hippocampus, space, and memory, *Behav. Brain Sci.*, 2 (1979) 313–322.
159. Orrenius, S., McConkey, D., Belloma, G., and Nicoterm, P. Role of Ca[2^+] in toxic killing, *Trends Pharmacol. Sci.*, 10 (1989) 281–285.
160. Ounsted, C., Lindsay, J., and Norman, R. *Biological Factors in Temporal Lobe Epilepsy. Clinics in Developmental Medicine, no. 22.*, The Spastics Society Medical Education and Information Unit, in association with William Heinemanh Medical Books, Ltd., London (1966) pp. 1–131.
161. Parent, J.M. and Lowenstein, D.H. Mossy fiber reorganization in the epileptic hippocampus, *Curr. Opin. Neurol.*, 10 (1997) 10–109.
162. Parent, J.M., Yu, T.W., Leibowitz, R.T., Geschwind, D.H., Sloviter, R.S., and Lowenstein, D.H. Dentate granule cell neurogenesis is increased by seizures and contributes to aberrant network reorganization in the adult rat, *J. Neurosci.*, 17 (1997) 9727–3738.
163. Pellegrini-Giampietro, D.E., Gorter, J.A., Bennett, M.V.L., and Zukin, R.S. The GluR2 (GluR-B) hypothesis: Ca^{2+}-permeable AMPA receptors in neurological disorders, *Trends Neurosci.*, 20 (1997) 464–470.
164. Pisa, M., Sanberg, M.R., Corcoran, M.E., and Fibiger, H.C. Spontaneously recurrent seizures after intracerebral injections of kainic acid in rat: a possible model of human temporal lobe epilepsy, *Brain Res.*, 200 (1980) 481–487.
165. Rauschecker, J.P., Egert, U., and Kossel, A. Effects of NMDA antagonists on developmental plasticity in kitten visual cortex, *Int. J. Development Neurosci.*, 8 (1990) 425–435.
166. Represa, A., Le Gall La Salle, G., and Ben-Ari, Y. Hippocampal plasticity in the kindling model of epilepsy in rats, *Neurosci. Lett.*, 99 (1989) 345–350.
167. Ribak, C.E. and Baram, T.Z. Selective death of hippocampal CA3 pyramidal cells with mossy fiber afferents after CRH-induced status epilepticus in infant rats, *Dev. Brain Res.*, 91 (1996) 245–251.
168. Ribak, C.E. and Navetta, M.S. An immature mossy fiber innervation of hilar neurons may explain their resistance to kainate-induced cell death in 15-day-old rats, *Dev. Brain Res.*, 79 (1994) 47–62.
169. Rocca, W.A., Sharbrough, F.W., Hauser, W.A., Annegers, J.F., and Schoenberg, B.S. Risk factors for complex partial seizures: A population-based case-control study, *Ann. Neurol.*, 21 (1987) 22–31.
170. Rodin, E.A., Schmaltz, S., and Twitty, G. Intellectual functions of patients with childhood-onset epilepsy, *Developmental Medicine and Child Neurology*, 28 (1986) 25–33.

171. Rothe, F., Schmidt, W., and Wolf, G. Postnatal changes in the activity of glutamate dehydrogenase and aspartate aminotransferase in the rat nervous system with special reference to the glutamate transmitter metabolism, *Dev. Brain Res.*, 11 (1983) 67–74.
172. Rutecki, P.A. and Yang, Y. Metabotropic glutamate receptor activation modulates epileptiform activity in the hippocampus, *Neuroscience*, 81 (1997) 927–935.
173. Saccan, A.I. and Schoepp, D.D. Activation of hippocampal metabotropic excitatory amino acid receptors leads to seizures and neuronal damage, *Neurosci. Lett.*, 139 (1992) 77–82.
174. Sanes, J.R. and Lichtman, J.W. Can molecules explain long-term potentiation? *Nature Neurosci.*, 2 (1999) 597–604.
175. Sankar, R., Shin, D.H., Liu, H., Mazarati, A., Pereira de Vasconcelos, A., and Wasterlain C.G. Patterns of status epilepticus-induced neuronal injury during development and long-term consequences, *J. Neurosci.*, 18 (1998) 8382–8393.
176. Sastry, B.R., Goh, J.W., and Auyeung, A. Associative induction of posttetanic and long-term potentiation in CA1 neurons of rat hippocampus, *Science*, 232 (1986) 988–990.
177. Scheyer, R.D. Involvement of glutamate in human epileptic activities, *Progr. Brain Res.*, 116 (1998) 359–369.
178. Schmidt, W. and Wolf, G. High-affinity uptake of L-[^{3}H]glutamate and D-[^{3}H]aspartate during postnatal development of the hippocampal formation: a quantitative autoradiographic study, *Exp. Brain Res.*, 70 (1988) 50–54.
179. Schoepfer, R., Monyer, H., Sommer, B., Wisden, W., Sprengel, R., Kuner, T., Lomeli, H., Herb, A., Kohler, M., Burnashev, N., et al. Molecular biology of glutamate receptors, *Prog. Neurobiol.*, 42 (1994) 353–357.
180. Schoepp, D.D., Bockaert, J., and Sladeczek, F. Pharmacological and functional characteristics of metabotropic excitatory amino acid receptors, *Trends Pharmacol. Sci.*, 11 (1990) 508–515.
181. Sherwin, A.L., Quesney, L.F., Gauthier, S., Olivier, A., Robitaille, Y., McQuaid. P., Harvey, C., and van Gelder, N. Enzyme changes in actively spiking areas of human epileptic cerebral cortex, *Neurology*, 34 (1984) 927–933.
182. Siesjo, B.K. Historical review: calcium, ischemia, and death of brain cells, *Ann. NY Acad. Sci.*, 522 (1988) 638–661.
183. Sloviter, R.S. Decreased hippocampal inhibition and a selective loss of interneurons in experimental epilepsy, *Science*, 253 (1987) 73–76.
184. Sloviter, R.S. Permanently altered hippocampal structure, excitability, and inhibition after experimental status epilepticus in the rat: The "Dormant Basket Cell" hypothesis and its possible relevance to temporal lobe epilepsy, *Hippocampus*, 1 (1991) 41–66.
185. Sloviter, R.S. The functional organization of the hippocampal dentate gyrus and its relevance to the pathogenesis of temporal lobe epilepsy, *Ann. Neurol.*, 35 (1994) 640–654.
186. Sloviter, R.S. and Dempster, D.W. "Epileptic" brain damage is replicated qualitatively in the rat hippocampus by central injection of glutamate or aspartate but not by GABA or acetylcholine, *Brain Res. Bull.*, 15 (1985) 39–60.
187. Sperber, E.F., German, I.M., Friedman, L.K., Veliskova, J., and Romero, M.T. The resiliency of the immature brain to seizure induced damage. In: A. Nehlig, J. Motte, S.L. Moshe, and P. Plounin (Eds.): *Childhood Epilepsies and Brain Development*, John Libbey, London (1999) pp. 255–262.
188. Sperber, E.F., Haas, K.Z., Stanton, P.K., and Moshé, S.L. Resistance of the immature hippocampus to seizure-induced synaptic reorganization, *Dev. Brain Res.*, 60 (1991) 88–93.
189. Stafstrom, C.E., Holmes, G.L., Chronopoulos, A., Thurber, S., and Thompson, J.L. Age-dependent cognitive and behavioral deficits following kainic acid-induced seizures, *Epilepsia*, 34 (1993) 420–432.
190. Stafstrom, C.E., Tandon, P., Hori, A., Liu, Z., Mikati, M.A., and Holmes, G.L. Acute effects of MK801 on kainic acid-induced seizures in neonatal rats, *Epilepsy Res.*, 26 (1997) 335–344.
191. Stafstrom, C.E., Thompson, J.L., and Holmes, G.L. Kainic acid seizures in the developing brain: status epilepticus and spontaneous recurrent seizures, *Dev. Brain Res.*, 65 (1992) 237–246.
192. Stanfield, B.B. Excessive intra- and supragranular mossy fibers in the dentate gyrus of tottering (tg/tg) mice, *Brain Res.*, 480 (1989) 294–299.
193. Sugiyama, H., Ito, I., and Hirono, C. A new type of glutamate receptor linked to inositol phospholipid metabolism, *Nature*, 325 (1987) 531–533.
194. Sugiyama, H., Ito, I., and Watanabe, M. Glutamate receptor subtypes may be classified into two major categories: a study on Xenopus oocytes injected with rat brain mRNA, *Neuron*, 3 (1989) 129–132.
195. Sutor, B. and Hablitz, J.J. Long-term potentiation in frontal cortex: role of NMDA-modulated polysynaptic excitatory pathways, *Neurosci. Lett.*, 97 (1989) 111–117.

196. Sutula, T., Cascino, G., Cavazos, J., Parada, I., and Ramirez, L. Mossy fiber synaptic reorganization in the epileptic human temporal lobe, *Ann. Neurol.*, 26 (1989) 321–330.
197. Sutula, T., Koch, J., Golarai, G., Watanabe, Y., and McNamara, J.O. NMDA receptor dependence of kindling and mossy fiber sprouting: evidence that the NMDA receptor regulates patterning of hippocampal circuits in the adult brain, *J. Neurosci.*, 16 (1996) 7398–7406.
198. Sutula, T., Lauersdorf, S., Lynch, M., Jurgella, C., and Woodard, A. Deficits in radial arm maze performance in kindled rats: evidence for long-lasting memory dysfunction induced by repeated brief seizures, *J. Neurosci.*, 15 (1995) 8295–8301.
199. Sutula, T., Xiao-Xian, H., Cavazos, J., and Scott, G. Synaptic reorganization in the hippocampus induced by abnormal functional activity, *Science*, 239 (1988) 1147–1150.
200. Sutula, T.P. Experimental models of temporal lobe epilepsy: new insights from the study of kindling and synaptic reorganization, *Epilepsia*, 31 (Suppl. 3) (1990) S45–S54.
201. Sutula, T.P. The pathology of the epilepsies: Insights into the causes and consequences of epileptic syndromes. In: W.E. Dodson and J.M. Pellock (Eds.): *Pediatric Epilepsy: Diagnosis and Treatment*, Demos Publications, New York (1993) pp. 37–44.
202. Tanabe, Y., Masu, M., Ishii, T., Shigemoto, R., and Nakanishi, S. A family of metabotropic glutamate receptors, *Neuron*, 8 (1992) 169–179.
203. Tandon, P., Liu, Z., Stafstrom, C.E., Sarksian, M., Werner, S.J., Mikati, M., Yang, Y., and Holmes, G.L. Long-term effects of excitatory amino acid antagonists NBQX and MK-801 on the developing brain, *Dev. Brain Res.*, 95 (1996) 256–262.
204. Tauck, D. and Nadler, J.V. Evidence of functional mossy fiber sprouting in the hippocampal formation of kainic acid-treated rats, *J. Neurosci.*, 5 (1985) 1016–1022.
205. Thurber, S., Chronopoulos, A., Stafstrom, C.E., and Holmes, G.L. Behavioral effects of continuous hippocampal stimulation in the developing rat, *Dev. Brain Res.*, 68 (1992) 35–40.
206. Tremblay, E., Represa, A., and Ben-Ari, Y. Autoradiographic localization of kainic acid binding sites in the human hippocampus, *Brain Res.*, 343 (1985) 378–382.
207. Tremblay, E., Roisin, M.P., Represa, A., Charriaut-Marlangue, C., and Ben-Ari, Y. Transient increased density of NMDA binding sites in the developing rat hippocampus, *Brain Res.*, 461 (1988) 393–396.
208. Turski, L., Ikonomidou, C., Turski, W.A., Bortolotto, Z.A., and Cavalheiro, E.A. Review: Cholinergic mechanisms and epileptogenesis. The seizures induced by pilocarpine: A novel experimental model of intractable epilepsy, *Synapse*, 3 (1989) 154–171.
209. Turski, W.A., Cavalheiro, E.A., Schwarz, M., Czuczwar, S.J., Kleinrok, Z., and Turski, L. Limbic seizures produced by pilocarpine in rats: behavioral, electroencephalographic and neuropathological study, *Behav. Brain Res.*, 9 (1983) 315–335.
210. Tursky, T., Lassanova, M., Sramka, M., and Nadvornik, P. Formation of glutamate and GABA in epileptogenic tissue from human hippocampus in vitro, *Acta Neurochir.*, 23 (1976) 111–118.
211. Ungerer, A., Mathis, C., Melan, C., and De Barry, J. Role of neuroexcitatory amino acids in memory processes. Study with gamma-L-glutamyl-L-aspartic acid, *Encephale*, 16 (1990) 423–429.
212. Ungerer, A., Mathis, C., Nélan, C., and De Barry, J. The NMDA receptor antagonists, CPP and gamma-L-glutamyl-L-aspartate, selectively block post-training improvement of performance in a Y-maze avoidance learning task, *Brain Res.*, 549 (1991) 59–65.
213. Van Landingham, K.E., Heinz, E.R., Cavazos, J.E., and Lewis, D.V. Magnetic resonance imaging evidence of hippocampal injury after prolonged febrile convulsions, *Ann. Neurol.*, 43 (1998) 413–426.
214. Wasterlain, C.G., Fujikawa, D.G., Penix, L., and Sankar, R. Pathophysiological mechanisms of brain damage from status epilepticus, *Epilepsia*, 34 (Suppl. 1) (1993) S37–S53.
215. Wasterlain, C.G. and Sankar, R. Excitotoxicity and the developing brain. In: G. Avanzini, R. Fariello, U. Heinemann, and R. Mutani (Eds.): *Epileptogenic and excitotoxic mechanisms*, John Libbey, London (1993) pp. 135–151.
216. Williamson, P.D., French, J.A., Thadani, V.M., Kim, J.H., Novelly, R.A., Spencer, S.S., Spencer, D.D., and Mattson, R.H. Characteristics of medial temporal lobe epilepsy: II. Interictal and ictal scalp electroencephalography, neuropsychological testing, neuroimaging, surgical results, and pathology, *Ann. Neurol.*, 34 (1993) 781–787.
217. Zimmerman, H.M. The histopathology of convulsive disorders in children, *J. Pediatr.*, 13 (1941) 859–890.

18

THE LANDAU-KLEFFNER SYNDROME

Hanna Kolski and Hiroshi Otsubo

Neurology
The Hospital for Sick Children
555 University Avenue
Toronto, Ontario, CANADA M5G 1X8
tel: (416) 813-6660
fax: (416) 813-6334

Landau-Kleffner Syndrome is a rare childhood disorder which involves seizures and acquired aphasia. Anticonvulsants, or the passage of time, may control the seizures, but speech recovery is variable, and the aphasia may persist.

INTRODUCTION

Landau and Kleffner, in 1957, were the first to describe an idiopathic syndrome consisting of acquired aphasia, seizures and paroxysmal EEG abnormalities.[72] Since their original description of "epileptic aphasia" in 6 children, over 200 cases of Landau-Kleffner Syndrome (LKS) have been reported[5,37,86,89] with various clinical and neurophysiological features.[31,121]

CLINICAL CHARACTERISTICS

The first signs occur between 3–8 years of age and consist of either language dysfunction (more than 50% of cases)[3] or clinical seizures (40–50%).[5,104] Aphasia is apparent before age 6 years in approximately 70% of cases,[5] but may appear after 8 years (in less than 10%).[37] Affected males outnumber females 2 : 1.[3,5,15]

Language Deficits

Disturbances occur eventually in all aspects of language, with comprehension generally affected earlier and more profoundly than speech production.[21,31,39,73,144] In rare cases, however, expressive rather than comprehensive deficits predominate.[12,31,32,40,47,127] The condition is often mistaken for acquired deafness.[31,114]

The typical progression initially involves receptive disturbances (85% of cases)[5] with loss of comprehension for both nonverbal (i.e., dog barking, doorbell ring) and verbal sounds[90] (so called "word deafness").[23,104,114] This has been considered a disorder of auditory processing[12] or verbal auditory agnosia.[5,23,113,114,144] Shortly thereafter, expressive speech is affected, with paraphasias, phonological errors, shortened sentences or articulation problems.[104,127,139] Eventually, the child may appear mute[21,127] and globally aphasic.[139] Reading and writing, if previously acquired, are lost in some[32,33,47,78,104,144] but not all cases.[28,114,127] Generally, the onset is subacute or fluctuating over days to months[3,55] with several exacerbations and remissions throughout its course.[9,29,34,54,104,114] If there is language recovery, receptive abilities improve first, followed by expressive speech.[32,70]

Clinical Seizures

Clinical seizures occur in 70–80% of patients with LKS.[5,60,92,104] They tend to be mild,[70] and their onset may occur before (20%)[5] during, or after (20–30%) the onset of the language disturbance.[3,55] Various seizure types have been reported in LKS. Focal motor, generalized tonic-clonic, atypical absence and atonic seizures are commonly cited.[5,31] Complex partial and myoclonic seizures are infrequent[5,72] while tonic seizures have never been recorded.[5,31,79,86] The frequency of clinical seizure occurrence varies,[104] with seizures tending to be few and nocturnal.[31,38,55] Most clinical seizures respond readily to conventional anticonvulsants[45,84,90,104] though refractory cases have been reported.[21,67,86]

EEG Patterns

EEG findings are consistently abnormal in LKS, regardless of the occurrence of clinical seizures.[55] In waking patients, EEG records may show a normal background,[60] and commonly, there are high amplitude rhythmic spikes or spike and slow-wave discharges (frequency of 1–3 Hz) present unilaterally or bilaterally over the temporal (greater than 1/2 cases), temporoparietal or parieto-occipital regions (1/3 cases).[3,5,21,31] Anterior and midtemporal foci are rare.[5] At times, the paroxysms appear multifocal[31,127] or generalized.[60,118] Both the waking and sleep EEG's tend to be highly variable;[31] the morphology, abundance and topography change frequently (even within an individual patient).[5,11,17,60] EEG abnormalities are invariably activated by non-REM sleep[38,60,67] particularly at sleep onset.[5,117] During the evolution of the syndrome, spike and slow-wave discharges become more bilateral, symmetrical, diffuse, and continuous.[3,5,31,38,60] This pattern may have the appearance of continuous spike waves during slow wave sleep (CSWS)[38,88,117,118] also known as subclinical "electrical status epilepticus induced by sleep"[108] or "electrical status epilepticus during slow sleep" (ESES)[51,68,117,132] (see below).

Table 1. Cardinal Symptoms of Landau-Kleffner Syndrome

Language Deficit (Acquired aphasia)
Clinical Seizures (Including focal motor, tonic-clonic atypical absence and atonic)
EEG Abnormalities (especially ESES during sleep)

CSWS/ ESES are characterized by a pattern of continuous, diffuse spikes and waves occupying at least 85% of slow-wave sleep.[88] Although in LKS discharges are usually bilateral, they may also be focal or unilaterally hemispheric during a given night.[60] Overall, EEG abnormalities tend to persist, but there may be silent periods with normal recordings.[29,64]

Interrelationships

The relationship between the language deficits, clinical seizures, and EEG abnormalities in LKS continues to be ambiguous.[32,86] Several reports cite aphasia progressing or fluctuating in conjunction with the appearance of spike-waves during slow-wave sleep on EEG[2,60,77,86,104,118,144] and resolving only when the electrographic disturbance disappears.[40,86,127,137,144] The temporal correlation, however, may not be strict[40,60,72,104,140] as language improvements may lag behind EEG changes by weeks to months.[86,144] At any rate, normalization of the EEG appears to be a prerequisite for, but does not guarantee, complete reacquisition of language.[11,127] In some cases, an association between aphasia and abnormal EEG is not found,[33,104] which may depend on factors including age of onset, duration, intensity, and localization.[10,56]

Other Features

It is generally thought that patients with LKS are neurodevelopmentally normal premorbidly,[5,72,78,89,104] with relative preservation of nonverbal cognitive capacities.[3,32,78,90,127,128] Some patients however, may have some degree of language disturbance prior to the onset of aphasia (13–75% of children).[5,84,114,127,128] In addition, there may be associated behavioral disturbances in up to 72% of patients, including attention deficit and hyperactivity,[5,23,34,49,78,127,128] aggressiveness, rage outbursts, stereotypies, and other autistic features.[3,5,10,12,45,55,60,118,122,127,135,142] It is uncertain whether this is a primary phenomenon, due to limbic or diencephalic dysfunction,[49,55] or secondary, due to frustration at not being able to communicate.[31,114] There are rare reports of associated apraxias,[2,45] visuospatial disorders,[9,33,128] psychotic behavior[60,118,145] and dystonia.[45] The diagnosis of Landau-Kleffner is contingent upon the exclusion of other diagnoses including deafness, elective mutism, acute behavioral disorder, autism, and postictal dysphasia.[55]

NEUROPHYSIOLOGY AND NEUROIMAGING

Localization of Abnormality

The etiology of LKS is not known.[104,121] Both the functional impairment (i.e., aphasia) and the EEG changes are thought to be of focal origin.[51] The anatomical localization is presumed to lie in the perisylvian speech area or auditory association cortices on the basis of several neurophysiological investigations. For instance, electrocorticography (Ecog) and EEG dipole source localization using scalp electrodes localize epileptic discharges to the dorsal surface of the superior temporal gyrus in LKS, the latter demonstrating negativity above the sylvian fissure and positivity at the temporal sites.[61,89,90,100] Magnetic source imaging (MSI), superimposing magnetoencephalography (MEG) on MRI, is particularly suited to detect potential generators in the depths of the fissure, originating from either the superior temporal or inferior frontal gyri.[89,90] Spikes near the auditory cortex may occur spontaneously or be elicited by pure tone audiometry.[102,103] In some, though not all, cases[69] auditory evoked early-, middle-[66] and long-[21] latency potentials have been found to be abnormal. These findings suggest dysfunction of central auditory pathways.[21,66] A defect in phonological decoding is presumed to be responsible[73,89,144] though this has not been confirmed by intraoperative event-related potentials.[14] Positron Emission Tomography (PET; FDG) studies reveal increased or decreased glucose utilization in LKS, particularly involving the temporal lobe(s)[24] during sleep[80,116] though metabolic patterns tend to be very variable.[21,79,119] Single photon emission computed tomography (SPECT; Tc HMPAO) may demonstrate abnormal cerebral perfusion in the temporoparietal areas, particularly the left temporal lobe,[58,101] middle frontal gyrus and mesiotemporal/ hippocampal regions. In some studies, however, perfusion has been found to be normal [221].[93]

The spikes in LKS are thought to be of unilateral focal cortical origin. This is substantiated by intracarotid amytal testing. Injection of sodium amytal into the hemisphere with the primary focus suppresses discharges bilaterally, whereas contralateral injection suppresses spikes only on the injected side.[89,90] Methohexital suppression has been used to suppress EEG activity of the entire brain in order to find the primary epileptic region.

Lesions

In the vast majority of cases, CT and MRI are normal.[5,32,47,56,60,113,119] Cerebrospinal fluid is characteristically negative.[47,49,84] Pathological specimens, if available, have revealed only mild subpial gliosis[21] or been normal.[84] The LKS phenotype however, has been reported in conjunction with specific infections, including neurocysticercosis[99] toxoplasma gondii[85] and *Mycoplasma pneumoniae.*[110] Chronic encephalitis,[49,84] due to a slow virus and subacute sclerosing panencephalitis,[8] has also been suspected, but not proven.[21] In addition, cerebral arteritis,[83,106] temporal lobe tumors[126] and demyelinating diseases[97,109] have been implicated. Some authors, however, postulate that these may be examples of symptomatic focal epilepsies with a particularly severe and persistent aphasic symptomatology.[29] Regardless of etiology, it appears that timing (early childhood) and location (perisylvian) are paramount in the evolution of this condition.[62,135]

Genetic predisposition does not appear to be a strong consideration:[5] Only 2 pairs of affected siblings have been reported.[72,95]

Pathophysiology

It remains controversial as to whether aphasia in LKS is due to an active epileptic focus in the area subserving language (akin to a focal epilepsy)[13,29,31,33,70,73,144] or whether the EEG abnormalities and aphasia are "epiphenomena", both being secondary to underlying pathology in the cortical areas responsible for speech.[60,63,130] Landau and Kleffner were the first proponents of the former theory, suggesting that "persistent convulsive discharges in brain tissue concerned with linguistic communication result in the functional ablation of these areas for normal linguistic behavior".[72] It has been suggested that LKS is a developmental disorder, occurring in a critical period of development (1–8 years).[90] It may be that "epileptic bombardment" results in excess activation of language circuits which would normally be "pruned" or eliminated.[65,98,112] This could lead to permanent, aberrant and extraneous synaptic connections in the cortex and hence disordered language.[19,79,90]

Other authors suggest that the fluctuating and relapsing nature of this condition, with or without treatment, is incompatible with a chronic lesion.[70] In addition, persistent aphasia may be due to a prolonged or recurrent postictal state.[29] On occasion, there is a spectacular clinical improvement with antiepileptic drugs or steroids which correlates with decreased or absent paroxysmal discharges.[25,29] Finally, comparisons have been made between LKS and benign epilepsy of childhood with centrotemporal spikes,[29,40] wherein epileptiform activity in anterior temporal regions correlates with episodic oromotor dysfunction and disordered speech.[30,48,123,126] Clinical and EEG features of typical benign partial epilepsy of childhood with rolandic spikes (BPEC-R) have been demonstrated to antedate the onset of aphasia in some patients[21] suggesting that the bilateral spike-and-wave discharges may be an associated phenomenon leading to functional deafferentation of language areas.[127] In addition, in both conditions seizures are self-limited, infrequent, and augmented by sleep.[29]

Those favoring the hypothesis that the language and epileptiform discharges are secondary to an unknown lesion point out that the onset, severity and course of the language disorder does not always temporally correlate with EEG findings.[60,63,86,125,142] With time, the EEG consistently remits, but language deficits are often permanent.[49,51] Furthermore, if epileptiform activity were the cause of speech abnormalities, as in other focal seizures, anticonvulsants would be expected to consistently improve the language dysfunction, which seldom occurs.[47,51,63,86] It may be, however, that appropriate antiepileptic drugs exist but have not yet been identified,[135] or that treatment duration has been too short. Considering the developmental theory above, it has been speculated that excess synapses and collaterals during the first 2–3 years of life create an imbalance between excitatory and inhibitory drives, thereby simultaneously impairing language/cognitive functions and predisposing to synchronous firing of neurons and seizures.[79] It has also been proposed that dysfunctional or immature subcortical connections could lead to deafferentation[5] perhaps of the temporal lobes, causing both aphasia and epilepsy.[49,109]

THERAPY

Medications

Corticosteroids have met with the greatest success,[74,81,84] albeit incomplete. Prednisone has been reported to result in normalization of EEG and recovery of language[50,59,70,86] even after long periods of aphasia.[18,74,81] Furthermore, its withdrawal has been associated with the recurrence or worsening of aphasia.[50,59,68,70,73,74,81,84,86] High initial doses (prednisone 1 mg/kg/day or prednisolone 2 mg/kg/day), given early in the course of LKS and for extended periods (>6 months), appear to be most effective.[18,59,74,75] Similar results have been associated with ACTH.[74,84] In general, conventional anticonvulsants fail to alleviate the language deficit,[68,81,84,138] or suppress the EEG abnormality for sustained periods[67] though clinical seizures often subside.[60] Valproate, ethosuximide and benzodiazepines have been reported to be transiently or partially effective.[25,27,34,47,79,81,94,115] In addition, vigabatrin[3] and felbamate[53] have improved seizure control and aphasia in certain cases. In contrast, phenobarbital and phenytoin have aggravated symptoms.[79,81] Carbamazepine can be helpful[11] particularly with prednisone[31,86] or deleterious.[57,79,81] Nicardipine, combined with conventional anticonvulsants, was reported to improve aphasia in 6 children.[106,107] Intravenous immunoglobulin administration has coincided with improvement in language function and EEG abnormalities.[46,70] In general, because of the potentially devastating outcome in LKS and cormorbidity with seizures, a trial of empiric anticonvulsants (especially valproic acid) followed by steroids has been recommended.[7]

Surgical Approaches

Multiple subpial transection has been used in intractable LKS.[14,62,90,96,116] This procedure interrupts intracortical horizontal fibers (to prevent hypersynchronization & seizure spread) while sparing vertically aligned fibers (to retain thalamocortical pathways). It has been successful in restoring functional speech in up to 79% of children, even after 2 years of profound language deficit.[91] The cases studied thus far however, had fairly strict inclusion criteria.[90]

Evaluating the effectiveness of various treatment modalities is difficult in view of the fluctuating nature of LKS.[60,78,81,138] Furthermore, no controlled studies have been published on any of these treatments.

PROGNOSIS

There are few long term follow-up studies of LKS into adulthood.[12,32,40,78,128] In the natural history of LKS, seizures and EEG abnormalities usually remit by late adolescence.[5,31,32,78,104,132] Speech reacquisition is variable, ranging from full recovery to severe impairment.[12,23,31,32,33,36,38,49,63,78,80,127,125,140] Comprehension-wise, auditory discrimination deficits can be very serious and lasting.[9] It is difficult to predict outcome[105] though the aphasia often stabilizes and improves.[29] In general, the earlier the age of onset (less than 4–5 years), the worse the prognosis for recovery of language and comprehension.[12,40,47,86] If language dysfunction persists for longer than 1 year, spontaneous recovery is rare, and deficits tend to be lifelong.[12,31,32,33,76,89,141] In contrast, fluctuation in the course of aphasia may be associated with a favorable outcome.[33,97] In those with the best language outcomes, visuoperceptive difficulties have been noted.[78] Factors not correlated with outcome include frequency and type of seizures[5,33,78] epileptic activity in an waking EEG recordings[104] length of aphasia and the use of anticonvulsants.[63]

DIFFERENTIAL DIAGNOSIS—OVERLAPPING SYNDROMES: CLINICAL AND ELECTROGRAPHIC

CSWS (Continuous Spikes and Waves during Slow Sleep Syndrome)

Depending on what inclusion criteria are used to define LKS, the spectrum of this condition may be quite broad.[89] Since its first description in 1957, there has been much debate centering on whether CSWS[89,117,118,131,143]syndromes are along the clinical and

Table 2. Therapy for Landau-Kleffner Syndrome

1) Drug Therapy: an anticonvulsant (especially valproate) followed by a corticosteroid (such as prednisone)
2) Surgical Therapy: Multiple subpial transection.

pathophysiological continuum of LKS[10,12,26,31,40,60,68,79,81,82,103,118] or whether they represent a distinct disorder.[13,22,50,111] There is some speculation that patients with LKS represent a subgroup of CSWS patients.[89] In both cases, children are normal premorbidly, regress in higher cortical functions, and are left with significant permanent deficits during adolescence despite the remission of seizures and normalization of the EEG.[60,79,103] In CSWS syndromes however, the deterioration tends to be more global and pronounced in the cognitive than language spheres.[4,38,55,67,79,118] Few patients with CSWS have an isolated language disorder.[51] Behavioral and intellectual disturbances, including dementia, are commonly associated with CSWS, but are relatively rare in LKS, even when severe aphasia persists into adulthood.[32,38] In comparison with LKS, CSWS patients are more likely to present with more severe epilepsy, demonstrate additional seizure types (especially atypical absences and nonconvulsive seizures with falls), have an increased number of symptomatic lesions and show more neuroimaging abnormalities.[15]

Electrographically, both conditions demonstrate a certain percentage of spike and slow waves during slow wave sleep. "Typical" CSWS syndrome consists of diffuse bilateral almost continuous (>85%) spike-wave complexes throughout all stages of slow-wave sleep.[87,88,108,117,132] They are usually absent during the awake state and REM sleep,[67] and tend to be frontally predominant.[38,51,52,67] Spike-wave indices >85% have also been reported with acquired epileptic aphasia,[9,60,73,86,104] though they are infrequent.[4,38,60] The spike-wave density in LKS varies widely between patients (10–100%),[11,38,86,113,130] and CSWS may be absent altogether.[38,104] Some argue however, that CSWS is not consistently seen with LKS due to inopportune timing of the EEG in relation to development of the syndrome[60] lack of overnight polysomnographic recordings to capture events or patients "outgrowing" the finding.[38] In contrast to classical CSWS, LKS discharges are usually predominantly in temporal and/or parietal areas and present in all sleep stages including REM.[4,38,50,51,52,133] Furthermore, apparent CSWS may occur in epileptic conditions other than LKS[116] including secondary or lesional syndromes,[15,117,130,132] Lennox-Gastaut syndrome,[5,67,86] "atypical" benign partial epilepsy with atonic seizures,[1,35] anterior opercular syndrome of oral motor dysfunction[20,123] and even benign epilepsy with centrotemporal/rolandic spikes (BECT/BECRs).[6,11,16,117,127,130,132]

Developmental Disorders

Pronounced language deficits are also found in developmental dysphasia and pervasive developmental disorder (PDD).[29,91,105] The former condition is characterized by delayed oral language acquisition, normal intelligence, fluctuating course and a high incidence of paroxysmal EEG abnormalities during waking[42,77,120,136] and sleep[41,42] records. On rare occasions, continuous spikes and waves during slow sleep have been reported.[29,31,41,42] There does not appear to be a direct relationship between the language and EEG abnormalities, however.[41,42] Clinical seizures and receptive dysphasia are occasionally associated.[42,77] There is speculation as to whether a subset children with developmental dysphasia have early onset LKS.[43] Of interest is the fact that, homozygotic twin girls, one with acquired epileptic aphasia and the other with developmental dysphasia, have been reported, suggesting that there is a common genetic predisposition to the two conditions.[31,44]

In autism (PDD), it has been proposed that "epilepsy or subclinical epileptiform EEG activity is responsible for regression and persistent deficits in behavior, language, and socio-communicative skills in a significant minority of children".[134] As such, it has been recommended that patients with autistic regression undergo overnight sleep EEG's to ensure that a potentially treatable epileptiform condition is not overlooked. Prednisone has been reported to reverse language and behavioral regression in children with PDD.[91,129] This has generated interest in the use of corticosteroids in the treatment of autistic syndromes.

CONCLUSIONS

Since 1957, LKS has remained an elusive, heterogenous disorder in terms of etiology, pathophysiology and management.[33,104] With time and continued research, it will cease to be, in W. Landau words, an "eponymic badge of ignorance".[71]

REFERENCES

1. Aicardi, J. *Epilepsy in children*, Raven Press, New York (1986) pp. 176–182.
2. Ansink, B.J., Sarphatie, H., and Van Dongen, H.R. The Landau-Kleffner syndrome: case report and theoretical considerations, *Neuropediatrics*, 20 (1989) 170–172.
3. Appleton, R. The Landau-Kleffner syndrome, *Arch. Dis. Childhood*, 72 (1995) 386–387.
4. Beaumanoir, A., Bureau, M., and Mira, L. Identification of the syndrome. In: A. Beaumanoir, M. Bureau, T. Deonna, L. Mira, and C.A. Tassinari (Eds.) *Continuous spikes and waves during slow sleep Electrical status epilepticus during slow sleep Acquired epileptic aphasia and related conditions*, John Libbey, London, (1995) pp. 243–249.
5. Beaumanoir, A. The Landau-Kleffner syndrome. In: J. Roger, M. Bureau, C.H. Dravet, F.E. Dreifuss, A. Pettet, and P. Wolf (Eds.) *Epileptic syndromes in infancy, childhood and adolescence*, 2nd Ed. John Libbey, London, (1992) pp. 231–43.
6. Beaumanoir, A. and Grandjean E. Continuous spike-and-wave discharges during sleep. Significance. *Electroencephalogr. Clin. Neurolphysiol.*, 55 (1983) 18–19.
7. Besag, F.M.C. Epilepsy, learning, and behaviour in childhood, *Epilepsia*, 36 (1995) (Suppl. 1) S58–63.
8. Bicknese, A.R., Preston, T., and Ettinger, A.B. Epileptic aphasia (Landau-Kleffner syndrome) secondary to progressive encephalitis, *Ann. Neurol.*, 40 (1996) 306–307.
9. Billard, C., Loisel, M.L., Gillet, P., Lucas, B., Autret, A., Degiovanni, E., Santini, J.J., Dulac, O., and Picard, A. Relationship between acquired neuropsychological deficit and nocturnal EEG abnormalities in a case of Landau-Kleffner syndrome, *ANAE*, 3 (1991) 39–43.
10. Billard, C., Autret, A., Lucas, B., Degiovanni, E., Gillet, P., Santini, J.J., and de Toffol, B. Are frequent spike-waves during non-REM sleep in relation with an acquired neuro-psychological deficit in epileptic children? *Neurophys. Clin.*, 20 (1990) 439–53.
11. Billard, C., Autret, A., Laffont, F., Lucas, B., and de Giovanni, E. Electrical status epilepticus during sleep in children: a reappraisal from eight new cases. In: M.B. Sterman, M.N. Shouse, P. Passouant (Eds.) *Sleep and epilepsy*, Academic Press, London (1982) pp. 481–494.
12. Bishop, D.V.M. Age of onset and outcome in "acquired aphasia with convulsive disorder" (Landau-Kleffner syndrome), *Dev. Med. Child Neurol.*, 27 (1985) 705–712.
13. Boel, M. and Casaer, P. Continuous spikes and waves during slow sleep: a 30 month follow-up study of neuropsychological recovery and EEG findings, *Neuropediatrics*, 20 (1989) 176–80.
14. Boyd, S.G., Rivera-Gaxiola, M., Towell, A.D., Harkness, W., and Neville, B.G.R. Discrimination of speech sounds in a boy with Landau-Kleffner syndrome: an intraoperative event-related potential study, *Neuropediatrics*, 27 (1996) 211–215.
15. Bureau, M. Outstanding cases of CSWS and LKS: analysis of the data sheets provided by the participants. In: A. Beaumanoir, M. Bureau, T. Deonna, L. Mira, and C.A. Tassinari (Eds.) *Continuous spikes and waves during slow sleep Electrical status epilepticus during slow sleep Acquired epileptic aphasia and related conditions*, John Libbey, London (1995) pp. 213–316.
16. Bureau, M. Continuous spikes and waves during slow sleep (CSWS): definition of the syndrome. In: A. Beaumanoir, M. Bureau, T. Deonna, L. Mira, and C.A. Tassinari (Eds.) *Continuous spikes and waves during slow sleep Electrical status epilepticus during slow sleep Acquired epileptic aphasia and related conditions*, John Libbey, London (1995) pp. 17–26.

17. Burr, W. and Stefan, H. Day by day variations of the amount of spike-wave activity in 24 hour cassette recordings, *Electroencephalogr. Clin. Neurophysiol.*, 67 (1987) 40–41.
18. Caraballo, R.H., Grillo, E., Soprano, A., and Fejerman, N. Long term steroid treatment in 2 patients with acquired epileptic aphasia, *Pediatr. Neurol.*, 8 (1992) 264.
19. Chugani, H.T., Phelps, M.E., and Mazziotta, J.C. Positron emission tomography study of human brain functional development, *Ann. Neurol.*, 22 (1987) 487–497.
20. Colamaria, V., Sgro, V., Caraballo, R., Simeone, M., Zullni, E., Fontana, E., Zanetti, R., Grimau-Merino, R., and Dalla Bernardina, B. Status epilepticus in benign rolandic epilepsy manifesting as anterior operculum syndrome, *Epilepsia*, 32 (1991) 329–334.
21. Cole, J., Andermann, F., Taylor, L., Olivier, A., Rasmussen, T., Robitaille, Y., and Spire, J.P. The Landau-Kleffner syndrome of acquired epileptic aphasia: unusual clinical outcome, surgical experience, and absence of encephalitis, *Neurology*, 38 (1988) 31–38.
22. Commission on the Classification and Terminology of the International League Against Epilepsy. Proposal for the classification of epilepsies and epileptic syndromes, *Epilepsia*, 26 (1985) 268–278.
23. Cooper, J.A. and Ferry, P.C. Acquired auditory verbal agnosia and seizures in childhood, *J. Speech Hear Disord.*, 43 (1978) 176–184.
24. da Silva, E.A., Chugani, D.C., Muzik, O., and Chugani, H.T. Landau-Kleffner syndrome: metabolic abnormalities in temporal lobe are a common feature, *J. of Child Neuro.*, 12 (1997) 489–495.
25. DeMarco, P. Electrical status epilepticus during slow sleep: one case with sensory aphasia, *Clin. Electroencephalogr.*, 19 (1988) 111–113.
26. De Negri, M. Electrical status epilepticus during sleep (ESES). Different clinical syndromes: towards a unifying view? *Brain Dev.*, 19 (1997) 447–451.
27. De Negri, M., Baglietto, M.G., and Biancheri, R. Electrical status epilepticus in childhood: treatment with short cycles of high dosage benzodiazepine (preliminary note), *Brain Dev.*, 15 (1993) 311–312.
28. Denes, G., Balliello, S., Pellegrini, A., and Volterra, V. Phonemic deafness in infancy and acquisition of written language. In: M. Coltheart, G. Sartori, and R. Job (Eds.) *The cognitive neuropsychology of language*, Erlbaum, London (1987) pp. 337–49.
29. Deonna, T. and Roulet, E. Acquired epileptic aphasia (AEA): definition of the syndrome and current problems. In: A. Beaumanoir, M. Bureau, T. Deonna, L. Mira, and C.A. Tassinari (Eds.) *Continuous spikes and waves during slow sleep Electrical status epilepticus during slow sleep Acquired epileptic aphasia and related conditions*, John Libbey, London (1995) pp. 37–45.
30. Deonna, T.W., Roulet, E., Fontan, D., and Marcoz, J.P. Speech and oromotor deficits of epileptic origin in benign partial epilepsy of childhood with rolandic spikes (BPERS). Relationship to the acquired aphasia-epilepsy syndrome, *Neuropediatrics*, 24 (1993) 83–7.
31. Deonna, T.W. Acquired epileptiform aphasia in children (Landau-Kleffner Syndrome), *J. Clin. Neurophysiol.*, 8 (1991) 288–298.
32. Deonna, T., Peter, C.L., and Ziegler, A.L. Adult follow-up of the acquired aphasia-epilepsy syndrome in childhood. Report of seven cases, *Neuropediatrics*, 20 (1989) 132–138.
33. Deonna, T., Beaumanoir, A., Gaillard, F., and Assal, G. Acquired aphasia in childhood with seizure disorder: a heterogeneous syndrome, *Neuropediatrie*, 8 (1977) 263–273.
34. Deuel, R.K. and Lenn, N.J. Treatment of acquired epileptic aphasia, *J. Pediatr.*, 90 (1977) 959–61.
35. Doose, H. and Baier, W.K. Benign partial epilepsy and related conditions: multifactorial pathogenesis with hereditary impairment of brain maturation, *Eur. Pediatr.*, 149 (1989) 152–158.
36. Drought, C. An unusual form of acquired aphasia in children, *Dev. Med. Child Neurol.*, 13 (1971) 563–571.
37. Dugas, M., Gerard, C.L., Franc, S., and Lecendreux, M. Late onset acquired epileptic aphasia. In: A. Beaumanoir, M. Bureau, T. Deonna, L. Mira, and C.A. Tassinari (Eds.) *Continuous spikes and waves during slow sleep Electrical status epilepticus during slow sleep Acquired epileptic aphasia and related conditions*, John Libbey, London (1995) pp. 143–147.
38. Dugas, M., Franc, S., Gerard, L.C., and Lecendreux, M. Evolution of acquired epileptic aphasia with or without continuous spikes and waves during slow sleep. In: A. Beaumanoir, M. Bureau, T. Deonna, L. Mira, and C.A. Tassinari (Eds.) *Continuous spikes and waves during slow sleep Electrical status epilepticus during slow sleep Acquired epileptic aphasia and related conditions*, John Libbey, London (1995) pp. 47–55.
39. Dugas, M. The Landau-Kleffner syndrome. Infantile "acquired" aphasia, paroxysmal electroencephalographic changes and epileptic seizures, *Nouvelle Presse Medicale*, 11 (1982) 1387–1391.
40. Dulac, O., Billard, C., and Arthuis, M. Electroclinical and developmental aspects of epilepsy in the aphasia-epilepsy syndrome, *Arch. Fr. Pediatr.*, 40 (1983) 299–308.

41. Duvelleroy-Hommet, C., Billard, C., Lucas, B., Gillet, P., Barthez, M.A., Santini, J.J., Degiovanni, E., Henry, F., De Toffol, B., and Autret, A. Sleep EEG and developmental dysphasia: lack of a consistent relationship with paroxysmal EEG activity during sleep, *Neuropediatrics*, 26 (1995) 14–18.
42. Echenne, B., Cheminal, R., Rivier, F., Negre, C., Touchon, J., and Billiard, M. Epileptic electroencephalographic abnormalities and developmental dysphasias: a study of 32 patients, *Brain Dev.*, 14 (1992) 216–225.
43. Echenne, B., Cheminal, R., Rivier, F., Touchon, J., and Billiard, M. Epileptic EEG abnormalities and developmental dysphasias: Results of repeated EEG investigations during sleep and waking. In: *Abstracts from Second International Symposium Specific Speech and Language Disorders in Children,* AFASIC, Harrogate (1991) p. 36.
44. Echenne, B. Les formes severes de retard de langage chez l'enfant, *Rev. Int. Pediatr.*, 201 (1990) 5–11.
45. Eslava-Cobos, J. and Mejia, L. Landau-Kleffner syndrome: much more than aphasia and epilepsy, *Brain & Lang*, 57 (1997) 215–224.
46. Fayad, M.N., Choueiri, R., and Mikati, M. Landau-Kleffner syndrome: consistent response to repeated intravenous gamma-globulin doses: a case report, *Epilepsia*, 38 (1997) 489–494.
47. Feekery, C.J., Parry-Fielder, B., and Hopkins, I.J. Landau-Kleffner syndrome: Six patients including discordant monozygotic twins, *Pediatr. Neurol.*, 9 (1993) 49–53.
48. Fejerman, N. and Di Blasi, A.M. Status epilepticus of benign partial epilepsies in children: report of 2 cases, *Epilepsia*, 28 (1987) 351–355.
49. Gascon, G., Victor, D., and Lombroso, C.T. Language disorder, convulsive disorder, and electroencephalographic abnormalities. Acquired syndrome in children, *Arch. Neurol.*, 28 (1973) 156–162.
50. Genton, P., Guerrini, R., Bureau, M., and Dravet, C. Continuous focal discharges during REM sleep in a case of Landau-Kleffner syndrome: a 3 year follow-up. In: A. Beaumanoir, M. Bureau, T. Deonna, L. Mira, and C.A. Tassinari (Eds.) *Continuous spikes and waves during slow sleep Electrical status epilepticus during slow sleep Acquired epileptic aphasia and related conditions*, John Libbey, London (1995) pp. 155–159.
51. Genton, P. and Guerrini, R. What differentiates Landau-Kleffner syndrome from the syndrome of continuous spikes and waves during slow sleep? *Arch. Neurol.*, 50 (1993) 1008–1009.
52. Genton, P., Maton, B., Ogihara, M., Samoggia, G., Guerrini, R., Medina, M.T., Dravet, C., and Roger, J. Continuous focal spikes during REM sleep in a case of acquired aphasia (Landau-Kleffner syndrome), *Sleep*, 15 (1992) 454–460.
53. Glauser, T.A., Olberding, L.S., Titanic, M.K., and Piccirillo, D.M. Felbamate in the treatment of acquired epileptic aphasia, *Epil. Res.*, 20 (1995) 85–89.
54. Gomez, M.R. and Klass, D.W. Epilepsies of infancy and childhood, *Ann. Neurol.*, 13 (1983) 113–124.
55. Gordon, N. The Landau-Kleffner syndrome: increased understanding, *Brain Dev.*, 19 (1997) 311–316.
56. Gordon, N. Acquired aphasia in childhood: the Landau-Kleffner syndrome, *Dev. Med. Child Neurol.*, 32 (1990) 270–274.
57. Gross-Selbeck, G. Treatment of 'benign' partial epilepsies of childhood, including atypical forms, *Neuropediatrics*, 26 (1995) 45–50.
58. Guerreiro, M.M., Camargo, E.E., Kato, M., Menezes Netto, J.R., Silva, E.A., Scotoni, A.E., Silveira, D.C., and Guerreiro, C.A.M. Brain single photon emission computed tomography imaging in Landau-Kleffner syndrome, *Epilepsia*, 37 (1996) 60–67.
59. Hirsch, E., Maquet, P., Metz-Lutz, M., Motte, J., Finck, S., and Marescaux, C. The eponym "Landau-Kleffner syndrome" should not be restricted to childhood-acquired aphasia with epilepsy. In: A. Beaumanoir, M. Bureau, T. Deonna, L. Mira, and C.A. Tassinari (Eds.) *Continuous spikes and waves during slow sleep Electrical status epilepticus during slow sleep Acquired epileptic aphasia and related conditions*, John Libbey, London (1995) pp. 57–62.
60. Hirsch, E., Marescaux, C., Maquet, P., Metz-Lutz, M.N., Kiesmann, M., Salmon, E., Franck, G., and Kurtz D. Landau-Kleffner syndrome: a clinical and EEG study of five cases, *Epilepsia*, 31 (1990) 756–767.
61. Hoeppner, J., Grote, C., Morrell, F., Whisler, W., and Brocken, C. Long term follow-up cognitive and behavioral function after surgery for Landau-Kleffner syndrome, *Epilepsia*, 34 (1993) (suppl 6) 72.
62. Hoeppner, T.J., Morrell, F., Smith, M.C., Chez, M., and Hasegawa, H. The Landau-Kleffner syndrome: a peri-sylvian epilepsy, *Epilepsia*, 33 (1992) (Suppl 3) 122.
63. Holmes, G.L., McKeever, M., and Saunders, Z. Epileptiform activity in aphasia of childhood: an epiphenomenon? *Epilepsia*, 22 (1981) 631–639.
64. Humpfrey, I.L., Knispstein, R., and Bumpass, E.R. Gradually developing aphasia. A diagnostic problem, *J. Am. Acad. Child Psychiatry*, 14 (1975) 652–655.

65. Huttenlocher, P.R. and de Courten, C. The development of synapses in striate cortex of man, *Hum. Neurobiol.*, 6 (1987) 1–9.
66. Isnard, J., Fischer, C., Bastuji, H., Badinand, N., and de Villard, R. Auditory early (BAEP) and middle-latency (MLAEP) evoked potentials in patients with CSWS and Landau-Kleffner syndrome. In: A. Beaumanoir, M. Bureau, T. Deonna, L. Mira, and C.A. Tassinari (Eds.) *Continuous spikes and waves during slow sleep Electrical status epilepticus during slow sleep Acquired epileptic aphasia and related conditions*, John Libbey, London (1995) pp. 99–103.
67. Jayakar, P.B. and Seshia, S.S. Electrical status epilepticus during slow-wave sleep: a review, *J. Clin. Neurophysiol.*, 8 (1991) 299–311.
68. Kellermann, K. Recurrent aphasia with subclinical bioelectric status epilepticus during sleep, *Eur. J. Pediatr.*, 128 (1978) 207–212.
69. Klein, S.K., Kurtzberg, D., Brattson, A., Kreuzer, J.A., Stapells, D.R., Dunn, M.A., Rapin. I., and Vaughan, H.G. Jr. Electrophysiologic manifestations of impaired temporal lobe auditory processing in verbal auditory agnosia, *Brain and Language*, 5 (1995) 383–405.
70. Lagae, L.G., Silberstein, J., Phillippe, L.G., and Casaer, P.J. Successful use of intravenous immunoglobulins in Landau-Kleffner Syndrome, *Pediatr. Neurol.*, 18 (1998) 165–168.
71. Landau, W.M. Landau-Kleffner syndrome: an eponymic badge of ignorance, *Arch. Neurol.*, 49 (1992) 353.
72. Landau, W.M. and Kleffner, F.R. Syndrome of acquired aphasia with convulsive disorder in children, *Neurology*, 7 (1957) 523–530.
73. Lanzi, G., Veggiotti, P., Conte, S., Partesana, E., and Resi, C. A correlated fluctuation of language & EEG abnormalities in a case of the Landau-Kleffner syndrome, *Brain Dev.*, 16 (1994) 329–334.
74. Lerman, P., Lerman-Sagie, T., and Kivity, S. Effect of early corticosteroid therapy for Landau-Kleffner syndrome, *Dev. Med. Child Neurol.*, 33 (1991) 257–266.
75. Lerman, P. and Lerman-Sagie, T. Early steroid therapy in Landau-Kleffner syndrome. In: J. Manelis, E. Bental, J.N. Loeber, and F.E. Dreifuss. *Advances in epileptology. Vol XVII,* Raven Press, New York (1989) pp. 330–332.
76. Loonan, M.C.B. and van Dongen, H.R. Acquired childhood aphasia. Outcome one year after onset, *Arch. Neurol.*, 49 (1990) 1324–1328.
77. Maccario, M., Hefferen, S.J., Keblusek, S.J., and Lipinski, K.A. Developmental dysphasia and electroencephalographic abnormalities, *Dev. Med. Child Neurol.*, 24 (1982) 141–155.
78. Mantovani, J.F. and Landau, W.M. Acquired aphasia with convulsive disorder: course and prognosis, *Neurology*, 30 (1980) 524–529.
79. Maquet, P., Hirsch, E., Metz-Lutz, M.N., Motte, J., Dive, D., Marescaux, C., and Franck, G. Regional cerebral glucose metabolism in children with deterioration of one or more cognitive functions and continuous spike-and wave discharges during sleep, *Brain*, 118 (1995) 1497–1520.
80. Maquet, P., Hirsch, E., Dive, D., Salmon, E., Marescaux, C., and Franck, G. Cerebral glucose utilization during sleep in Landau-Kleffner syndrome: a PET study, *Epilepsia*, 31 (1990) 778–783.
81. Marescaux, C., Hirsch, E., Finck, S., Maquet, P., Schlumberger, E., Sellal, F., Metz-Lutz, M.N., Alembik, Y., Salmon, E., and Franck, G. Landau-Kleffner syndrome: a pharmacologic study of five cases, *Epilepsia*, 31 (1990) 768–777.
82. Marescaux, C., Kiesmann, M., and Hirsch, E. Are the Landau-Kleffner syndrome and the syndrome of continuous spike-wave discharges during sleep one and the same? *Epilepsia*, 30 (1989) 693.
83. Martinez Bermejo A., Castroviejo, I., Lopez Martin, V., Arcas, J., and Perez Higueras, A. Acquired aphasia syndrome with epilepsy (Landau-Kleffner syndrome) secondary to cerebral arteritis, *Neurologia*, 4 (1989) 296–299.
84. McKinney, W. and McGreal, D.A. An aphasic syndrome in children, *Can. Med. Assoc. J.*, 110 (1974) 637–639.
85. Michaoowicz, R., Jozwiak, S., Ignatowicz, R., and Szwabowska-Orzeszko, E. Landau-Kleffner syndrome-epileptic aphasia in children-possible role of toxoplasma gondii infection, *Acta Paediatr. Hung.*, 29 (1989) 337–342.
86. Ming, L., Xiao-yu, H., Jiong, Q., and Xi-ru, W. Correlation between CSWS and aphasia in Landau-Kleffner syndrome: a study of three cases, *Brain Dev.*, 18 (1996) 197–200.
87. Morikawa, T., Seino, M., Watanabe, Y., Watanabe, M., and Yagi, K. Clinical relevance of continuous spike-waves during slow wave sleep. In: J. Manelis, E. Bental, J.N. Loeber, and F.E. Dreifuss (Eds.) *The XVIIth Epilepsy International Symposium*, Raven Press, New York (1989) pp. 359–363. (Advances in epileptology: vol 17).

88. Morikawa, T., Seino, M., Osawa, T., and Yagi, K. Five children with continuous spike-wave discharges during sleep. In: J. Roger, C. Dravet, M. Bureau, F. Dreifuss, and P. Wolf (Eds.) *Epileptic syndromes in infancy, childhood and adolescence,* John Libbey, London (1985) pp. 205–12.
89. Morrell, F. Electrophysiology of CSWS in Landau-Kleffner syndrome. In: A. Beaumanoir, M. Bureau, T. Deonna, L. Mira, and C.A. Tassinari (Eds.) *Continuous spikes and waves during slow sleep Electrical status epilepticus during slow sleep Acquired epileptic aphasia and related conditions*, John Libbey, London (1995) 77–90.
90. Morrell, F., Whisler, W.W., Smith, M.C., Hoeppner, T.J., de Toledo-Morrell, L., Pierre-Louis, S.J.C., Kanner, A.M., Buelow, J.M., Ristanovic, R., and Bergen, D. Landau-Kleffner syndrome: treatment with subpial intracortical transection, *Brain*, 118 (1995) 1529–1546.
91. Mott, S.H., Weinstein, S.L., Conry, J.A., Kenworthy, L.E., Lockwood, S., Wagner, A., and Packer, R.J. Pervasive developmental disorder/autism versus Landau-Kleffner syndrome: Steroid-responsive encephalopathy characterized by language and social interactive impairment, *Ann. Neurol.*, 40 (1996) 332.
92. Mouridsen, S.E. The Landau-Kleffner Syndrome: a review, *Eur. Adolesc. Psychiatry*, 4 (1995) 223–228.
93. Mouridsen, S.E., Videbaek, C., Sogaard, H., and Andersen, A.R. Regional cerebral blood-flow measured by HMPAO and SPECT in a 5-year-old boy with Landau-Kleffner syndrome, *Neuropediatrics*, 24 (1993) 47–50.
94. Nagafuchi, M., Iinuma, K., Yamamoto, K., and Kitahara, T. Diazepam therapy of verbal auditory agnosia, *Brain and Language*, 45 (1993) 180–188.
95. Nakano, S., Okuno, T., and Mikawa, H. Landau-Kleffner syndrome: EEG topographic studies, *Brain Dev.*, 11 (1989) 43–50.
96. Neville, B.G.R., Harkness, W.F.J., Cross, J.H., Cass, H.C., Burch, V.C., Lees, J.A., and Taylor, D.C. Surgical treatment of severe autistic regression in childhood epilepsy, *Pediatr. Neurol.*, 16 (1997) 137–140.
97. Nevsimalowva, S., Tauberova, A., Doutlik, S., Kucera, V., and Dlouha, O. A role of autoimmunity in the etiopathogenesis of Landau-Kleffner syndrome? *Brain Dev.*, 14 (1992) 342–345.
98. O'Leary, D.D. Development of connectional diversity and specificity in the mammalian brain by the pruning of collateral projections, *Curr. Opin. Neurobiol.*, 2 (1992) 70–77.
99. Otero, E., Cordova, S., Diaz, F., Garcia-Teruel, I., and Del Brutto, O.H. Acquired epileptic aphasia (the Landau-Kleffner syndrome) due to neurocysticercosis, *Epilepsia*, 30 (1989) 569–572.
100. Otsubo, H., Hunjan, A., Hwang, P.A., Snead, O.C., Mosskin, M., and Chuang, S.H. Dipole localization of epileptiform discharges of patients with Landau-Kleffner syndrome II: magnetic source imaging and EEG mapping, *Epilepsia*, 3 (1996) (Suppl 5) S90.
101. O'Tuama, L.A., Urion, D.K., Janicek, M.J., Treves, S.T., Bjornson, B., and Moriarty, J.M. Regional cerebral perfusion in Landau-Kleffner syndrome and related childhood aphasias, *J. Nucl. Med.*, 33 (1992) 1758–1765.
102. Paetau, R. Sounds trigger spikes in the Landau-Kleffner syndrome, *J. Clin. Neurophysiol.*, 11 (1994) 231–234.
103. Paetau, R., Kajola, M., Korkman, M., Hamalainen, M., Granstrom, M.L., and Hari, R. Landau-Kleffner syndrome: epileptic activity in the auditory cortex, *NeuroReport*, 2 (1991) 201–204.
104. Paquier, P.F., Van Dongen, H.R., and Loonen, M.C.B. The Landau-Kleffner syndrome or 'acquired aphasia with convulsive disorder'. Long-term follow-up of six children and review of the recent literature, *Arch. Neurol.*, 49 (1992) 354–359.
105. Parry-Fielder, B., Nolan, T.M., Collins, K.J., and Stojcevski, Z. Review Article: developmental language disorders and epilepsy, *J. Paediatr. Child Health*, 33 (1997) 277–280.
106. Pascual Castraviefo, I., Lopez-Martin, V., MartinezBermefo, A., and Gigueras, A. Is cerbral arteritis the cause of the Landau-Kleffner syndrome? Four cases in childhood with angiographic study, *Can. J. Neurol. Sciences*, 19 (1992) 46–52.
107. Pascual Castroviedo, I. Nicardipine in the treatment of acquired aphasia and epilepsy, *Dev. Med. Child Neurol.*, 32 (1990) 930.
108. Patry, G., Lyagoubi, S., and Tassinari, C.A. Sub-clinical "electrical status epilepticus" induced by sleep in children. A clinical and electroencephalographic study of six cases, *Arch. Neurol.*, 24 (1971) 242–252.
109. Perniola, T., Margari, L., Buttiglione, M., Anreula, C., Simone, I.L., and Santostasi, R. A case of Landau-Kleffner syndrome secondary to inflammatory demyelinating disease, *Epilepsia*, 39 (1993) 551–556.
110. Pouplard, F. and Pasquier, C. Le syndrome d'aphasie acquise avec epilepsie est-il une complication du Mycoplasma pneumoniae? *Nouv. Presse Med.*, 7 (1978) 2970.

111. Proposal for revised classification of epilepsies and epileptic syndromes, *Epilepsia*, 30 (1989) 389–99.
112. Purves, D. and Lichtman, J.W. Elimination of synapses in the developing nervous system, *Science*, 210 (1980) (4466) 153–157.
113. Rapin, I. Verbal auditory agnosia in children, *Dev. Med. Child Neurol.*, 30 (1988) 685.
114. Rapin, I., Mattis, S., Rowan, A.F., and Goldern, G.G. Verbal auditory agnosia in children, *Dev. Med. Child Neurol.*, 19 (1977) 192–207.
115. Ravnik, I. A case of Landau-Kleffner syndrome: effect of intravenous diazepam. In: J. Roger, C. Dravet, M. Bureau, R.E. Dreifuss, and P. Wolf (Eds.) *Epileptic syndromes in infancy, childhood and adolescence*, John Libbey, London (1985) 192–193.
116. Rintahaka, P.J., Chugani, H.T., and Sankar, R. Landau-Kleffner syndrome with continuous spikes and waves during slow-wave sleep, *J. Child Neurol.*, 10 (1994) 127–133.
117. Rodriquez, I. and Niedermeyer, E. The aphasia-epilepsy syndrome in children: electroencephalographic aspects, *Clin. Electroencephalogr.*, 13 (1982) 23–35.
118. Roulet, E., Deonna, T., Gaillard, F., Peter-Favre, C., and Despland, P.A. Acquired aphasia, dementia, and behavior disorder with epilepsy and continuous spike and waves during sleep in a child, *Epilepsia*, 32 (1991) 495–503.
119. Sankar, R., Chugani, H.T., Lubens, P., Philippart, M., and Phelps, M. Heterogeneity in the patterns of cerebral glucose utilization in children with the Landau-Kleffner syndrome, *Neurology*, 40 (1990) (Suppl 1) 257.
120. Sato, S. and Dreifuss, F.E. Electroencephalographic findings in a patient with developmental expressive aphasia, *Neurology*, 23 (1973) 181–185.
121. Sawhney, I.M.S., Robertson, I.J.A., Polkey, C.E., Binnie, C.D., and Elwes, R.D.C. Multiple subpial transection: a review of 21 cases, *J. Neurol., Neurosurg., and Psych.*, 58 (1995) 344–349.
122. Sawhney, I.M., Suresh, N., Dhand, U.K., and Chopra, J.S. Acquired aphasia with epilepsy: Landau-Kleffner syndrome, *Epilepsia*, 29 (1988) 283–287.
123. Shafrir, Y. and Prensky, A.L. Acquired epileptiform opercular syndrome: a second case report, review of the literature, and comparison to the Landau-Kleffner syndrome, *Epilepsia*, 36 (1995) 1050–1057.
124. Sheffer, I.E., Jones, L., Pozzebon, M., Howell, R.A., Saling, M.M., and Berkovic, S.F. Autosomal dominant rolandic epilepsy and speech dyspraxia: a new syndrome with anticipation, *Ann. Neurol.*, 38 (1995) 633–642.
125. Shoumaker, R.D., Bennett, D.R., Bray, P.F., and Curless, R.G. Clinical and EEG manifestations of an unusual aphasic syndrome in children, *Neurology*, 24 (1974) 10–16.
126. Solomon, G.E., Carson, D., Pavlakis, S., Fraser, R., and Labar, D. Intracranial EEG monitoring in Landau-Kleffner syndrome associated with left temporal lobe astrocytoma, *Epilepsia*, 34 (1993) 557–560.
127. Soprano, A.M., Garcia, E.F., Caraballo, R., and Fejerman, N. Acquired epileptic aphasia: neuropsychologic follow-up of 12 patients, *Pediatr. Neurol.*, 11 (1994) 230–235.
128. Soprano, A., Garcia, E., Caraballo, R., and Fejerman, N. Neuropsychological and neurolinguistic aspects in a group of patients with acquired epileptic aphasia or Landau-Kleffner syndrome, *Pediatr. Neurol.*, 8 (1992) 398–399.
129. Stefanos, G.A., Grover, W., and Geller, E. Case study: corticosteroid treatment of language regression in pervasive developmental disorder, *J. Am. Acad. Child Adolesc. Psychiatry*, 34 (1995) 1107–1111.
130. Tassinari, C.A. The problems of "continuous spikes and waves during slow sleep" or "electrical status epilepticus during slow sleep" today. In: A. Beaumanoir, M. Bureau, T. Deonna, L. Mira and C.A. Tassinari (Eds.) *Continuous spikes and waves during slow sleep Electrical status epilepticus during slow sleep Acquired epileptic aphasia and related conditions*, John Libbey, London (1995) 251–255.
131. Tassinari, C.A., Bureau, M., Dravet, C., Dalla Bernardina, D., and Roger, J. Epilepsy with continuous spikes and waves during slow sleep, otherwise described as ESES (epilepsy with electrical status epilepticus during slow sleep). In: J. Roger, C. Dravet, M. Bureau, F.E. Dreifuss, and P. Wolf (Eds.) *Epileptic Syndromes in Infancy, childhood, and adolescence*, John Libbey Eurotext, London, England (1985) 194–212.
132. Tassinari, C., Bureau, M., Dravet, C., and Roger, J. Electrical status epilepticus during sleep in children (ESES). In: M. Sterman, M. Shouse, and P. Passouant (Eds.) *Sleep and epilepsy*, Academic Press, London (1982) 465–479.
133. Tiberge, M., Calvet, U., Soubiran, C., and Arbus, L. Landau-Kleffner syndrome with continuous sharp waves during REM sleep, *Electroencephalogr. Clin. Neurophysiol.*, 70 (1988) 11.
134. Tuchman, R.F. and Rapin, I. Regression in pervasive developmental disorders: seizures and epileptiform electroencephalogram correlates, *Pediatrics*, 99 (1997) 560–566.

135. Tuchman, R.F. Epilepsy, language, and behavior: clinical models in childhood, *J. Child Neurol.*, 9 (1994) 95–102.
136. Tuchman, R.F., Rapin, I., and Shinnar S. Autistic and dysphasic children. II: Epilepsy, *Pediatrics*, 88 (1991) 1219–1225.
137. Urion, D.K., Kleinman, S.A., and Maister, B. correlation between language status and electroencephalographic normalization in the Landau-Kleffner syndrome (verbal auditory agnosia): report of 3 cases, *Ann. Neurol.*, 24 (1988) 325.
138. van de Sandt-Koenderman, W.M., Smit, I.A., van Dongen, H.R., and van Hest, J.B. A cases of acquired aphasia and convulsive disorder: some linguistic aspects of recovery and breakdown, *Brain Lang.*, 21 (1984) 174–183.
139. van Dongen, H.R., Meulstee, J., and Paquier, P.F. 2 contrasting cases of children with acquired epiletpic aphasia and continuous spikes and waves during sleep. In: A. Beaumanoir, M. Bureau, T. Deonna, L. Mira, and C.A. Tassinari (Eds.) *Continuous spikes and waves during slow sleep electrical status epilepticus during slow sleep Acquired epileptic aphasia and related conditions*, John Libbey, London (1995) 137–141.
140. van Dongen, H., Meulstee, J., Blauw-van Mourik, M., and van Harskamp, F. Landau-Kleffner syndrome: A case study with fourteen-year follow-up, *Eur. Neurol.*, 29 (1989) 109–114.
141. van Harskamp, F., van Dongen, H.R., and Loonen, M.C.B. Acquired aphasia with convulsive disorder in children: a case study with a seven-year follow-up, *Brain Lang.*, 6 (1978) 141–148.
142. White, H. and Sreenivasan, V. Epilepsy-aphasia syndrome in children: an unusual presentation to psychiatry, *Can. J. Psychiatry*, 32 (1987) 599–601.
143. Yasuhara, A., Yoshida, H., Hatanaka, T., Sugimoto, T., Kobayashi, Y., and Dyken, E. Epilepsy with continuous spike-waves during slow sleep and its treatment, *Epilepsia*, 32 (1991) 59–62.
144. Zardini, G., Molteni, B., Nardocci, N., Sarti, D., Avanzini, G., and Granata, T. Linguistic development in a patient with Landau Kleffner syndrome: a nine-year follow-up, *Neuropediatrics*, 26 (1995) 19–25.
145. Zivi, A., Broussaud, G., Daymas, S., Hazard, J., and Sicard, C. Epilepsy-acquired aphasia syndrome with psychosis: Report of a case, *Annal de Pediatrie*, 37 (1990) 391–394.

19

NEUROPROTECTIVE STRATEGIES IN EPILEPSY

Peter L. Carlen, Marc R. Pelletier, Aviv Ouanounou,
Michael Tymianski, and Liang Zhang

The Toronto Western Hospital
399 Bathurst Street
Toronto, Ontario, CANADA M5T 2S8
tel: (416) 603-5040
fax: (416) 603-5745

Dr. Carlen reviews the evidence that seizures may cause cell death and discusses possible strategies for preventing seizure-induced brain damage.

INTRODUCTION

It is well known that brain damage can cause epilepsy. Post-traumatic epilepsy (PTE), for instance, is a consequence of traumatic brain injury. The risk of PTE ranges from less than 5 percent of patients following blunt head trauma (e.g., simple skull fracture) to 50 percent in patients following penetrating wounds or those involving intracranial hematoma.[11] One of the accepted causes of temporal-lobe epilepsy is perinatal brain injury, which can also lead to other epileptic syndromes, such as West's syndrome and Lennox-Gastaut syndrome. Epilepsy can also be a presenting symptom of both brain tumors and strokes.

Conversely, it has become clear that epilepsy can also cause brain damage, and that seizure-induced brain damage may in turn cause further seizures. The present paper will discuss a variety of pathogenic mechanisms related to seizure-induced brain damage, with a focus on the role of unregulated increases in intracellular calcium (Ca^{2+}).

BRAIN DAMAGE FROM SEIZURES: CLINICAL STUDIES

Pathology of the hippocampus (cell death, gliosis) in humans with temporal-lobe epilepsy—referred to as "Ammon's horn sclerosis", "hippocampal sclerosis", or "mesial temporal sclerosis"—was first described over 100 years ago.[160] It has since been documented in many studies.[10,32] Pathophysiological features associated with epilepsy include excessive neuronal excitation,[111] disinhibition, or both.[76,77,171] These processes can lead to cell death in susceptible brain regions (see below), and ectopic synaptic connections.[166] Hypersynchrony, the hallmark of epilepsy, can also result from enhanced inhibition (discussed by Engel[39]). In the extreme case of convulsive status epilepticus (SE), profound systemic derangements and altered brain metabolism can present a life-threatening situation.[18,85,187]

The immature brain is particularly permissive in supporting epileptiform activity,[31,60,61,62,147] and approximately 85 percent of cases of SE occur before the age of 5 years.[4,92] This has a profound import for prognosis because, under certain circumstances (e.g., number and duration of seizures), SE itself may promote epileptogenesis.[86] Interestingly, and for unknown reasons, the immature brain is more resistant to seizure-induced damage compared to the adult brain.[5,71,115]

Sysemic Facors: Hypoxia

Clearly, when patients have prolonged SE and are apneic, controlling the airway and ensuring adequate oxygenation are essential. The multiple effects of hypoxia are complex and under intense study. SE is associated with a marked increase in catecholamines, which causes many systemic changes (e.g., hypertension, tachycardia, cardiac arrhythmias, hyperglycaemia, acidosis, and fever). These are initially reversible, but later can cause neuropathological changes.[186]

Initially there was controversy about whether brain damage was the effect of seizures or of systemic factors such as hypoxia. Studies by Meldrum suggest that the damage is done by the seizures themselves. In a model of convulsive SE, induced by the intravenous administration in baboons of the type A γ-aminobutyric acid ($GABA_A$) receptor antagonist bicuculline, distinct phases of systemic perturbations (e.g., blood pressure, lactate, glucose, and pH) and brain damage to the neocortex, cerebellum, and hippocampus were observed.[101] Damage to the hippocampus was evident even after the baboons had been paralysed and artificially ventilated, demonstrating that, while systemic perturbations increased the severity of brain damage, non-convulsive SE is sufficient to produce brain damage.[100] Similar results were observed when the experiments were repeated using rats.[19,98] Meldrum and colleagues concluded that asynchrony between blood flow and brain metabolism failed to explain the brain damage observed under these conditions. An important hypothesis emerged from these studies—that the mechanism of brain damage might relate to an accumulation of intracellular Ca^{2+}.

Sustained Electrical Stimulation

Compelling support for the excitotoxicity/seizure-induced brain damage hypothesis has been provided by experimental models utilizing sustained electrical stimulation of an afferent pathway.[25,73,158,162,185] Unlike the administration of convulsive drugs, electrical stimulation provides the advantage of isolating an excitatory afferent pathway. Sustained stimulation of the afferents produces excitotoxic damage in susceptible postsynaptic neurons.

Perhaps the most extensive experiments have been done by Sloviter and colleagues, who demonstrated, via sustained electrical stimulation of the perforant path, hippocampal neuropathology identical to that induced by kainic acid.[154,158] After Sloviter's stimulation protocol (or Lothman's continuous hippocampal stimulation paradigm), the inhibitory GABAergic basket interneurons of the hippocampus appeared relatively resistant to seizure-induced brain damage. It was discovered that the basket cells were "disconnected" from their presynaptic excitatory input (damaged hilar somatostatin-positive interneurons, hilar mossy cells), which results in a net loss of inhibition. These observations contributed to the "dormancy" hypothesis of hippocampal interneurons.[13,155,157]

More germane to this discussion was the demonstration that the neurons that were resistant to seizure-induced death (dentate granule cells, GABA-positive basket cells, CA2 pyramidal cells) stained positively for Ca^{2+} binding protein calbindin-D28k and parvalbumin-like immunoreactivity.[155,156] Consistent with these data are the following observations: 1) neurons containing parvalbumin-like and calbindin-D28k immunoreactivity in human neocortical epileptic foci are relatively preserved[41] (but see DeFelipe et al.[34]); 2) dentate hilar cells in rats that do not contain Ca^{2+} binding proteins, and which demonstrate electrophysiological deterioration after prolonged afferent stimulation, are afforded protection after chelation of intracellular Ca^{2+} with BAPTA;[142] 3) chelation of intracellular Ca^{2+} with BAPTA-AM, which is cell-permeant, reduced early excitotoxic and ischemic neuronal death both *in vitro* and *in vivo*;[180] and 4) a neuroprotective role for calretinin, a Ca^{2+} binding protein similar to calbindin, has been reported recently in rat neocortical cultures treated with either the Ca^{2+} ionophore A23187, NMDA, or kainate.[89] Therefore, these data suggest that the vulnerability of neurons may be inversely related to their capacity to sequester free intracellular Ca^{2+} (although see Freund et al.;[48] Weiss et al.[190]).

Kindling

Kindling is a phenomenon in which repeated presentation of an initially subconvulsant electrical stimulus to various brain regions results in a progressive and permanent enhancement of seizure susceptibility.[56] Kindling is considered to be a valid model of human temporal lobe epilepsy.[42,141,144] The amygdala and piriform cortex are considered to be critical structures in the initiation and propagation of epileptiform activity, and interictal discharges appear first in these structures, regardless of the site of kindling stimulation.[74,128] Utilizing a model of partial SE based upon kindling of seizures in the rat amygdala, McIntyre et al.[96] reported extensive neuronal pathology (degeneration, gliosis), which was most marked in the ipsilateral piriform cortex. In addition, hippocampal cell death has been reported after kindling,[15,127] and markers of apoptosis (below) are elevated after even a single discharge.[15]

Apoptosis

There are two morphological patterns of cell death, necrosis and apoptosis (programmed cell death) seen in both human and animal models. Necrosis, the more common form of cell death, occurs after an exogenous insult (e.g., ischemia, chemical injury) and is manifested by cell swelling or rupture, denaturation and coagulation of cytoplasmic proteins and breakdown of cell organelles. It is accompanied by a significant inflammatory response. A sudden and large increase in intracellular free Ca^{2+} can lead to necrosis.

Apoptosis, described first by Kerr, Wyllie and Currie,[79] a feature of normal development in both the central and the peripheral nervous system, requires *de novo* gene transcription, and drives the elimination of unwanted cell populations during embryogenesis.

The biology of apoptosis has been characterized most thoroughly in the nematode *Caenorhabditis elegans* (for reviews see Stewart;[163] White[191]), and involves members of the *ced* gene family. The most important members of this family include the death genes, *ced*-3 and *ced*-4[197] and the survival gene, *ced*-9,[67] which is homologous to the mammalian *bcl*-2 gene.[86]

NEUROPROTECTION TARGETS IN EPILEPSY

Accepting that seizures can damage the brain, it is important to look for interventions which can prevent that damage. A number of possible targets exist for pharmacological intervention. These include intracellular Ca^{2+}, glutamatergic neurotransmission and excessive neuronal discharge.

Calcium

Ca^{2+} is critical for many cellular processes. For normal functioning, the intracellular Ca^{2+} concentration must be maintained in the nM range. This feat is significant, considering that the cell contains large intracellular Ca^{2+} stores and that the extracellular Ca^{2+} concentration is greater than the free intracellular concentration by four orders of magnitude.[27,121] The best characterized mechanisms of Ca^{2+} influx are voltage-dependent Ca^{2+} channels (L, N, T, P, R), ligand-operated Ca^{2+} permeable channels (non-selective cation channels, e.g., NMDA), second-messenger-activated Ca^{2+}-permeable ion channels (e.g., metabotropic GluR1) and Ca^{2+}-release-activated currents (capacitative Ca^{2+} influx). The critical elements participating in neuronal intracellular Ca^{2+} homeostasis are: mitochondria, organelles associated with the endoplasmic reticulum (e.g., calciosomes), Na^+/Ca^{2+} exchange, Ca^{2+} binding proteins, and Ca^{2+} pumps.[17,97,106,107]

Perturbation of the intracellular Ca^{2+} concentration—with excess accumulation—can result in the initiation of a variety of Ca^{2+}-dependent cascades that can result in cell death.[27,46,80,107] The therapeutic efficacy of the antagonism of Ca^{2+} channels (e.g., dihydropyridine-sensitive, T-type) has been demonstrated previously.[105,150,151]

The current hypothesis of excitotoxicity describes both a rapid and a delayed component in cell death.[104,130] The early damage is the result of an influx of Na^+, Cl^-, and H_2O with osmolysis. The delayed cell death is secondary to an increase in intracellular Ca^{2+}, which is thought to initiate a variety of deleterious consequences including: 1) activation of phospholipase A2, which leads to the production of arachadonic acid and free radicals; 2) endonuclease fragmentation of DNA; and 3) production of nitric oxide, which contributes to inhibition of mitochondrial oxidative phosphorylation.[27,99]

Glutamatergic Transmission

Glutamate is considered to be the major excitatory neurotransmitter in the mammalian central nervous system. Based upon signal transduction mechanisms, glutamate receptors have been broadly categorized as either "ionotropic" or "metabotropic".[91,111,188] The ionotropic receptors are ligand-gated, non-specific cation channels that mediate synaptic transmission. Based upon their affinities for structural analogues of glutamate, this category has been subdivided further into the NMDA and the non-NMDA receptors. The non-NMDA receptors bind α-amino-3-hydroxy-5-methyl-4-isoxazolepropionate

(AMPA) and kainate. Metabotropic glutamate receptors (mGluRs) activate second-messenger signalling cascades via G-proteins.[30,145,174] At least seven distinct cDNAs encoding mGluR subtypes (mGluR1–7) have been cloned.[51,113,146,168] Three major groups of mGluRs have been identified based upon amino-acid sequence, coupling to second-messenger cascades and pharmacology. Activation of mGluR1 and mGluR5 (Group I) stimulates the phosphokinase C-mediated formation of inositol-1,4,5-trisphosphate (IP_3) and mobilization of intracellular Ca^{2+}, and also arachadonic acid release.[3,8] In contrast, mGluR2 and 3 (Group II) appear to be negatively coupled to cyclic adenosine monophosphate (cAMP).[113,168] Group III comprises mGluR4, 6, and 7 (for review see Miller et al.;[108] Pin and Duvoisin;[125] Recasens and Vignes[134]).

The prominent role of Ca^{2+} influx through NMDA receptors in hypoxic-ischemic injury and excitotoxicity,[47,50,99,138,180] and the neuroprotection afforded by antagonists of NMDA receptors,[28,50,104] is well established. Neurotoxicity of Ca^{2+} influx via non-NMDA receptors,[21,45] and the neuroprotection by antagonists of non-NMDA receptors,[120] has also been reported.

Less clear is the role of mGluRs in epileptogenesis and neurotoxicity. Stimulation of mGluRs with the active stereoisomer of 1-aminocyclopentane-1,3-dicarboxylic acid (trans-ACPD), 1S,3R-ACPD has been reported to produce effects consistent with enhanced excitability,[36,59,129] epileptiform discharges[102,170,176] and seizure-related brain injury.[94,95,139] Conversely, stimulation of mGluRs has also been reported to produce a decrease in epileptiform activity[23,148] and to provide neuroprotection against NMDA-induced neurotoxicity.[16,22,29,81,90,152] The heterogeneity of mGluRs, differences in their developmental and regional expression in the brain, neuronal localization[93,149,169,183] and lack of specific antagonists likely explain these opposing effects.

Excessive Neuronal Discharge

As indicated above, excessive neuronal discharge is associated with brain damage. This is thought to be due to excitotoxicity resulting from excessive glutamate release, which activates both N-methyl-D-aspartate (NMDA) and non-NMDA receptors to cause increased Ca^{2+} influx and subsequent neurotoxicity. Hence, a rational approach to prevent seizure-induced brain damage is to control epileptiform discharges by administering anticonvulsants, which would be expected to decrease the epileptic bursting discharges. In addition to stopping seizures *per se*, other neuroprotective mechanisms have been attributed to the anticonvulsants, based on *in vitro* and *in vivo* animal models.

There is evidence that blockade of voltage-dependent sodium (Na^+) channels is a reasonable target for neuroprotection.[173] Phenytoin has been shown to be neuroprotective in a model of focal brain ischemia in the rat,[20] and the local anaesthetic and Na^+ channel blocker, lidocaine, accelerated neuroelectrical recovery after incomplete global ischemia in rabbits.[132] Tetrodotoxin, a specific Na^+ channel blocker, when topically applied to the hippocampus prior to the induction of transient forebrain ischemia in rats, decreased post-ischemic CA1 cell loss.[196] BW619C89 pyrimidine, a use-dependent blocker of voltage-dependent Na^+ channels, was neuroprotective in a rat model of focal cerebral ischaemia, possibly by inhibiting glutamate release.[57] Global ischemia-induced hippocampal neuronal loss in gerbils was significantly attenuated by phenytoin.[167] Na^+ channel blocking agents protect against *in vitro* hypoxia-induced impairment of hippocampal synaptic transmission,[161] hypoglycaemia-induced neurotoxicity in hippocampal culture,[172] and combined oxygen and glucose deprivation-induced neurotoxicity in hippocampal slices.[189]

Another putative neuroprotective effect of Na^+ channel blockers is to prevent the influx of Na^+ into neurons. Acutely, this influx could cause neuronal swelling with increased ephaptic coupling and synchronization, and also could reverse the Na^+/Ca^{2+} exchange gradient giving rise to increased and potentially neurotoxic Ca^{2+} entry. In CNS white matter (optic nerve), it has been shown that both Na^+ channel blockers and reduction of extracellular Na^+ protect against anoxic injury.[164] Whether reversed Na^+/Ca^{2+} exchange plays a role in grey matter injury is unclear.

BRAIN DAMAGE IN EXPERIMENTAL MODELS OF EPILEPSY

Experiments with animal models of epilepsy have provided further evidence that seizures can produce brain damage:

Kainic Acid Model

In experimental conditions, epileptic syndromes are commonly induced by the administration of convulsive drugs. One of the most investigated models has been the kainic acid model. Kainic acid is structurally similar to the EAA neurotransmitter glutamate. Documentation of its powerful convulsive property provided a major impetus for the excitotoxin hypothesis.[116] Although the administration of convulsants such as kainic acid also produces direct neurotoxic effects, and the limits of their distribution are difficult to assess, the kainic acid model provides evidence in support of the hypothesis that seizures themselves, independent of any systemic sequelae, can produce brain damage.[14,87,112]

In addition to its physiological role, apoptosis has also been observed under pathological conditions (reviewed in Thompson[175]), including ischemia,[83,84] hypertension,[63] Huntington's and Alzheimer's disease[37] and epilepsy.[26,135] Perturbation of intracellular Ca^{2+} homeostasis has been demonstrated to also trigger apoptosis.[114] Because both necrosis and apoptosis are often found in pathology, distinguishing between these types of cell death is becoming progressively more difficult. Indeed, some investigators describe the phenotype of cell death in terms of variations upon an apoptosis-necrosis continuum.[126]

Neurotrophins

The induction of apoptosis (programmed cell death) can be prevented by the administration of neurotrophins, nerve growth factor (NGF) proteins which activate members of the *trk* family of tyrosine kinase receptors (for review see Barbacid;[12] Kaplan and Stephens[75]). The regulation of cell fate by NGF has been demonstrated recently to be receptor-dependent, i.e., activation of *trk*A receptors promotes survival. Activation of the non-tyrosine kinase receptor, p75, however, causes cell death.[44] Seizure-induced cell death of hilar neurons is associated with axonal sprouting of dentate granule cells.[165,166] The axonal sprouting is promoted by the activity-dependent activation of genes encoding neurotrophic factors.[117] Consistent with these data are the following observations: 1) a progressive and persistent increase in mRNA levels for both NGF and brain-derived nerve growth factor (BDNF) when seizures are kindled with electrical stimulation of the amygdala;[140] 2) prolonged increases in neurotrophic activity following kainate-induced hippocampal synaptic reorganization;[88] and 3) the retardation of kindling and blockade of mossy fiber sprouting following the intraventricular infusion of either anti-NGF IgGs[182] or a synthetic inhibitory NGF peptide.[131] Drugs which activate NGF receptors, therefore, might be developed for use in neuroprotection.

Gap Junctions

Gap junctions are composed of proteins called "connexins" that form a channel between adjacent cells creating a direct cytoplasmic connection between them (reviewed by Dermietzel and Spray[35]). Gap junctions are now recognized as playing a critical role in neural synchronization in some seizure models.[24,123,133,181] Gap junctions not only transfer electrical charge when open, but can also transfer small molecules. It has been hypothesized that they play a role in diseases such as the X-linked dominant Charcot-Marie-Tooth disease[159] and the development of spreading depression[69]—a phenomenon associated with seizures and ischemia. Therefore, modulating gap junctional function might prevent not only seizures but also affect other neuropathological processes.

Free Radicals

Production of free radicals has been implicated in several models of neurotoxicity. Intracortical injection of iron compounds (e.g., $FeCl_2$, $FeCl_3$) produces focal epileptiform activity, formation of superoxide radicals, cell death and gliosis.[153,178,192,193] Corticosteroid[195] or antiperoxidant[194] pretreatment reduces iron-induced seizures, free radical formation, and brain injury. Interestingly, phenytoin prevents the seizures, but is without effect on lipid peroxidation.[195]

Chemically different antioxidants, in doses known to scavenge free radicals, markedly suppressed the development of kindling and were shown to have anticonvulsant properties in rats.[143] In tottering mice, a genetic model of human generalized epilepsy, reduced glutathione (GSH), the most important free radical scavenging compound *in vivo*, has been found to be significantly lower in the occipital cortex, and total GSH is lower in hippocampus and occipital cortex.[2] These data suggest that strategies to decrease free radical formation might play an important role in treating seizure syndromes.

Osmotic Shifts

Reducing extracellular space increases the extracellular resistance, thereby enhancing electrical interactions. This occurs in hyposmolar states, such as water intoxication and hypoxia-ischemia. Hyposmolar states can cause clinical seizures and have been used in *in vitro* models of epilepsy.[6,38,138,177] Dudek et al.[38] showed that the epileptiform bursting in hippocampal CA1 neurons provoked by zero Ca^{2+} perfusate, which also blocks synaptic transmission, was enhanced by decreasing further the osmolality, thereby reducing the extracellular volume. Conversely, membrane-impermeant solutes, which increase the extracellular volume, reduced or blocked epileptiform discharges both in the zero-Ca^{2+}seizure model[38] and in an *in vitro* model of high potassium-induced seizures.[177] Furosemide also blocked epileptiform activity in hippocampal slices, probably by blocking a decrease in the extracellular space.[70] It is possible that, with cell swelling, increased neuronal apposition could lead to increased field coupling between adjacent neurons.[24] Whether blocking cellular swelling during seizures will have neuroprotective effects above and beyond diminishing neural synchrony and seizures is not known.

Cytokines

Injury to the CNS elicits a complex cellular response involving both astrocytes and microglia. Reactive glial populations make up the so called "glial scar" that has long been implicated as a barrier to axonal regeneration and as a causal factor in the genesis of epilepsy. Cytokines, a class of protein immunomodulators, are important signals that help control structural and functional responses to CNS injury.[53,58] Cytokines comprise a large number of diverse molecules including the interleukins (ILs), tumour necrosis factors (TNFs), the interferons (INFs) and the colony-stimulating factors (CSFs).[103] Some of these immunomodulators are produced within and act upon cells of the CNS.[53] In addition, clinical and pathological observations have shown increased cytokine levels during the course of stroke,[103] acquired immunodeficiency syndrome,[109] trauma[52] and epilepsy.[54] One long-standing hypothesis held by epileptologists[118] is that seizures are a consequence of reactive gliosis ("epileptogenic glial scar"). Giulian et al.,[54] employing cytokines as tools to manipulate reactive gliosis demonstrated that IL-1 supported neuronal growth by an action upon astroglia, while granulocyte-macrophage CSF (GM-CSF) initiated epileptic-like discharges through mechanisms involving reactive microglia. Perhaps IL-1 promoted neuronal growth and survival by driving astroglial release of neurotrophic factors, as reported by Fagan and Gage.[40] The importance of cytokine research is that it advances the understanding of neuroimmunomodulators in the diseased brain—modulators that might offer alternative therapeutic strategies to control reactive gliosis and, thus, to improve recovery from CNS injury.

Autoimmune Mechanisms

It is possible to induce epilepsy in animals not only by electrical or pharmacological stimulation but also through autoimmune mechanisms.[78] Analogous to other autoimmune diseases, in which high doses of gamma-globulin are used therapeutically with good success, Laffont et al.[82] and Pechadre et al.[119] have used gamma-globulin in children with epilepsy. More recently, altered immunoglobulin subtypes have been tried,[1,9] especially immunoglobulin light chains.[64,65] Ariizumi et al.[9] showed that, of eight patients who had had attacks for two years or less, four showed complete clinical and EEG remission, two showed marked improvement and the remaining two showed a moderated improvement in electrical seizures after high dose non-treated gamma-globulin (NTGG) therapy. Although there is as yet no consensus concerning the clinical relevance of immune mechanisms in epilepsy, there are some suggestive findings. For example, asialomonoganglioside antibody is raised in serum and cerebrospinal fluid,[72] brain nicotinic acetylcholine receptor antibody is increased in serum,[43] and Daute and co-workers[33] found that immunosuppression greatly benefited children with intractable epilepsy. Moreover, Ramussen's encephalitis, a progressive childhood disorder characterized by intractable focal seizures and inflammatory histopathology, has recently been suggested to be caused, at least in some cases, by an autoimmune mechanism: autoantigen activation of GluR3.[136,179] Consistent with this hypothesis is the observation of a reduction in seizure frequency in patients with Rasmussen's following administration of immunoglobulins, steroids[66] or plasmaphoresis.[7]

Summary

The principal points that emerge from the preceding discussion are that a) seizures can produce brain damage, b) there are numerous factors that contribute to seizure-induced neuronal death, and c) altered intracellular Ca^{2+} homeostasis is a prominent mechanism.

REFERENCES

1. Aarli, J.A. Immunological aspects of epilepsy, *Brain Development*, 15 (1993) 41–50.
2. Abbott, L.C., Nejad, H.H., Bottje, W.G., and Hassan, A.S. Glutathione levels in specific brain regions of genetically epileptic (tg/tg) mice, *Brain Research Bulletin*, 25 (1990) 629–631.
3. Abe, T., Sugihara, H., Shigemoto, R., Mizuno, N., and Nakanishi, S. Molecular characterization of a novel metabotropic glutamate receptor mGluR5 coupled to inositol phosphate/Ca^{2+} signal transduction, *Journal of Biological Chemistry*, 267 (1992) 13361–13368.
4. Aicardi, J. and Chevrie, J.J. Convulsive status epilepticus in infants and children: a study of 239 cases, *Epilepsia*, 11 (1971) 187–197.
5. Albala, B.J., Moshe, S.L., and Okada, R. Kainic acid-induced seizures: a developmental study, *Developmental Brain Research*, 13 (1984) 139–148.
6. Andrew, R.D. Seizure and acute osmotic change: clinical and neurophysiological aspects, *Journal of Neurology*, 101 (1991) 7–18.
7. Andrews, P.I., Dichter, M.A., Berkovic, S.F., Newton, M.R., and McNamara, J.O. Plasmapheresis in Rasmussen's encephalitis, *Neurology*, 46 (1996) 242–246.
8. Aramori, I. and Nakanishi, S. Signal transduction and pharmacological characteristics of a metabotropic glutamate receptor, mGluR1, in transfected CHO cells, *Neuron*, 8 (1992) 757–765.
9. Arriizumi, M., Baba, K., Shiihara, H., Ogawa, K., Hibio, S., Hinib, Y., Ryo, S., and Momoki, T. High dose gamma-globulin for intractable childhood epilepsy, *Lancet*, ii (1983) 162–163.
10. Babb, T.L., Brown, W.J., Pretorius, J., Davenport, C., Lieb, J.P., and Crandall, P.H. Temporal lobe volumetric cell densities in temporal lobe epilepsy, *Epilepsia*, 25 (1984) 729–740.
11. Babb, T.L. and Pretorious, J.K. Pathologic substrates of epilepsy. In: E. Wylie (Ed.) *The treatment of epilepsy: Principles and practice*, Lea and Febiger, Philadelphia (1993) pp. 55–70.
12. Barbacid, M. The Trk family of neurotrophin receptors, *Journal of Neurobiology*, 25 (1994) 1386–1403.
13. Bekenstein, J.W. and Lothman, E.W. Dormancy of inhibitory interneurons in a model of temporal lobe epilepsy, *Science*, 259 (1993) 97–100.
14. Ben-Ari, Y. Limbic seizure and brain damage produced by kainic acid: mechanisms and relevance to human temporal lobe epilepsy, *Neuroscience*, 14 (1985) 375–403.
15. Bengzon, J., Kokaia, Z., Elmer, E., Nanobashvili, A., Kokaia, M., and Lindvall, O. Apoptosis and proliferation of dentate gyrus neurons after single and intermittent limbic seizures. *Proc. Natl. Acad. Sci. USA*, 94 (1997) 10432–10437.
16. Birrell, G.J., Gordon, M.P., and Marcoux, F.W. (1S,3R)-1-aminocyclopentane-1,3-dicarboxylic acid attenuates N-methyl-D-aspartate-induced neuronal cell death in cortical cultures via a reduction in delayed Ca^{2+} accumulation, *Neuropharmacology*, 32 (1993) 1351–1358.
17. Blaustein, M.P. Calcium transport and buffering in neurons, *Trends in Neuroscience*, 11 (1988) 438–443.
18. Bleck, T.P. Convulsive disorders: Status epilepticus, *Clinical Neuropharmacology*, 14 (1991) 191–198.
19. Blennow, G., Brierly, J.B., Meldrum, B.S., and Siesjo, B.K. Epileptic brain damage. The role of systemic factors that modify cerebral energy metabolism, *Brain*, 101 (1978) 687–700.
20. Boxer, P.A., Cordon, J.J., Mann, M.E., Rodolosi, L.C., Vartanian, M.G., Rock, D.M., Taylor, C.P., and Marcoux, F.W. Comparison of phenytoin with non-competitive N-Methyl-D-Aspartate antagonists in model of focal brain ischemia in rat, *Stroke*, 21 (1990) 47–51.
21. Brorson, J.R., Manzolillo, P.A., and Miller, R.J. Ca^{2+} entry via AMPA/KA receptors and excitotoxicity in cultured cerebellar purkinje cells, *Journal of Neuroscience*, 14 (1994) 187–197.
22. Bruno, V., Battaglia, G., Copani, A., Giffard, R.G., Raciti, G., Raffaele, R., Shinozaki, H., and Nicoletti, F. Activation of class II or III metabotropic glutamate receptors protects cultured cortical neurons against excitotoxic degeneration, *Eur. J. Neurosci.*, 7 (1995) 1906–1913.
23. Burke, J.P. and Hablitz, J.J. Metabotropic glutamate receptor activation decreases epileptiform activity in rat neocortex, *Neuroscience Letters*, 174 (1994) 29–33.
24. Carlen, P.L., Perez-Velazquez, J.L., Valiante, T.A., Jahromi, J.J., and Bardakjian, B.L. Electric coupling in epileptogenesis. In: R. Dermietzel and D.C. Spray (Eds.) *Gap Junctions in the Nervous System*, R.G. Landes Co., Austin [i.e., Georgetown] Tex., (1996).
25. Cavazos, J.E., Das, I., and Sutula, T.P. Neuronal loss induced in limbic pathways by kindling: Evidence for induction of hippocampal sclerosis by repeated brief seizures, *J. Neurosci.*, 14 (1994) 3106–3121.

26. Charriaut-Marlangue, C., Aggoun-Zouaoui, D., Represa, A., and Ben-Ari, Y. Apoptotic features of selective neuronal death in ischemia, epilepsy and gp120 toxicity, *Trends Neurosci.*, 19 (1996) 109–114.
27. Choi, D.W. Calcium-mediated neurotoxicity: relationship to specific channel types and role in ischemic damage, *Trends Neurosci.*, 11 (1988) 465–469.
28. Clifford, D.B., Zorumski, C.F., and Olney, J.W. Ketamine and MK-801 prevent degeneration of thalamic neurons induced by focal cortical seizures, *Experimental Neurology*, 105 (1989) 272–279.
29. Colwell, C.S., Altemus, K.L., Cepeda, C., and Levine, M.S. Regulation of N-methyl-D-aspartate-induced toxicity in the neostriatum: a role for metabotropic glutamate receptors? *Proc. Natl Acad. Sci. USA*, 489 (1996) 115–125.
30. Conn, P.J. and Desai, M.A. Pharmacology and physiology of metabotropic glutamate receptors in mammalian central nervous system, *Drug Development Research*, 24 (1991) 207–229.
31. Connors, B.W., Bernardo, L.S., and Prince, D.A. Coupling between neurons of the developing rat neocortex, *J. Neurosci.*, 3 (1983) 773–782.
32. Dam, A.M. Epilepsy and neuron loss in the hippocampus, *Epilepsia*, 21 (1980) 617–629.
33. Daute, K.H., Hassler, A., and Klust, E. Myoklonisch-Astatisches Petit Mal: Immunologische Aspekte in Atiopathogenese und Therapie, *IVth Prague International Symposium of Child Neurology*, (1978) abstr. p. 16.
34. DeFelipe, J., Sola, R.G., Marco, P., del Rosario del Rio, M., Pulido, P., and Cajal, S.R. Selective changes in the microorganization of the human epileptogenic neocortex revealed by parvalbumin immunoreactivity, *Cerebral Cortex*, 3 (1993) 39–48.
35. Dermietzel, R. and Spray, D.C. Gap junctions in the brain: where, what type, how many and why? *Trends Neurosci.*, 16 (1993) 186–192.
36. Desai, M.A., McBain, C.J., Kauer, J.A., and Conn, P.J. Metabotropic glutamate receptor-induced disinhibition is mediated by reduced transmission at excitatory synapses onto interneurons and inhibitory synapses onto pyramidal cells, *Neuroscience Letters*, 181 (1994) 78–82.
37. Dragunow, M., Faull, R.L., Lalor, P., Beilharz, E.J., Singleton, K., Walker, E.B., and Mee, E. In situ evidence for DNA fragmentation in Huntington's disease striatum and Alzheimer's disease temporal lobes, *Neuroreport*, 6 (1995) 1053–1057.
38. Dudek, F.E., Obenaus, A., and Tasker, J.G. Osmolality-induced changes in extracellular volume alter epileptiform bursts independent of chemical synapses in the rat: importance of non-synaptic mechanisms in hippocampal epileptogenesis, *Neuroscience*, 120 (1990) 267–270.
39. Engel, J., Jr. Excitation and inhibition in epilepsy, *Can. J. Neurol. Sci.*, 23 (1996) 167–174.
40. Fagan, A.M. and Gage, F.H. Cholinergic sprouting in the hippocampus: A proposed role for IL-I, *Experimental Neurology*, 100 (1990) 105–120.
41. Ferrer, I., Oliver, B., Russi, A., Casas, R., and Rivera, R. Parvalbumin and calbindin-D28k immunocytochemistry in human neocortical epileptic foci, *J. Neurol. Sci.*, 123 (1994) 18–25.
42. Fisher, R.S. Animal models of the epilepsies, *Brain Res. Reviews*, 14 (1989) 245–278.
43. Fontana, A., Fulpius, B.W., and Cuenoud, S. Antibodies against muscle and brain nicotinic acetylcholine receptors in IgA deficient patients with epilepsy, *Adt. Cytopharmacology*, 3 (1979) 287–292.
44. Frade, J.M., Rodriguez-Tébar, A., and Barde, Y.-A. Induction of cell death by endogenous nerve growth factor through its p75 receptor, *Nature*, 383 (1996) 166–168.
45. Frandsen, A., Drejer, J., and Schousboe, A. Direct evidence that excitotoxicity in cultured neurons is mediated via N-methyl-D-aspartate (NMDA) as well as non-NMDA receptors. *J. Neurochemistry*, 53 (1989) 297–299.
46. Frandsen, A. and Schousboe, A. Dantrolene prevents glutamate cytotoxicity and Ca^{2+} release from intracellular stores in cultured cerebral cortical neurons, *J. Neurochemistry*, 56 (1991) 1075–1078.
47. Frandsen, A. and Schousboe, A. Excitatory amino acid-mediated cytotoxicity and calcium homeostasis in cultured neurons, *J. Neurochemistry*, 60 (1993) 1202–1210.
48. Freund, T.F., Buzsaki, G., Leon, A., Baimbridge, K.G., and Somogyi, P. Relationship of neuronal vulnerability and calcium binding protein immunoreactivity in ischemia, *Exp. Brain Res.*, 83 (1990) 55–66.
49. Garthwaite, G. and Garthwaite, J. Neurotoxicity of excitatory amino acid receptor agonists in rat cerebellar slices: Dependence on calcium concentration, *Neuroscience Letters*, 66 (1986) 193–198.
50. Garthwaite, G. and Garthwaite, J. Differential dependence on Ca^{2+} of N-methyl-D-aspartate and quisqualate neurotoxicity in young rat hippocampal slices, *Neuroscience Letters*, 97 (1989) 316–322.
51. Gereau, R.W. IV. and Conn, P.J. Roles of specific metabotropic receptor subtypes in regulation of hippocampal CA1 pyramidal cell excitability, *J. Neurophysiology*, 74 (1995) 122–129.
52. Giulian, D., Chen, J., George, J., and Noponen, M. The role of mononuclear phagocytes in wound healing after traumatic injury to the adult mammalian brain, *J. Neuroscience*, 9 (1989) 4416–4429.

53. Giulian, D., Johnson, B., Krebs, J.F., George, J., and Tapscott, M. Microglial mitogens are produced in the developing and injured mammalian brain, *J. Cell Biology*, 112 (1991) 323–333.
54. Giulian, D., Li, J., Leara, B., and Keenen, C., Phagocytic microglia release cytokines and cytotoxins that regulate the survival of astrocytes and neurons in culture, *Neurochem. Int.*, 25 (1994) 227–233.
55. Giulian, D., Li, J., Li, X., George, J., and Rutecki, P.A. The impact of microglia-derived cytokines upon gliosis in the CNS, *Dev. Neurosci.*, 16 (1994) 128–136.
56. Goddard, G.V., McIntyre, D.C., and Leech, C.K. A permanent change in brain function resulting from daily electrical stimulation, *Experimental Neurology*, 25 (1969) 295–330.
57. Graham, S.H., Chen, J., Lan, J., Leach, M.J., and Simon, R.P. Neuroprotective effects of a use-dependent blocker of voltage-dependent sodium channels, BW619C89, in rat middle cerebral artery occlusion, *J. Pharmacol. Exp. Ther.*, 269 (1994) 854–859.
58. Grau, G.E., Fajardo, L., Piguet, P.F., Allet, B., Lambert, P.H., and Vassalli, P. Tumor necrosis factor (cachectin) as an essential mediator in murine cerebral malaria, *Science*, 237 (1987) 1210–1212.
59. Greene, C., Schwindt, P., and Crill, W. Metabotropic receptor mediated afterdepolarization in neocortical neurons, *Eur. J. Pharmacol.*, 226 (1992) 279–280.
60. Haas, K., Sperber, E.F., and Moshe, S.L. Kindling in developing animals: expression of severe seizures and enhanced development of bilateral foci, *Developmental Brain Research*, 56 (1990) 275–280.
61. Hablitz, J.J. and Heinemann, U. Extracellular K^+ and Ca^{2+} changes during epileptiform discharges in the immature rat neocortex, *Brain Res.*, 433 (1987) 299–303.
62. Hablitz, J.J. and Heinemann, U. Alterations in the microenvironment during spreading depression associated with epileptiform discharges in the immature neocortex, *Developmental Brain Research*, 46 (1989) 243–252.
63. Hamet, P., Richard, L., Dam, T.-V., Teiger, E., Orlov, S.N., Gaboury, L., Gossard, F., and Tremblay, J. Apoptosis in target organs of hypertension, *Hypertension*, 26 (1995) 642–648.
64. Haraldsson, A., van Engelen, B.G.M., Reiner, W.O., Bakkeren, A.J.M., and Weemaes, C.M.R. Light chain ratios and concentrations of serum immunoglobulins in children with epilepsy, *Epilepsy Res.*, 13 (1992b) 255–260.
65. Haraldsson, A., Weemaes, C.M.R., Koch-Jansen, M.J.H., van Eck-Arts, P.B.J.M., de Boo, T., Bakkeren, J.A.J.M., and Stoelinga, G.B.A. Immunoglobulin G, A and M light chain ration in children. *Annals of Clinical Biochemistry*, 29 (1992a) 270–274.
66. Hart, Y.M., Cortez, M., Andermann, F., Hwang, P., Fish, D.R., Dulac, O., Silver, K., Fejerman, N., Cross, H., Sherwin, A. et al. Medical treatment of Rasmussen's syndrome (chronic encephalitis and epilepsy): effect of high-dose steroids or immunoglobulins in 19 patients, *Neurology*, 44 (1995) 1030–1036.
67. Hengartner, M.O., Ellis, R.E., and Horvitz, H.R. *Caenorhabditis elegans* gene ced-9 protects cells from programmed cell death, *Nature*, 356 (1992) 494–499.
68. Hengartner, M.O. and Horvitz, H.R. *C. elegans* cell survival gene ced-9 encodes a functional homolog of the mammalian proto-oncogene bcl-2, *Cell*, 76 (1994) 665–676.
69. Herreras, O., Largo, C., Ibarz, J.M., Somjen, G.G., and Martin del Rio, R. Role of neuronal synchronizing mechanisms in the propagation of spreading depression in the in vivo hippocampus, *J. Neuroscience*, 14 (1994) 7087–7098.
70. Hochman, D.W., Baraban, S.C., Owens, J.W., and Schwartzkroin, P.A. Dissociation of synchronization and excitability in furosemide blockade of epileptiform activity, *Science*, 270 (1995) 99–102.
71. Holmes, G.L. and Thompson, J.L. Effects of kainic acid on seizure susceptibility in the developing brain, *Developmental Brain Research*, 39 (1988) 51–59.
72. Hoshino, C., Yotsumiya, N., and Yata, J. Relationship between anti-GA1 antibody and CNS destructions in nervous diseases, *Acta Pediatrica Japon*, 86 (1982) 1694.
73. Inoue, K., Morimoto, K., Sato, K., Yamada, N., and Otsuki, S. Mechanisms in the development of limbic status epilepticus and hippocampal neuronal loss: An experimental study in a model of status epilepticus induced by kindling-like electrical stimulation of the deep prepyriform cortex in rats, *Acta Medica Okayama*, 46 (1992) 129–139.
74. Kairiss, E.W., Racine, R.J., and Smith, G.K. The development of the interictal spike during kindling in the rat, *Brain Res.*, 322 (1984) 101–110.
75. Kaplan, D.R. and Stephens, R.M. Neurotrophin signal transduction by the Trk receptor, *J. Neurobiology*, 25 (1994) 1404–1417.
76. Kapur, J. and Lothman, E.W. Loss of inhibition precedes delayed spontaneous seizures in the hippocampus after tetanic electrical stimulation, *J. Neurophysiology*, 61 (1989) 427–434.

77. Kapur, J., Stringer, J.L., and Lothman, E.W. Evidence that repetitive seizures in the hippocampus cause a lasting reduction of GABAergic inhibition, *J. Neurophysiology*, 61 (1989) 417–426.
78. Karpiak, S.E., Graf, L., and Rapport, M.M. Antiserum to brain gangliosides produces recurrent epileptiform activity, *Science*, 194 (1976) 735–737.
79. Kerr, J.F.R., Wyllie, A.H., and Currie, A.R. Apoptosis: a basic biological phenomenon with wide-ranging implications in tissue kinetics, *Br. J. Cancer*, 26 (1972) 239–257.
80. Kiedrowski, L., Brooker, G., Costa, E., and Wroblewski, J.T. Glutamate impairs neuronal calcium extrusion while reducing sodium gradient, *Neuron*, 12 (1994) 295–300.
81. Koh, J., Palmer, E., and Cotman, C.W. Activation of the metabotropic glutamate receptor attenuates N-methyl-D-aspartate neurotoxicity in cortical cultures, *Proc. Natl Acad. Sci.*, 88 (1991) 9431–9437.
82. Laffont, F., Esnault, S., Gilbert, A., Peytour, M.A., Cathala, H.P., and Eygonnet, J.P. Effet des gammaglobulines sur des epelepsies rebelles, *Ann. Med. Int.* 130 (1979) 307–312.
83. Li, Y., Sharov, V.G., Jiang, N., Zaloga, C., Sabbah, H.N., and Chopp, M. Ultrastructural and light microscopic evidence of apoptosis after middle cerebral artery occlusion in the rat, *Am. J. Pathology*, 146 (1995) 1045–1051.
84. Linnik, M.D., Zobrist, R.H., and Hatfield, M.D. Evidence supporting a role for programmed cell death in focal cerebral ischemia in rats, *Stroke*, 24 (1993) 2002–2008.
85. Lothman, E. The biochemical basis and pathophysiology of status epilepticus, *Neurology*, 40 (Suppl.2) (1990) 13–23.
86. Lothman, E.W. and Bertram, E.H. III. Epileptogenic effects of status epilepticus, *Epilepsia*, 34 (Suppl.1) (1993) S59–S70.
87. Lothman, E.W. and Collins, R.C. Kainic acid induced limbic seizures: Metabolic, behavioral, electroencephalographic and neuropathological correlates, *Brain Res.*, 218 (1981) 299–318.
88. Lowenstein, D.H., Seren, M.S., and Longo, F.M. Prolonged increases in neurotrophic activity associated with kainate-induced hippocampal synaptic reorganization, *Neuroscience*, 56 (1993) 597–604.
89. Lukas, W. and Jones, K.A. Cortical neurons containing calretinin are selectively resistant to calcium overload and excitotoxicity in vitro, *Neuroscience*, 61 (1994) 307–316.
90. Maiese, K., Swiriduk, M., and TenBroeke, M. Cellular mechanisms of protection by metabotropic glutamate receptors during anoxia and nitric oxide toxicity, *J. Neurochemistry*, 66 (1996) 2419–2428.
91. Mayer, M.L. and Westbrook, G.L. The physiology of excitatory amino acids in the vertebrate central nervous system, *Progress in Neurobiology*, 28 (1987) 197–276.
92. Maytal, J., Shinnar, S., Moshe, S.L., and Alvarez, L.A. Low morbidity and mortality of status epilepticus in children, *Pediatrics*, 83 (1989) 323–331.
93. McBain, C.J., DiChiara, T.J., and Kauer, J.A. Activation of metabotropic glutamate receptors differentially affects two classes of hippocampal interneurons and potentiates excitatory synaptic transmission, *J. Neuroscience*, 14 (1994) 4433–4445.
94. McDonald, J.W., Fix, A.S., Tizzano, J.P., and Schoepp, D.D. Seizures and brain injury in neonatal rats induced by 1S,3R-ACPD, a metabotropic glutamate receptor agonist, *J. Neuroscience*, 13 (1993) 4445–4455.
95. McDonald, J.W. and Schoepp, D.D. The metabotropic excitatory amino acid receptor agonist 1S,3R-ACPD selectively potentiates N-methyl-D-aspartate-induced brain injury, *Eur. J. Pharmacology*, 215 (1992) 353–354.
96. McIntyre, D.C., Nathanson, D., and Edson, N. A new model of partial status epilepticus based on kindling, *Brain Res.*, 250 (1982) 53–63.
97. Meldolesi, J., Volpe, P., and Pozzan, T. The intracellular distribution of calcium, *Trends Neurosci.*, 11 (1988) 449–452.
98. Meldrum, B.S. Metabolic factors during prolonged seizures and their relation to nerve cell death. In: A.V. Delgado-Escueta, C.G. Wasterlain, D.M. Treiman, and R.J. Porter (Eds.) *Advances in Neurology. Status epilepticus: mechanisms of brain damage and treatment*, vol 34. Raven, New York (1983) pp. 261–275.
99. Meldrum, B. and Garthwaite, J. Excitatory amino acid neurotoxicity and neurodegenerative disease, *Trends Pharmacol. Sci.*, 11 (1990) 379–387.
100. Meldrum, B., Vigouroux, R., and Brierley, J. Systemic factors and epileptic brain damage: prolonged seizures in paralyzed artificially ventilated baboons, *Arch. Neurology Psychology*, 29 (1973) 82–87.
101. Meldrum, B.S. and Brierly, J.B. Prolonged epileptic seizures in primates: ischemic cell damage and its relation to ictal physiological events, *Arch. Neurology*, 28 (1973) 10–17.
102. Merlin, L.R., Taylor, G.W., and Wong, R.K. Role of metabotropic glutamate receptor subtypes in the patterning of epileptiform activities in vitro, *J. Neurophysiology*, 74 (1995) 896–900.

103. Merril, J.E. Tumour necrosis factor alpha, interleukin 1 and related cytokines in brain development: Normal and pathological, *Developmental Neuroscience*, 14 (1992) 1–10.
104. Meyer, F.B. Calcium, neuronal hyperexcitability and ischemic injury. *Brain Res. Reviews*, 14 (1989) 227–243.
105. Meyer, F.B., Tally, P.W., Anderson, R.E., Sundt, T.M. Jr., Yaksh, T.L., and Sharbrough, F.W. Inhibition of electrically induced seizures by a dihydropyridine calcium channel blocker, *Brain Res.*, 384 (1986) 180–183.
106. Miller, R.J. Calcium signalling in neurons, *Trends Neurosci.*, 11 (1988) 415–419.
107. Miller, R.J. The control of neuronal Ca^{2+} homeostasis, *Progress in Neurobiology*, 37 (1991) 255–285.
108. Miller, S., Kesslack, J.P., Romano, C., and Cotman, C.W. Roles of metabotropic glutamate receptors in brain plasticity and pathology, *Annals of the New York Academy of Sciences*, 757 (1995) 460–474.
109. Mintz, M., Rapoport, R., Oleske, J.M., Connor, E.M., Koenigsberger, M.R., Denny, T., and Epstein, L.G. Elevated serum levels of tumour necrosis factor reassociated with progressive encephalopathy in children with acquired immunodeficiency syndrome, *Am. J. Diseases in Childhood*, 143 (1989) 771–774.
110. Mody, I. and Heinemann, U. NMDA receptors of dentate gyrus granule cells participate in synaptic transmission following kindling, *Nature*, 326 (1987) 701–704.
111. Monaghan, D.T., Bridges, R.J., and Cotman, C.W. The excitatory amino acid receptors: their classes, pharmacology and distinct properties in the function of the central nervous system, *Annual Review of Pharmacology and Toxicology*, 29 (1989) 365–402.
112. Nadler, J.V. Kainic acid as a tool for the study of temporal lobe epilepsy, *Life Sciences*, 29 (1981) 2031–2042.
113. Nakanishi, S. Molecular diversity of glutamate receptors and implications for brain function, *Science*, 258 (1992) 597–603.
114. Nicotera, P., McConkey, D.J., Dypbukt, J.M., Jones, D.P., and Orrenius, S. Ca^{2+}- activated mechanisms in cell killing, *Drug Metabolism Review*, 20 (1992) 193–201.
115. Nitecka, L., Tremblay, E., Charton, G., Bouillot, J.P., Berger, M.L., and Ben-Ari, Y. Maturation of kainic acid seizure-brain damage syndrome in the rat. II. histopathological sequelae, *Neuroscience*, 13 (1984) 1073–1094.
116. Olney, J., Rhee, V., and Ho, O. Kainic acid: a powerful neurotoxic analogue of glutamate, *Brain Res.*, 77 (1974) 507–512.
117. Patel, M.N. and McNamara, J.O. Selective enhancement of axonal branching of cultured dentate gyrus neurons by neurotrophic factors, Neuroscience, 69 (1995) 763–770.
118. Payan, J. Anterior transposition of the ulnar nerve: an electrophysiological study, *J. Neurol. Neruosurg. Psychiatry*, 33 (1970) 157–165.
119. Pechadre, J.C., Sauvezie, B., Osier, C., and Gilbert, J. Traitement des encephalopathies epileptiques de l'enfant par les gammaglobulines, *Rev EEG Neurophysiol*, 7 (1979) 443–447.
120. Penix, L.P. and Wasterlain, C.G. Selective protection of neuropeptide containing dentate hilar interneurons by non-NMDA receptor blockade in an animal model of status epilepticus, *Brain Res.*, 644 (1994) 19–24.
121. Penner, R., Fasolato, C., and Hoth, M. Calcium influx and its control by calcium release, *Current Opinion in Neurobiology*, 3 (1993) 368–374.
122. Perez-Velazquez, J.L., Valiante, T.A., and Carlen, P.L. Changes in calcium currents during ethanol withdrawal in a genetic mouse model, *Brain Res.*, 649 (1994) 305–309.
123. Perez-Velazquez, J.L., Valiante, T.A., and Carlen, P.L. Modulation of gap junctional mechanisms during calcium-free induced field burst activity: a possible role for electrotonin coupling in epileptogenesis, *J. Neurosci.*, 14 (1994) 4308–4317.
124. Perez-Velazquez, J.L. and Zhang, L. In vitro hypoxia induces expression of the NR2C subunit of the NMDA receptor in rat cortex and hippocampus, *J. Neurochem.*, 63 (1994) 1171–1173.
125. Pin, J.P. and Duvoisin, R. The metabotropic glutamate receptors: structure and functions. *Neuropharmacology*, 34 (1995) 1–26.
126. Portera-Cailliau, C., Price, D.L., and Martin, L. Brain maturity and selective glutamate receptor subtype activation influence excitotoxic neuronal death morphology along an apoptosis-necrosis continuum, *Society of Neuroscience abstract*, Washington, DC 589 (1996) 1.
127. Pretel, S., Applegate, C.D., and Piekut, D. Apoptotic and necrotic cell death following kindling induced seizures. *Acta Histochem.*, 99 (1997) 71–79.
128. Racine, R.J., Mosher, M., and Kairiss, E.W. The role of the piriform cortex in the generation of interictal spikes in the kindled preparation, *Brain Res.*, 454 (1988) 251–263.

129. Rahman, S. and Neuman, R.S. Characterization of metabotropic glutamate receptor-mediated facilitation of N-methyl-D-aspartate depolarization of neocortical neurones, *Br. J. Pharmacology*, 117 (1996) 675–683.
130. Randall, R.D. and Thayer, S.A. Glutamate-induced calcium transient triggers delayed calcium overload and neurotoxicity in rat hippocampal neurons, *J. Neuroscience*, 12 (1992) 1882–1895.
131. Rashid, K., Van der Zee, C.E., Ross, G.M., Chapman, C.A., Stanisz, J., Riopelle, R.J., Racine, R.J., and Fahnestock, M. A nerve growth factor peptide retards seizure development and inhibits neuronal sprouting in a rat model of epilepsy, *Proc. Natl Acad. Sci. USA*, 92 (1995) 9495–9499.
132. Rasool, N., Muzzafar, M.D., and Rubinstein, E.H. Lidocaine accelerates neuroelectrical recovery after incomplete global ischemia in rabbits, *Stroke*, 21 (1990) 929–935.
133. Rayport, S.G. and Kandel, E.R. Epileptogenic agents enhance transmission at an identified weak electrical synapse in aplysia, *Science*, 213 (1981) 462–464.
134. Recasens, M. and Vignes, M. Excitatory amino acid metabotropic receptor subtypes and calcium regulation, *Annals of the New York Academy of Sciences*, 757 (1995) 418–429.
135. Represa, A., Niquet, J., Pollard, H., and Ben-Ari, Y. Cell death, gliosis, and synaptic remodelling in the hippocampus of epileptic rats, *J. Neurobiology*, 26 (1995) 413–425.
136. Rogers, S.W., Andrews, P.I., Gahring, L.C., Whisenand, T., Cauley, K., Crain, B., Hughes, T.E., Heinemann, S.F., and McNamara, J.O. Autoantibodies to glutamate receptor GluR3 in Rasmussen's encephalitis, *Science*, 265 (1994) 648–651.
137. Roper, S.N., Obenaus, A., and Dudek, F.E. Osmolality and nonsynaptic epileptiform bursts in rat CA1 and dentate gyrus, *Annals of Neurology*, 31 (1992) 81–85.
138. Rothman, S.M. and Olney, J.W. Glutamate and the pathophysiology of hypoxic-ischemic brain damage, *Annals of Neurology*, 19 (1986) 105–111.
139. Sacaan, A.I. and Schoepp, D.D. Activation of hippocampal metabotropic excitatory amino acid receptors leads to seizures and neuronal damage, *Neuroscience Letters*, 139 (1992) 77–82.
140. Sato, K., Kashihara, K., Morimoto, K., and Hayabara, T. Regional increases in brain-derived neurotrophic factor and nerve growth factor mRNAs during amygdaloid kindling, but not in acidic and basist growth factor mRNAs, *Epilepsia*, 37 (1996) 6–14.
141. Sato, M., Racine, R.J., and McIntyre, D.C. Kindling: basic mechanisms and clinical validity, *Electroencephalography Clin. Neurophysiol.*, 76 (1990) 459–472.
142. Scharfman, H.E. and Schwartzkroin, P.A. Protection of dentate hilar cells from prolonged stimulation by intracellular calcium chelation, *Science*, 246 (1989) 257–260.
143. Schmidt, J. New aspects in the pharmacology of kindling implications for mechanism of action in kindling, *Polish J. Pharmacol. Pharmacy*, 39 (1987) 527–536.
144. Schmutz, M. Relevance of kindling and related processes to human epileptogenesis, *Progress in Neuropsychopharmacology and Biological Psychiatry*, 11 (1987) 505–525.
145. Schoepp, D.D. The biochemical pharmacology of metabotropic glutamate receptors, *Biochemical Society Transactions*, 21 (1993) 97–102.
146. Schoepp, D.D. and Conn, P.J. Metabotropic glutamate receptors in brain function and pathology, *Trends Pharmacol. Sci.*, 14 (1993) 13–20.
147. Schwartzkroin, P.A. Epileptogenesis in the immature central nervous system. In: P.A. Schwartzkroin and H.V. Wheal (Eds.) *Electrophysiology of epilepsy.*: Academic Press, London (1984) pp. 389–412.
148. Sheardown, M.J. Metabotropic glutamate receptor agonists reduce epileptiform activity in the rat neocortex, *NeuroReport*, 3 (1992) 916–918.
149. Shigemoto, R., Nakanishi, S., and Mizuno, N. Distribution of the mRNA for a metabotropic glutamate receptor (mGluR1) in the central nervous system: An in situ hybridization study in adult and developing rat, *J. Comp. Neurology*, 322 (1992) 121–135.
150. Siesjo, B.K. Basis for the application of calcium antagonists in neurology: calcium in the brain under physiological and pathological conditions, *Eur. Neurology*, 30 (1990) 3–9.
151. Siesjo, B.K. and Bengtsson, F. Calcium fluxes, calcium antagonists, and calcium-related pathology in brain ischemia, hypoglycemia, and spreading depression: a unifying hypothesis, *Journal of Cerebral Blood Flow and Metabolism*, 9 (1989) 127–140.
152. Siliprandi, R., Lipartiti, M., Fadda, E., Sautter, J., and Manev, H. Activation of the glutamate metabotropic receptor protects retina against N-methyl-D-aspartate toxicity, *Eur. J. Pharmacology*, 219 (1992) 173–174.
153. Singh, R. and Pathak, D.N. Lipid peroxidation and glutathione peroxidase, glutathione reductase, superoxide dismutase, catalase, and glucose-6-phosphate dehydrogenase activities in $FeC1_3$-induced epileptogenic foci in the rat brain, *Epilepsia*, 31 (1990) 15–26.

154. Sloviter, R.S. "Epileptic" brain damage in rats induced by sustained electrical stimulation of the perforant path. I. Acute electrophysiological and light microscopic studies, *Brain Res. Bulletin*, 10 (1983) 675–697.
155. Sloviter, R.S. Decreased hippocampal inhibition and a selective loss of interneurons in experimental epilepsy, *Science*, 235 (1987) 73–76.
156. Sloviter, R.S. Calcium-binding protein (Calbindin-D28K) and parvalbumin immunocytochemistry: Localization in the rat hippocampus with specific reference to the selective vulnerability of hippocampal neurons to seizure activity, *J. Comp. Neurology*, 280 (1989) 183–196.
157. Sloviter, R.S. Permanently altered hippocampal structure, excitability and inhibition after experimental status epilepticus in the rat: the "dormant basket cell" hypothesis and its possible relevance to temporal lobe epilepsy, *Hippocampus*, 1 (1991) 41–46.
158. Sloviter, R.S. and Damiano, B.P. Sustained electrical stimulation of the perforant path duplicates kainate-induced electrophysiological effects and hippocampal damage in rats, *Neuroscience Letters*, 24 (1981) 279–284.
159. Spray, D.C. and Dermietzel, R. X-linked dominant Charcot-Marie-Tooth disease and other potential gap-junction diseases of the nervous system, *Trends Neurosci.*, 18 (1995) 256–262.
160. Sommer, W. Erkrankung des Ammonshorns als aetiologisches moment der epilepsien, *Arch. Psychiatric Nervenkr.*, 10 (1880) 631–675.
161. Stanton, P.K. and Moskal, J.R. Diphenylhydantoin protects against hypoxia-induced impairment of hippocampal synaptic transmission, *Brain Res.*, 546 (1991) 351–354.
162. Stelzer, A., Slater, N.T., and ten Bruggencate, G. Activation of NMDA receptors blocks GABAergic inhibition in an in vitro model of epilepsy, *Nature*, 326 (1987) 698–701.
163. Stewart, B.W. Mechanisms of apoptosis: Integration of genetic, biochemical, and cellular indicators, *Journal of the National Cancer Institute*, 86 (1994) 1286–1296.
164. Stys, P.K., Waxman, S.G., and Ransom, B.R. Ionic mechanisms of anoxic injury in mammalian CNS white matter: role of Na^+-Ca^{2+} exchanger, *J. Neuroscience*, 12 (1992) 430–439.
165. Sutula, T., Cascino, G., Cavazos, J., Parada, I., and Ramirez, L. Mossy fiber synaptic reorganization in the epileptic human temporal lobe, *Annals of Neurology*, 26 (1989) 321–330.
166. Sutula, T., He, X.X., Cavazos, J., and Scott, G. Synaptic reorganization in the hippocampus induced by abnormal functional activity, *Science*, 239 (1988) 1147–1150.
167. Taft, W.C., Clifton, G.L., Blair, R.E., and DeLorenzo, R.J. Phenytoin protects against ischemia-produced neuronal cell death, *Brain Res.*, 483 (1989) 143–148.
168. Tanabe, Y., Masu, M., Ishii, T., Shigemoto, R., and Nakanishi, S. A family of metabotropic glutamate receptors, *Neuron*, 8 (1992) 169–179.
169. Tanabe, Y., Nomura, A., Masu, M., Shigemoto, R., Mizuno, N., and Nakanishi, S. Signal transduction, pharmacological properties, and expression patterns of two metabotropic glutamate receptors, mGluR3 and mGluR4, *J. Neuroscience*, 13 (1993) 1372–1378.
170. Taschenberger, H., Roy, B.L., and Lowe, D.A. Effects of a metabotropic glutamate agonist, trans-ACPD, on cortical epileptiform activity, *NeuroReport*, 3 (1992) 629–632.
171. Tasker, J.G. and Dudek, F.E. Electrophysiology of GABA-mediated synaptic transmission and possible roles in epilepsy, *Neurochemical Res.*, 16 (1991) 251–262.
172. Tasker, R.C., Coyle, J.T., and Vornov, J.J. The regional vulnerability to hypoglycemia-induced neurotoxicity in organotypic hippocampal culture: protection by early tetrodotoxin or delayed MK-801, *J. Neuroscience*, 12 (1992) 4298–4308.
173. Taylor, C.P. and Meldrum, B.S. Na^+ channels as targets for neuroprotective drugs, *TiPS*, 16 (1995) 309–316.
174. Thomas, R.J. Excitatory amino acids in health and disease, *J. Am. Geriatrics Society*, 43 (1995) 1279–1289.
175. Thompson, C.B. Apoptosis in the pathogenesis and treatment of disease, *Science*, 267 (1995) 1456–1462.
176. Tizzano, J.P., Griffey, K.I., Johnson, J.A., Fix, A.S., Helton, D.R., and Schoepp, D.D. Intracerebral 1S,3R-dicarboxylic acid (1S,3R-ACPD) produces limbic seizures that are not blocked by ionotropic glutamate receptor antagonists, *Neuroscience Letters*, 162 (1993) 12–16.
177. Traynelis, S.F. and Dingledine, R. Role of extracellular space in hyperosmotic suppression of potassium-induced electrographic seizures, *J. Neurophysiology*, 61 (1989) 927–938.
178. Triggs, W.J. and Willmore, L.J. In vivo lipid peroxidation in rat brain following intracortical Fe^{2+} injection, *J. Neurochemistry*, 42 (1984) 976–980.
179. Twyman, R.E., Gahring, L.C., Spiess, J., and Rogers, S.W. Glutamate receptor antibodies activate a subset of receptors and reveal an agonist binding site, *Neuron*, 14 (1995) 755–762.

180. Tymianski, M., Wallace, M.C., Spigelman, I., Uno, M., Carlen, P.L., Tator, C.H., and Charlton, M.P. Cell-permeant Ca^{2+} chelators reduce early excitotoxic and ischemic neuronal injury in vitro and in vivo, *Neuron*, 11 (1993) 221–235.
181. Valiante, T.A., Perez-Velazquez, J.L., Jahromi, S.S., and Carlen, P.L. Coupling potential in CA1 neurons during calcium-free-induced field burst activity, *J. Neurosci.*, 15 (1995) 6946–6956.
182. Van der Zee, C.E., Rashid, K., Le, K., Moore, K.A., Stanisz, J., Diamond, J., Racine, R.J., and Fahnestock, M. Intraventricular administration of antibodies to nerve growth factor retards kindling and blocks mossy fiber sprouting in adult rats, *J. Neuroscience*, 15 (1995) 5316–5323.
183. Vecil, G.G., Li, P.P., and Warsh, J.J. Evidence for metabotropic excitatory amino acid receptor heterogeneity: Developmental and brain regional studies, *J. Neurochemistry*, 59 (1992) 252–258.
184. Vicedomini, J.P. and Nadler, J.V. A model of status epilepticus based on electrical stimulation of hippocampal afferent pathways, *Experimental Neurology*, 96 (1987) 681–691.
185. Vicedomini, J.P. and Nadler, J.V. Stimulation-induced status epilepticus: role of hippocampal mossy fibers in the seizures and associated neuropathology, *Brain Res.*, 521 (1990) 70–74.
186. Walton, N.Y. Systemic effects of generalized convulsive status epilepticus, *Epilepsia*, 34 (1993) S54–S58.
187. Wasterlain, C.G., Fujikawa, D.G., Penix, L., and Sankar, R. Pathophysiological mechanisms of brain damage from status epilepticus, *Epilepsia*, 34 (Suppl.1) (1993) S37-S53.
188. Watkins, J.C., Krogsgaard-Larsen, P., and Honore, T. Structure-activity relationships in the development of excitatory amino acid receptor agonists and competitive antagonists, *Trends Pharmacol. Sci.*, 11 (1990) 25–33.
189. Weber, M.L. and Taylor, C.P. Damage from oxygen and glucose deprivation in hippocampal slices is prevented by tetrodotoxin, lidocaine and phenytoin without blockade of action potentials, *Brain Res.*, 664 (1994) 167–177.
190. Weiss, J.H., Koh, J.Y., Baimbridge, K.G., and Choi, D.W. Cortical neurons containing somatostatin- or parvalbumin-like immunoreactivity are atypically vulnerable to excitotoxic injury in vitro, *Neurology*, 40 (1990) 1288–1292.
191. White, E. Life, death, and the pursuit of apoptosis, *Genes and Development*, 10 (1996) 1–15.
192. Willmore, L.J., Ballinger, W.E. Jr., Boggs, W., Sypert, G.W., and Rubin, J.J. Dendritic alterations in rat isocortex within an iron-induced chronic epileptic focus, *Neurosurgery*, 7 (1980) 142–146.
193. Willmore, L.J., Hiramatsu, M., Kochi, H., and Mori, A. Formation of superoxide radicals after $FeCl_3$ injection into rat isocortex, *Brain Res.*, 277 (1983) 393–396.
194. Willmore, L.J. and Rubin, J.J. Antiperoxidant pretreatment and iron-induced epileptiform discharges in the rat: EEG and histopathologic studies, *Neurology*, 31 (1981) 63–69.
195. Willmore, L.J. and Triggs, W.J. Effect of phenytoin and corticosteroids on seizures and lipid peroxidation in experimental posttraumatic epilepsy, *J. Neurosurgery*, 60 (1984) 467–472.
196. Yamasaki, Y., Kogure, K., Hara, H., Ban, H., and Akaike, N. The possible involvement of tetrodotoxin-sensitive ion channels in ischemic neuronal damage in the rat hippocampus, *Neuroscience Letters*, 121 (1991) 251–254.
197. Yuan, J.Y. and Horvitz, H.R. The *Caenorhabditis elegans* genes ced-3 and ced-4 act autonomously to cause programmed cell death, *Developmental Biology*, 138 (1990) 33–41.

20

THE KETOGENIC DIET

Eileen P. G. Vining

Pediatric Epilepsy
600 North Wolfe Street
Meyer 2-147
Baltimore, MD
21287
tel: (410) 955-9100
fax: (410) 614-0373

One of the most significant trends in therapy has been the renewed popularity of the ketogenic diet. Dr. Vining discusses the history of the diet, its possible mechanisms of action and its clinical efficacy.

EARLY HISTORY

The concept of treating seizures through dietary management has existed throughout much of recorded history. The underpinnings of the ketogenic diet (fasting and starvation) are referred to in the Bible in the story of the possessed boy being freed from the demon (seizures). When Jesus was questioned as to how he had been able to accomplish it, he said, "This kind does not leave but by prayer and fasting" (Matthew 17:14–21). Raphael's glorious *Transfiguration* depicts this scene. References are again made in church literature, where John of Gaddesden in medieval times recommended that not only the person, but the family, submit to prayer and fasting.

The scientific literature begins in 1911 and reflects the belief that still was prevalent: epilepsy was caused by the excesses in life (sexual, food, drink, etc.). Guelpa and Marie referred to their patients with intractable seizures as "trop alimentes," (too well fed) and placed 20 patients on a four-day fast with a reported modest effect on the control of their seizures.[9]

The story of the beginnings of dietary management in the United States involves the successful use of prayer and fasting by Bernarr Macfadden and his associate, Dr. Hugh Conklin, an osteopathic physician in Michigan. The son of a prominent New York attorney was cured using this method, and the family prevailed upon an eminent physician-friend in New York to try this approach. In 1921, Geyelin reported using a protocol of 3 weeks of fasting in which 26 patients were treated. Eighteen showed marked improvement, usually very early in the fast.[7]

In that same year, Wilder proposed the use of what would ultimately become a traditional ketogenic diet.[24] His suggestions were based on a knowledge of metabolism in diabetes—knowledge that had only recently been discovered. He suggested that the biochemical aspects of fasting (ketosis) could be achieved by utilizing a diet that was rich in ketogenic foods (fats) and low in antiketogenic foods (carbohydrates and protein). At this early stage, Wilder recognized that the ratio of ketogenic: antiketogenic foods needed to maintain ketosis had to be at least 3:1.

Wilder's colleagues at the Mayo Clinic proceeded to refine the ketogenic diet and reported on its success. In 1924, Peterman described a diet involving 1 gram of protein/kilogram of body weight and 10–15 gram of carbohydrate with the remainder of calories as fats.[17] His 1925 publication reported that 60% of patients were seizure free and 35% were markedly improved.[18] Helmholz later reported a somewhat less optimistic, yet realistic, rate of 31% virtually cured and 23% markedly improved.[10]

A number of studies followed throughout the 1920's and into the early 1950's, although interest clearly seemed to diminish after the introduction of diphenylhydantoin in 1938. These studies showed that 60–75% of children achieved better than 50% control of their seizures with 30–40% achieving more than 90% control. Livingston reported the largest series of 300 patients in 1954 and provided a thorough discussion of its use at that time.[16]

MECHANISM OF ACTION

While clinicians reported on its efficacy, the fledgling pediatric research community tried to unravel the mechanism of the diet's action. The major areas of investigation involved: acidosis, water balance/dehydration, the direct effect of lipids/ketones, and altered energy substrates. Several early investigators believed acidosis was the major mechanism involved.[15] Acidosis produced by administration of inorganic acids, however, did not produce seizure control, and several investigators have shown that initial acidosis in the ketogenic diet is quickly compensated. Children chronically on the diet have a normal pH. More recently, the ketogenic diet has been studied in animal models, and brain pH is normal.[1]

Not unexpectedly, many attributed the positive effects of the diet to dehydration, relating it to the action of acetazolamide and a negative balance of sodium and potassium. Again, in animal models, brain electrolytes and water content were not affected by the ketogenic diet.[2,3]

Several investigators believed that the lipidemia produced by the diet was the critical aspect of efficacy. There is no evidence, however, that children with higher lipid levels and more abnormal profiles achieve better seizure control.[13,20]

Alterations in the energy substrate used by the brain and in cerebral metabolism have been studied more recently (1970's), without conclusive evidence that the explanation for the diet's efficacy lies in that domain. Uhlemann and Neims[22] were able to develop a ketogenic diet for mice and demonstrated that the level of ketonemia appeared to be related to the degree of protection from bicuculline-induced seizures, and also that this ketonemia was better established in younger mice. Appleton and DeVivo studied the effects of a high-fat diet in rats and found that there was a rise in electroconvulsive threshold after 10 days on the diet that could be rapidly reversed. They did not see significant changes in brain glutamic acid, glutamine, gamma-aminobutyric acid (GABA), alanine or aspartic acid. They suggested that the effect was due to increased ketone body utilization and, in a subsequent study, found that ratios of ATP to ADP were significantly increased in rats on a high-fat diet, concluding that the mechanism of action of the diet might lie in increased cerebral energy reserves.[2]

Still more recent work, engendered by a recent increase in the clinical use of the diet, has shown that brain pH does not change in animals[1] and has suggested that there may be an effect on neurotransmitters.[4]

Despite its clinical use for over seventy years, we have little understanding of the mechanism by which the diet may work. Continued efforts to unravel this puzzle are critical both to elucidating potentially more effective and less life-altering therapy, and perhaps to understanding epilepsy. We also have no rational explanation for why the diet's effect (like most antiepileptic drugs) is so variable. Why do some children achieve virtually immediate relief while others may take a full month before good control is established? Why do some children achieve good control and then lose it while seeming to continue in excellent ketosis? These are some of the clinical questions that should promote fascinating basic science research.

THE BASICS OF THE DIET

The diet is calculated based upon a proscribed number of calories per kilogram at a specific ratio of fats to carbohydrate and protein. The number of calories allowed per day varies depending on a number of factors, including the age of the child, the current nutritional status of the child and the expected energy needs of the child. The ratio prescribed varies from 3:1 to 5:1. Lower ratios usually fail to produce good ketosis and the higher ratio usually fails to provide enough protein.

In calculating the diet for a four-year-old child who weighed 15 kilograms, we would estimate caloric needs at about 68 kcal/kg/day or a total of 1,000 calories/day. In prescribing a 4:1 diet, we would know that for every 4 grams of fat we would be able to allow 1 gram of combined protein and carbohydrate. Since a gram of fat contains 9 calories, our four grams would contain 36 calories, while our one gram of protein or carbohydrate would contain 4 calories. This "unit" of the diet, therefore, contains 40 calories and we would need 25 of these "units" to provide 1,000 kcal/day. Thus, the child would receive a total of 100 grams of fat (900 calories) and 25 grams of carbohydrate and protein (100 calories), with 15 grams set aside to protein (1 gram/kg/day), leaving 10 grams for the very limited carbohydrates allowed in this diet. This would then be divided into three equally proportioned meals, and the individual meals would be constructed for the child. Children also need to receive a carbohydrate-free, fat-soluble vitamin supplement and additional calcium. Fluids are routinely limited to 60–75 ml/kg of body weight, with a maximum under usual circumstances of 1,200 ml/day.

The initiation of the diet usually involves a brief hospitalization during which children are fasted for 36–48 hours. They are followed closely for evidence of hypoglycemia. Over the next several days, the diet is gradually introduced using a ketogenic egg-nog first at 1/3 of the daily calories, then 2/3 of the daily calories and then full calories with regular food. Throughout this period of time, extensive family teaching is undertaken so that the family can become independent in preparing and managing the various aspects of the ketogenic diet. This includes information dealing with intercurrent illness.[5,21]

OTHER DIET ISSUES

Many children begin the ketogenic diet on multiple medications. Our goal always is to minimize these medications and, where possible, to discontinue them. This should be done, however, after the diet has been stabilized, and in a very gradual manner. It is important to maintain confidence in the diet at the start, and rapid weaning from medication may result in loss of seizure control with the family believing that the diet will not be effective.

Theoretically, two medications pose some increased risk to children beginning the ketogenic diet. Acetazolamide may compound the problems of acidosis. Topiramate may also do this, as well as increasing the risk of kidney stones. Children on barbiturates and benzodiazepines often become more sedated as the diet is being initiated, and sometimes these medications need to be carefully decreased. Concerns have also been raised about the concomitant use of valproic acid.[14] However, we have not seen evidence of increased toxicity. We rarely use supplemental carnitine.

The ketogenic diet is not without complications of its own. Kidney stones occur in about 5% of children on the diet.[11] A family history of renal stones should alert the clinician to the need for careful monitoring of the urine and renal function in these children, and more liberal fluid use is warranted. Children are also at risk for disordered bone metabolism and for vitamin D deficiency osteomalacia. Supplementation of the diet with vitamin D is critical. There is some evidence in the literature of increased risk of infection due to abnormal white blood cell function.[26]

The diet is rigorous, and children, as well as parents, sometimes feel deprived. The children feel that they do not "fit in" with their friends at birthday parties, holidays, or more routine trips to fast-food chains. This cannot always be avoided, but some accommodations can be made. Special ketogenic cheese-cakes can be prepared, and alternative trick-or-treat strategies can be devised, with parents "buying back" the candy so the child can buy a more appropriate "treat". Parents can be creative in how they present the food, and with advance planning, can still eat out with their children. The use of "keto-coaches" and contact with other families is often important in helping families overcome these barriers.

There are several variations of the diet including the medium-chain triglyceride (MCT) diet.[12] MCT oil is more "ketogenic" and it was originally believed that this would yield a more effective, as well as more palatable, diet. The MCT diet does not actually appear more effective than the classic diet, and many children have problems with diarrhea and bloating when it is used. At Johns Hopkins we often add some MCT oil to the classic diet to alleviate constipation. Another variation is the corn oil diet suggested by Woody in 1988.[25] It is certainly cheaper to use than MCT, and it is similarly effective and seems to have similar disadvantages if large percentages of fats are administered simply as these oils.

CLINICAL STUDIES OF EFFICACY

Early studies of the diet's efficacy, prior to the implementation of the more modern antiepileptic drugs, suggested that 60–75% of children had better than 50% improvement in seizure control, with 30–40% being more than 90% improved.

Two modern studies were published in the 1989 and 1992. In a well-designed, although brief study of efficacy, Schwartz and colleagues compared the clinical effects of the classical 4:1 ketogenic diet, as used by Livingston, to the MCT diet of Huttenlocher and a modified MCT diet that had been developed at the Radcliffe Hospital.[19] In a companion study they compared the biochemical and endocrine consequences of the diet.[20] They accumulated a 3-week prospective baseline of seizure frequency, then admitted the patients to the hospital for studies and a 24-hour fast. In a non-random assignment, 15 were given the classical diet, 22 the MCT diet (adults included) and 13 the Radcliffe diet. Seizure frequency was examined at 6 weeks, although some patients continued the diets for up to 4 years. Eighty-one percent had >50% reduction. All three diets appeared to be equally effective, although large quantities of MCT were described as unpalatable. They found that all seizure types appeared to respond, that in general there was no EEG change, and that even older children did well.

In a retrospective study from Hopkins, Kinsman et al. examined the outcome of 58 consecutive patients placed on the diet between 1980 and 1985.[14] They were treated with the classical (4:1) ketogenic diet. This was a significantly handicapped population (84% mentally retarded, 45% cerebral palsy) and 80% had multiple seizure types. The median time on the diet for those who had improved was 24 months and only 4 months for those whose seizures did not improve. Virtually complete control was obtained in almost 30% and an additional 38% showed greater than 50% improvement, yielding a greater than 50% improvement in almost 70%. The authors noted than medications were reduced in almost 2/3 of the children. They also reported renal stones in three children and significant acidosis leading to diet discontinuation in one child.

As a result of a resurgence of interest in the ketogenic diet after the publicity surrounding its success in a young child from California, studies looking at its efficacy in even more modern and varied settings have been done. Data from a prospective 7-center study of 51 children placed on a classical 4:1 ketogenic diet after training at Johns Hopkins showed that half of the children remained on the diet at one year and 40% (N = 20) of those starting the diet (N = 51) had a >50% decrease in seizures at one-year follow-up.[23] A prospective evaluation of intervention with the diet in 150 consecutively treated children at Hopkins found that 55% remained on the diet at one year and 50% of those initiating the diet had a >50% decrease in seizures at one year.[6] Although this rate of response is less than what was seen in the older studies—or even in the more recent prospective British study—these numbers represent success in a population of children who had not been controlled by aggressive therapy with the many medications currently available. These studies present prospective data and provide a long-term picture of efficacy.

CONCLUSIONS

The ketogenic diet remains a very effective form of therapy for children with difficult to control seizures. If the diet is successful—providing better than 50% improvement in seizure control—children and their parents will continue to adhere to it. If it is not successful in controlling seizures, the benefits do not outweigh the disadvantages of such a restrictive life-style and it is abandoned.

There are numerous clinical questions that are being addressed currently. Although most studies have not indicated that there are subgroups in which the diet will be more effective, clinical bias suggests there may be groups of children who are more appropriate, more sensitive to it, and in whom it should be used earlier in the clinical course. We remain concerned about renal stones, and need to be able to monitor these children carefully and to discover treatments that will minimize this adverse event. We need to examine the effect of the diet on lipids, and to understand the impact of this over the longer-term use of the diet, since it is now being used (when effective) for more extended periods of time. We need to determine whether there are better ways of monitoring the diet so that efficacy can be maximized. Early attempts at measuring serum beta-hydroxy butyrate appear promising.[8]

Laboratory research is beginning once again, and, with more modern and sophisticated methodology, we may actually discover the mechanism of action of the diet. When this occurs, we may have better insights into the mechanisms of epilepsy, and we may be able to propose more specific therapies and relieve the burdens imposed by this very rigid, but effective, diet.

REFERENCES

1. Al-Mudallal, A.S., LaManna, J.C., Lust, D.W., and Harik, S.I. Diet-induced ketosis does not cause cereral acidosis, *Epilepsia*, 37 (1996) 35–53.
2. Appleton, D.B. and DeVivo, D.C. An animal model for the ketogenic diet, *Epilepsia*, 15 (1974) 211–227.
3. DeVivo, D.C., Leckie, M.P., Ferrendelli, J.S., and McDougal Jr., D.B. Chronic ketosis and cerebral metabolism, *Ann. Neurol.*, 3 (1978) 331–337.
4. Erecinska, M., Nelson, D., Daikhin, Y., and Yudkoff, M. Regulation of GABA level in rat brain synaptosomes: fluxes through enzymes of the GABA shunt and effects of glutamate, calcium and ketone bodies, *J. Neurochem.*, 67 (1996) 2325–2334.
5. Freeman, J.M., Kelly, M.T., and Freeman, J.M. *The Epilepsy Diet Treatment: An Introduction to the Ketogenic Diet*, New York, Demos (1994).
6. Freeman, J.M., Vining, E.P.G., Pillas, D.J., Pyzik, P.L., Casey, J.C., and Kelly, M.T. The efficacy of the ketogenic diet-1998: a prospective evaluation of intervention in 150 children, *Pediatrics*, 102 (1998) 1358–1363.
7. Geyelin, H.R. Fasting as a method for treating epilepsy, *Medical Record*, 99 (1921) 1037–1039.
8. Gilbert, D.L., Swink, T.D., Casey, J., Pyzik, P., and Freeman, J.M. Blood B-OH butyrate levels and management of the ketogenic diet, *Ann. Nuerol*, (1997) (Abstract).
9. Guelpa, G. and Marie, A. La Lutte contre l'epilepsie par la desintoxication et par la reeducation alimentaire, *Revue de Therapie Medico-Chirugicale*, 78 (1911) 8–13.
10. Helmholz, H.F. Treatment of epilepsy in childhood: five years experience with ketogenic diet, *JAMA*, 88 (1927) 2028.
11. Hertzbery, G.Z., Fivush, B.A., Kinsman, S.L., and Gearhart, J.P. Urolithiasis associated with the ketogenic diet, *J. Peds.*, 117 (1990) 743–745.
12. Huttenlocher, P.R., Wilbourn, A.J., and Signore, J.M. Medium-chain triglycerides as a therapy for intractable childhood epilepsy, *Neurology*, 21 (1971) 1097–1103.
13. Ide, T., Steinke, J., and Cahill, G. Metabolic interactions of glucose, lactate and B-hydroxybutyrate in rat brain slices, *Am. J. Physiol.*, 217 (1969) 784–791.
14. Kinsman, S.L., Vining, E.P., Quaskey, S.A., Mellits, D., and Freeman, J.M. Efficacy of the ketogenic diet for intractable seizure disorders: review of 58 cases, *Epilepsia*, 33 (1992) 1132–1136.
15. Lennox, W.G. and Cobb, S. Epilepsy from the standpoint of physiology and treatment, *Medicine (Baltimore)*, 7 (1928) 105–290.
16. Livingston, S. *The Diagnosis and Treatment of Convulsive Disorders in Children*. Charles C. Thomas (Ed.), Springfield, Illinois (1954).
17. Peterman, M.G. The ketogenic diet in the treatment of epilepsy: a preliminary report, *Am. J. Dis. Child*, 28 (1924) 28–33.
18. Peterman, M.G. The ketogenic diet in epilepsy, *JAMA*, 84 (1925) 1979–1983.
19. Schwartz, R.M., Boyes, S., and Aynsley-Green, A. Metabolic effects of three ketogenic diets in the treatment of severe epilepsy, *Dev. Med. Child Neurol.*, 31 (1989) 152–160.
20. Schwartz, R.H., Eaton, J., Bower, B.D., and Aynsley-Green, A. Ketogenic diets in the treatment of epilepsy: short-term clinical effects, *Dev. Med. Child Neurol.*, 31 (1989) 145–151.

21. Swink, T.D., Vining, E.P.G., and Freeman, J.M. The ketogenic diet, *Advances in Pediatrics*, 44 (1997) 297–329.
22. Uhlemann, E.R. and Neims, A.H. Anticonvulsant properties of the ketogenic diet in mice, *J. Pharmacol. Exp. Ther.*, 180 (1972) 231–238.
23. Vining, E.P.G., Freeman, J.M., Ballaban-Gil, K., Camfield, C.S., Cafield, P.R. Holmes, G.L., Shinnar, S., Shuman, R., Trevathan, E., and Wheless, J.W. A multi-center study of the efficacy of the ketogenic diet, *Arch. Neurol.*, 55 (1998) 1433–1437.
24. Wilder, R.M. The effect of ketonemia on the course of epilepsy, *Mayo Clinic Bulletin*, 2 (1921) 307.
25. Woody, R.C., Brodie, M., Hampton, D.K., and Piser Jr., R.H. Corn oil ketogenic diet for children with intractable seizures, *J. Child Neurol.*, 3 (1988) 21–24.
26. Woody, R.C., Steele, R.W, Knapple, W.L., and Pilkington Jr., N.S. Impaired neutrophilic function in children with seizures treated with the ketogenic diet, *J. Peds.*, 115 (1989) 427–430.

21

VAGAL STIMULATION FOR INTRACTABLE SEIZURES

Adrian Upton

1200 Main Street West
P.O. Box 2000, Station "A"
Hamilton, Ontario
L8N 3Z5
tel: (905) 521-2100 ext.6372
fax: (905) 521-2656

Vagal stimulation has recently been approved for use in North America. Dr. Upton discusses the findings of a study conducted at the McMaster Medical Centre.

INTRODUCTION

At the present time, the conventional management of seizures includes a trial of multiple medications, followed by consideration of surgical ablation of part of the brain. Since seizures are related to paroxysmal depolarization shifts, more specific therapies are possible. These include the use of electrical stimulation to the brain.

It is known that sensory stimulation can arrest seizures in about 30% of cases.[18] It is possible, therefore, that more precisely applied *electrical* stimulation might also alter the spread of abnormal electrical discharges in the brain. Electrical stimulation can be localized and lateralized, and it is adjustable and reversible. Anti-seizure medications cannot be lateralized or localized (although there has been no adequate trial of drug pumps). Surgical procedures are not adjustable or reversible. The potential use of electrical stimulation, therefore, offers a number of advantages over conventional surgical or medical therapies for the control of intractable seizures. Furthermore, neurostimulation can be used as a supplement to conventional treatments in patients with intractable seizures.

Despite these potential advantages of neurostimulation in the control of epilepsy, neurostimulation techniques have not met with wide acceptance. One possible reason is that early studies of cerebellar or thalamic stimulation did not involve blinded, controlled studies.[8,9,10] Blinded and controlled trials of vagal stimulation, however, have resulted in

Table 1. Indications for Vagal Stimulation

Indications
Intractable Epilepsy
Frequent Seizures
Failure of Anticonvulsant Drugs

acceptance of the technique as a clinical procedure in Europe, and, more recently, in Canada and the USA. Table 1 presents suggested indications for vagal stimulation.

A BRIEF HISTORY OF NEUROSTIMULATION

There is a long history of the application of electrical stimulation to the nervous system of animals and man, including stimulation of the brain, nerves (e.g., phrenic, acoustic, peroneal), and autonomic pathways.[23] There is a similar long history for the use of vagal stimulation.[26,2,3,25,16,24]

The common factor in much of this stimulation work is the *thalamus*. Cerebellar stimulation, for instance, alters thalamic evoked potentials.[21,8,9] Direct thalamic stimulation, particularly stimulation of the anterior thalamus, can result in complete control of seizures.[9,22] Vagal stimulation may well also act through hypothalamic, anterior thalamic and cingulate pathways. This provides some rationale for the use of vagal stimulation for the control of seizures.

Modification of autonomic pathways does not seem likely as a potential treatment for seizures. Some conventional drugs such as phenytoin, however, have some anti-sympathetic effects. Other anticonvulsants (e.g., ethosuximide) act on thalamic relay neurons. Such effects may overlap with those of direct vagal stimulation.

The advantage of vagal stimulation over other methods of stimulation of the central nervous system is that the surgical procedure to implant the electrodes is much simpler for vagal stimulation. There is, therefore, far less risk of complications (e.g., infection, hemorrhage).

The main concerns about vagal stimulation have related to its potential effects on the recurrent laryngeal nerve, the heart and the gastrointestinal system. We have demonstrated, however, that there are no significant adverse effects on cardiac or gastrointestinal function.[20] Furthermore, we have demonstrated no adverse effects of vagal stimulation on cognition or balance.[7,4,5,6]

THE McMASTER STUDY

Indications/Choice of Subjects

Our own studies have involved 10 patients with intractable seizures (simple or complex partial that may be secondarily generalized) that have failed to respond to medical or surgical therapy. They have involved patients with normal intelligence, but this should not be a long-term limitation. Other inclusion criteria have been ages 12–60, males or females, more than 6 seizures per month over a 3-month period, ability to understand and consent, women using acceptable birth control and the use of investigational drugs that have been taken for 5 times the elimination half-life plus 2 weeks.

Exclusion criteria have been progressive neurological disease or progressive medical conditions, prior cervical vagotomy, pregnancy, current use of more than 3 anticonvulsant medications or any medical condition that was likely to deteriorate over the next year.

Implantation of Vagal Stimulator

A helical bipolar stimulation electrode (Hungtington electrode—first 2 patients, then Cyberonics electrode licensed from HMR1) has been stimulated by an Itrel I or II (Medtronic-first 2 patients) or the Neurocybernetic prosthesis (NCP) system (Cyberonics Inc. Webster Texas, USA).

The implantation procedure takes about an hour. Under local or general anaesthetic, a 10-cm skin incision is placed midway between the left clavicle and the mastoid process, with division of the platysma muscle anterior to the sternomastoid muscle. After blunt dissection, the carotid sheath is incised, the vagus nerve is exposed and the nerve is lifted by two soft vessel loops. The proximal of the electrode coils is placed around the vagus nerve and then the distal coil is placed around the nerve. Sutures are used to secure the electrode.

In most operations, the generator is placed in the left chest with tunnelling of the electrode from the neck to the chest incision. The pacemaker is placed in a thoracic subcutaneous pocket and sewed to the pectoralis facia. At McMaster, the pacemaker is placed in the right chest by a similar procedure. Placement in the right chest reduces stress on the diagonally placed electrode. Furthermore, any defibrillation procedures would not be hampered by the right-sided pacemaker. The neck and chest wounds are sutured.

The electrodes are assessed at the time of operation by lead voltage measurements. The generator output is measured across a 1,000 Ohm resistor. A test cable allows recording of wave forms and voltage as the vagus nerve is stimulated. Voltages are measured at the leading and trailing edges of the wave form.[14] Parameters are indicated in Table 2.

Over 18 months, the average voltages measured 1.7 volts (leading edge) and 1.9 volts (trailing edge).

Stimulation Parameters

Our initial two patients received continuous trains of pulses of 330 pulses/sec at 30 Hz at 2 to 6 volts (Itrel I. Medtronic). Patients involved in the controlled study (Cyberonics) were given a baseline (no stimulation) of 12 weeks, randomization to *high* or *low* stimulation groups for 12 weeks, and then continuation at high stimulation. Parameters are indicated in Table 3.

Effects of Vagal Stimulation

At McMaster, we have assessed afferent and efferent effects of left vagal stimulation. Effects on evoked cortical potentials are indicated in Table 4: 1) *EEG changes: EEG,*

Table 2. Intra-Operative Assessment of Stimulator Function

Testing of stimulation	Pulse voltage/leading edge	Tracking edge
0.5 MA 500 pulses/sec 30 Hz	0.25 → 1 volts	0.6 → 2.2 volts
1 MA 250 pulses/sec 30 Hz	0.25 → 1 volts	0.5 → 1.5 volts

Table 3. Stimulation Parameters

High		Low
Frequency (20 to 50 Herz) 30 Hz		1–2 Hz
Pulse width 500 μsec		130 μsec
On 30 to 90 seconds		30 seconds
Off 5 to 10 minutes		60 to 80 minutes
Magnet		
Amplitude	same as OMA	intermittent
On time	30 to 90 seconds	30 seconds
Pulse width	500 μsec	130 μsec

24 hour EEG, and topographic EEG studies. There were no immediate effects on EEG, but there have been slow and progressive decreases in paroxysmal activity over 5 years. 2) *Metabolic changes—PET scan studies.* There are changes in metabolism of radioactive oxygen in the anterior thalamus and cingulate gyrus.[13] 3) *Clinical effects.* There is no effect on reaction time, cognition or balance. 4) *Efferent: effects on heart rate variability.* The low-frequency component of RR variability (0.05 to 0.15 Hz) is enhanced (the sympathetic component), and the high-frequency (0.15 to 0.5 Hz) component is reduced (the vagal component) by vagal stimulation.[15]

Efficacy of Vagal Stimulation

Our series consisted of 10 subjects with intractable complex partial seizures and secondary generalization. There were 8 males, 2 females, mean age 34 ± 7.8 years. Seizure records were obtained 4 to 6 months before and continuously after stimulation.

Mean seizure frequency was reduced by 47.6% in 7 subjects and 15% in 3 subjects. One patient progressed from 150 seizures per month to none for more than 2 years. Failure of the battery resulted in an increase in seizures in 4 subjects prior to stimulation change.

These results are similar to those from the full controlled study, presented in Table 5.

Table 4. Some Unwanted Effects of Anticonvulsant Medications

Neurological	Non-neurological
reduced memory	reduced white count
fatigue	low fibrogen
ataxia	hirsutism
movement disorders	firm hypertrophy, weight gain, rashes
Teratogenic effects	
cleft palate	
epicanthic folds	
atrial septal defects	
spina bifida	
dysmorphic features	

Table 5. Results of the Large International Controlled Trial

Seizure frequency	N = 31 (percent change re baseline)	(base vs. stim.)
Mean	−30.9	P < .0001
Median	−37.5	P < .0001
Seizures per 12 wks.		
Mean	−79	P < .036
Median	−16	P < .0002
% > 50% reduction 38.7 (Ben Menachem et al.[1])		

UNTOWARD EFFECTS OF VAGAL STIMULATION

The main concerns about the effects of neurostimulation have been: 1) *Infection*. We have not encountered this. 2) *Damage to neural tissue*. We have some patients where threshold levels have not changed in 22 years (cerebellar) or 13 years (anterior thalamic). 3) *Adverse effects of vagal stimulation*. The only significant adverse effects are hoarseness (35.5%), coughing (13.9%) and throat pain (12.9%).[17] 4) *Electrode breakage*. We have had 1 lead breakage in 10 patients over 5 years. 5) *Life of the battery*. The initial vagal stimulators had stimulation lives of about 18 months. One pacemaker (Itrel I Medtronic[R]), is still working well after 5 years. New pacemakers and batteries should have predicted lives of over 5 years. The life of the battery will be greatly prolonged by the development of pacemakers that are automatically triggered by seizure activity. Such contingent pacemakers have been developed for cardiological applications and we have been working on this problem.[11,12] 6) *Failure of the stimulator*. One patient in the first 67 patients suffered vocal cord paralysis. This resulted from a short circuit in the generator and about 4 hours of direct current to the left vagus nerve.[17] The stimulator has been modified to prevent this problem.

COST OF VAGAL STIMULATION

The costs of neurostimulation may well be comparable with the costs of chronic use of multiple anticonvulsant medication, despite the costs of stimulators (£5,000.00 to £8,000.00) and surgical implantation. The chronic use of anticonvulsant medications can produce many unwanted effects (Table 6).

Table 6. Effects on Cortical Evoked Potentials

Afferent	Latency (msec)
Evoked cortical potentials	N1 71.7 ± 12.7
Vagal stimulation	P1 127.43 ± 34.1
	N2 194.4 ± 32.9
	P2 261.0 ± 47.4
	N3 328.3 ± 45.8
200 μsec square waves recorded CZ to average 14 MA at 2 Hz average 25 stimuli[19]	

DISCUSSION

Stimulation of the left vagus nerve in man has proven to be a safe and effective treatment of untreatable seizures. In our series of 10 patients, one patient has improved from 150 seizures a month to being seizure free for over 2 years. About 40% of patients have experienced a greater than 50% improvement in seizure frequency. This same figure has applied to the results of a large international controlled trial.[1]

Adverse effects of the procedure have been hoarseness, throat irritation and voice change in 13 to 14% of patients.[17] Lead breakage has been a rare problem. Some initial stimulators had battery lives of only 18 months, but this has been improved to about 5 years. Battery life could be prolonged by the development of contingent pacemakers that are triggered by seizure activity in the EEG. Such contingent pacemakers have been developed for cardiological applications.

The cost of stimulators compares with the cost of the chronic use of 3 anticonvulsant drugs over the replacement life of the stimulator. However, most studies have involved the use of vagal pacemakers as a supplement to the use of medications.

Neurostimulation has not received wide acceptance as a treatment for intractable epilepsy despite encouraging results from the use of cerebellar, thalamic and vagal stimulation. The results of well designed international controlled trials of vagal stimulation have resulted in acceptance of vagal stimulation as a clinical procedure in European countries and, more recently, in North America. Wider use of vagal stimulation or other forms of neurostimulation should result in a reduction in the cost of the pacemaker and the procedure.

Initial controlled studies have shown that vagal stimulation is a safe and well tolerated procedure which can be used as a supplement to conventional medical management of intractable seizures. In the future, the procedure may be applied to paediatric cases as a primary treatment and the use of contingent stimulation should greatly prolong the battery life of the pacemaker.

REFERENCES

1. Ben Menachem, E., Manon-Espaillat, R., Ristanovic, R., Wilder, B.J., Stefan, H., Mirza, W., Tarver, W.B., and Wernicke, J.F. Vagus nerve stimulation for treatment of partial seizures: 1. A controlled study of effect on seizures. *Epilepsia* 35 (1994) 614–624.
2. Chase, M.H., Nakamura, Y., Clemente, C.D., and Sterman, M.B. Afferent vagal stimulation: Neurographic correlates of induced EEG synchronisation and desynchronisation. *Brain Res.* 5 (1967) 236–249.
3. Chase, M.H. and Nakamura, Y. Cortical and subcortical EEG patterns of response to afferent abdominal vagal stimulation: Neurographic correlates. *Physiol. Behav.* 3 (1968) 605–610.
4. Clarke, B., Upton, A.R.M., Kamath, M., Griffin, H., Fitzpatrick, D., and Denardis, M. Vagus nerve stimulation reduces seizure frequencies in man. *American EEG Society Annual Meeting*, Chicago, Sept. 1994.
5. Clarke, B., Upton, A.R.M., Kamath, M. et al. Vagus nerve stimulation reduces seizure frequencies in man. Abstract. Submitted to *North American Society of Pacing and Electrophysiology, 15th Annual Scientific Sessions*, 1994.
6. Clarke, B.M., Upton, A.R.M., and Griffin, H. Cognitive motor function after electrical stimulation of the vagus nerve. *PACE* 15 (II)(1992b) 1603–1607.
7. Clarke, B.M., Upton, A.R.M., Kamath, M., and Griffin, H. Electrostimulation effects of the vagus nerve on balance in epilepsy. *PACE* 15 (II)(1992a) 1614–1630.
8. Cooper, I.S. and Upton, A.R.M. Use of chronic cerebellar stimulation (CCS) for disorders of disinhibition in man. *Lancet* 1 (1978) 595–600.

9. Cooper, I.S., Upton, A.R.M., and Amin, I. Reversibility of chronic neurological deficits. Some effects of stimulation of the thalamus and internal capsule. *Applic. Neurophysiol.* 43 (1980) 244–258.
10. Cooper, I.S., Upton, A.R.M., and Amin, I. Chronic cerebellar stimulation (CCS) and deep brain stimulation (DBS) in voluntary movement disorders. *Appl. Neurophysiol.* 45 (1982) 209–217.
11. Dahanayake, B.W. and Upton, A.R.M. A novel approach for epileptic seizure detection. *7th IEEE Symposium*, Winston, Salem, N.C. June 10–11, 1994.
12. Dahanayake, B.W. and Upton, A.R.M. A novel approach to fast learning: smart neural nets. In: *Library of Congress IEEE Proceedings*, 188–193, 1994.
13. Garnet, E.S., Nahmias, C., Scheffel, A., Firnau, G., and Upton, A.R.M. Regional cerebral blood flow in man manipulated by direct vagal stimulation. *PACE* 15 (II)(1992) 1579–1582.
14. Holder, L.K., Wernicke, J.F., and Tarver, W.B. Treatment of refractory partial seizures: preliminary results of a controlled study. *PACE* 15 (10)(II)(1992) 1557–1571.
15. Kamath, M.V., Upton, A.R.M., Talalla, A., and Fallen, E.L. Neurocardiac responses to vagoafferent electrostimulation in humans. *PACE* 15 (II)(1992) 1581–1587.
16. Lockard, J.S. and Congdon, W.C. Effects of vagal stimulation on seizure rate monkey model. *Epilepsia* 27 (1986) 626.
17. Ramsay, E., Uthman, B.M, Augustinsson, L.E., Upton, A.R., Naritoku, D., Willis, J., Treig, T., Barolat, G., and Wernicke, J.F. Vagus nerve stimulation (VNS) for treatment of partial seizures: 2. Safety, side-effects and tolerability. *Epilepsia* 35 (3)(1994) 627–636.
18. Symonds, C. Sensory stimulation in centrencephalic and focal cortical epilepsy. *Electroenceph. Clin. Neurophysiol.* 13 (1961) 309.
19. Tougas, G., Hudoba, P., Fitzpatrick, D., Hunt, R.H., and Upton, A.R. Cerebral evoked potential responses following direct vagal and oesophageal electrical stimulation in humans. *Am. J. Physiol.* 264 (1993) 486–491.
20. Upton, A.R.M., Tougas, G., Talalla, A., White, A., Hudoba, P., Fitzpatrick, D., Clarke, B., and Hunt, R. Neurophysiological effects of left vagal stimulation in man. *PACE* 14 (1)(1991) 70–76.
21. Upton, A.R.M. Neurophysiological mechanisms in modification of seizures. In: I.S. Cooper (Ed.) *Cerebellar Stimulation in Man*, Raven Press, New York (1978) pp. 39–57.
22. Upton, A.R.M., Amin, I., Garnett, S., Springman, M., Nahmias, C., and Cooper, I.S. Evoked metabolic responses in the limbic-striate system produced by stimulation of anterior thalamic nucleus in man. *PACE* 10 (1987) 217–255.
23. Upton, A.R.M. and White, A. Autonomic stimulation. *PACE* 14 (1991) 50–70.
24. Woodbury, J.W. and Woodbury, D.M. Vagal stimulation reduces the severity of maximal electroshock seizures in intact rats: Use of a cuff electrode for stimulating and recording. *PACE* 14 (1991) 94–107.
25. Zabara, J. Time course of seizure control to brief repetitive stimuli. *Epilepsia* 26 (1985) 518.
26. Zanchetti, A., Wang, S.C., and Moruzzi, G. The effect of vagal afferent stimulation on the EEG pattern of the cat. *Electroenceph. Am. Neurophysiol.* 4 (1952) 357–361.

INDEX